Skyrmions
Topological Structures, Properties, and Applications

Series in Materials Science and Engineering

Series in Materials Science and Engineering

Skyrmions
Topological Structures, Properties, and Applications

Edited by
J Ping Liu
Zhidong Zhang
Guoping Zhao

CRC Press
Taylor & Francis Group
Boca Raton London New York

CRC Press is an imprint of the
Taylor & Francis Group, an **informa** business

CRC Press
Taylor & Francis Group
6000 Broken Sound Parkway NW, Suite 300
Boca Raton, FL 33487-2742

First issued in paperback 2020

© 2017 by Taylor & Francis Group, LLC
CRC Press is an imprint of Taylor & Francis Group, an Informa business

No claim to original U.S. Government works

ISBN 13: 978-1-4987-5388-3 (hbk)
ISBN 13: 978-0-367-49666-1 (pbk)

Visit the Taylor & Francis Web site at
http://www.taylorandfrancis.com

and the CRC Press Web site at
http://www.crcpress.com

Contents

Series Preface

The series publishes cutting-edge monographs and foundational text-books for interdisciplinary materials science and engineering.

Its purpose is to address the connections between properties, structure, synthesis, processing, characterization, and performance of materials. The subject matter of individual volumes spans fundamental theory, computational modeling, and experimental methods used for design, modeling, and practical applications. The series encompasses thin films, surfaces, and interfaces, and the full spectrum of material types, including biomaterials, energy materials, metals, semiconductors, optoelectronic materials, ceramics, magnetic materials, superconductors, nanomaterials, composites, and polymers.

It is aimed at undergraduate-level and graduate-level students, as well as practicing scientists and engineers.

Foreword

The book ventures into an important emerging area in magnetism and spintronics: the study of skyrmions. These new types of magnetic textures, skyrmions lattices, or individual skyrmions are topologically protected spin configurations, which are of great interest for their unprecedented physical properties. In particular, they represent the ultimate small achievable size for a nonvolatile magnetic memory element.

While the existence of magnetic skyrmions had been theoretically predicted more than 20 years ago, the first experimental observations date back to only 2009 when skyrmion lattices were found in noncentrosymmetric magnetic compounds and ascribed to Dzyaloshinskii–Moriya interactions (DMI) induced by spin–orbit coupling (SOC) in the absence of inversion symmetry. It is amazing to see how quickly these first experimental results have been followed by a series of observations of skyrmions not only in various exotic magnetic compounds but also, more importantly, in thin films of classical magnetic metals (Fe, Co) deposited on heavy metals such as platinum or iridium. In thin films, the presence of an interface breaks the inversion symmetry, and the SOC is brought about by the heavy metal. In the initial experiments on films, the skyrmions were observed at low temperatures in epitaxially grown ultrathin layers, and this type of skyrmion was not directly relevant for technological applications. But one now knows from the most recent works that skyrmions can also be found at room temperature in magnetic layers and multilayers fabricated by standard methods of the spintronic technologies, which clearly pave the way for the development of applied devices.

What is amazing is the number of remarkable original properties demonstrated by skyrmions. A skyrmion lattice can be the ground state of the magnetic material, but isolated skyrmions can also be stabilized in a metastable state by their topological protection and manipulated as quasi particles. These skyrmions can be put into motion by remarkably small electrical currents, and they can also be created or annihilated by electrical currents. Among the other original properties revealed by the skyrmions, one can also cite the topological Hall effect, specific excitation modes, their interaction with magnons, and so on. It is now clear that all of these remarkable properties raise great expectations for their use in spintronic devices, race-track type storage devices, nonvolatile memories, logic gates, oscillators, and so on.

This book reviews all aspects of recent developments in research on skyrmions, from the presentation of the observation and characterization techniques to the description of physical properties and expected applications. It will be of great use for all scientists working in this field.

Albert Fert
2007 Nobel Laureate in Physics

Preface

First proposed by Tony Skyrme in the 1960s, skyrmions—among the most important and exciting physical phenomena discovered in the past 50 years—have been studied both theoretically and experimentally for decades, starting from particle physics research. The experimental observation of magnetic skyrmions a few years ago, an important milestone on the road to an improved understanding of skyrmions, has triggered great attention from the community of magnetism and magnetic materials. From 2008 to the spring of 2016, there were already roughly 600 published articles and 10,000 citations on magnetic skyrmions, as indexed by the Web of Science.

In our opinion, there are strong reasons behind this explosion of interest in skyrmions. It is a highly interdisciplinary subject between condensed matter physics, chemistry, and topology. Although topology is one of the most fundamental and important fields in mathematics, it clearly has a close relationship with the physical properties of condensed matter, such as the Aharonov–Bohm effect, Berry phase effect, quantum Hall effect, and topological insulators. There has been a new trend in condensed matter physics and materials science to use topological properties to develop novel functional materials. People from the field of magnetism and magnetic materials have realized the importance of understanding the electronic spin states and magnetic (and electric) domain structures in ferromagnetic and ferroelectric substances. The topology is linked to the physical properties of condensed matter in many ways, including, for instance, the following:

1. Topological phases in eigenvalues and eigenvectors of a many-body system may be generated via the interchange of particles.
2. The Fermi surfaces of a material (for instance, topological insulators and Weyl Fermions) may possess different topological behaviors in momentum space.
3. The spin configurations (such as vortex, meron, and skyrmions) in real space may have different topological properties and are of significant interest in understanding magnetism and transport.

It is also obvious to us that investigation in magnetic skyrmions is not only important for fundamental research but also for advanced technological applications. Skyrmions, as a new type of magnetic domains with special topological structures, have outstanding magnetic and transport properties that may lead to applications in data storage and other advanced spintronic devices, as readers will find in our book.

This book had its genesis in the *Symposium on Magnetic Domains* that we organized on July 21, 2014, in Ningbo, People's Republic of China. Focused on magnetic skyrmions, the symposium was probably the earliest gathering of magnetic skyrmion researchers in China, with participation from most research groups in the country who had been investigating magnetic skyrmions. In view of the rapid developments in the field, we felt strongly the need for a book that introduces the fundamental aspects and possible applications of magnetic skyrmions to the community to benefit researchers in condensed matter physics, chemistry, and materials science, especially young scientists who are interested in skyrmions. Thanks are due to Dr. Luna Han at CRC Press/Taylor & Francis Group, who coincidentally sent us an invitation in early 2015 that initiated the book plan; we subsequently started to search for authors to write the book. Editing a comprehensive textbook such as this one, covering a research topic that is still rapidly developing, tends to be a huge challenge. We extended the author group from symposium participants to several other researchers globally who have carried out outstanding work on magnetic skyrmions, as readers can find from the Contents and the Contributors list. We are proud of all authors and grateful for their contributions, which have made this book a reality.

We also thank Prof. Albert Fert, the 2007 Nobel Prize winner in Physics and an active researcher in magnetic skyrmions, for his great support and insightful Foreword to this book.

J. Ping Liu
University of Texas at Arlington

Zhidong Zhang
Institute of Metal Research

Guoping Zhao
Sichuan Normal University

About the Editors

J. Ping Liu received his PhD degree in physics at the Van der Waals— Zeeman Institute, University of Amsterdam, the Netherlands. He is currently a Distinguished University Professor in the Department of Physics, University of Texas at Arlington, USA. His current research work is focused on nanostructured magnetic materials. He has organized and led several joint research programs in nanocomposite magnets and has authored or co-authored more than 260 peer-reviewed scholarly papers, including review articles, book chapters, and a book (*Nanoscale Magnetic Materials and Applications*). Dr. Liu is a fellow of the American Physical Society.

Zhidong Zhang received his PhD degree from the Institute of Metal Research (IMR), Chinese Academy of Sciences, Shenyang, People's Republic of China. He is a professor at IMR and the head of Magnetism and Magnetic Materials Division. Dr. Zhang has been engaged in research in the fields of magnetism and magnetic materials, including in the light rare earth giant magnetostrictive materials, magnetic nanocapsules, nanocomposite magnetic thin films, magnetocaloric materials, and theoretical work on Bose–Einstein condensates and superfluids. His recent focus is on topological structures in magnetic materials, such as vortex and skyrmions. He has published more than 600 papers in international journals, including 15 invited review articles.

Guoping Zhao received his PhD degree in physics from the National University of Singapore (NUS). He is a distinguished professor in Sichuan Normal University (SICNU), Chengdu, People's Republic of China. Prof. Zhao is the leader of the Scientific Research Innovation Team for Physics and the founder and director of the Magnetic Materials Lab, SICNU. He is the author or the co-author of more than 80 publications in international journals and referred conferences. He is an active researcher in fundamental issues in magnetism and magnetic materials, including coercivity mechanisms, magnetic domain structures (including skyrmions), and exchange interactions.

Contributors

Felix Büttner
Department of Materials
  Science and Engineering
Massachusetts Institute of
  Technology
Cambridge, Massachusetts

Gong Chen
National Center for Electron
  Microscopy (NCEM),
  Molecular Foundry
Lawrence Berkeley National
  Laboratory
Berkeley, California

Chia-Ling Chien
Department of Physics and
  Astronomy
Johns Hopkins University
Baltimore, Maryland

C. P. Chui
Department of Physics
The University of Hong Kong
Pokfulam, Hong Kong

Yingying Dai
Shenyang National Laboratory
  for Materials Science
Institute of Metal Research
Chinese Academy of Sciences
Shenyang, People's Republic
  of China

Haifeng Ding
National Laboratory
  of Solid State
  Microstructures
Department of Physics
and
Collaborative Innovation
  Center of Advanced
  Microstructures
Nanjing University
Nanjing, People's Republic
  of China

Haifeng Du
High Magnetic Field
  Laboratory
Chinese Academy
  of Sciences
Hefei, People's Republic
  of China
and
Collaborative
  Innovation Center
  of Advanced
  Microstructures
Nanjing University
Nanjing, People's Republic
  of China

Motohiko Ezawa
Department of Applied Physics
University of Tokyo
Hongo, Japan

Hans Fangohr
Engineering and the
    Environment
University of Southampton
Southampton, United Kingdom

Wei Liang Gan
Division of Physics and
    Applied Physics
School of Physical and
    Mathematical Sciences
College of Science
Nanyang Techonological
    University
Singapore

Sunxiang Huang
Department of Physics
University of Miami
Coral Gables, Florida

Chiming Jin
High Magnetic Field Laboratory
Chinese Academy of Sciences
Hefei, People's Republic of
    China
and
Collaborative Innovation
    Center of Advanced
    Microstructures
Nanjing University
Nanjing, People's Republic of
    China

Mathias Kläui
Institute of Physics
Johannes Gutenberg Universität
    Mainz
and
Graduate School of Excellence
    Materials Science in Mainz
Mainz, Germany

Ping Lai
College of Physics and
    Electronic Engineering
Sichuan Normal University
Chengdu, People's Republic of
    China

Matthew Langner
Materials Science Division
Lawrence Berkeley National
    Laboratory
Berkeley, California

James Lee
Advanced Light Source
and
Materials Science Division
Lawrence Berkeley National
    Laboratory
Berkeley, California

Wen Siang Lew
Division of Physics and
    Applied Physics
School of Physical and
    Mathematical Sciences
College of Science
Nanyang Technological
    University
Singapore

Shi-Zeng Lin
Theoretical Division
Los Alamos National Laboratory
Los Alamos, New Mexico

J. Ping Liu
Department of Physics
University of Texas at
    Arlington
Arlington, Texas

Wenqing Liu
York-Nanjing Joint Center
    for Spintronics and Nano
    Engineering
School of Electronic Science
    and Engineering
Nanjing University
Nanjing, People's Republic
    of China

Hubin Luo
Key Laboratory of Magnetic
    Materials and Devices
Ningbo Institute of Materials
    Technology and Engineering
Chinese Academy of Sciences
Ningbo, People's Republic
    of China

Fusheng Ma
Division of Physics and
    Applied Physics
School of Physical and
    Mathematical Sciences
College of Science
Nanyang Techonological
    University
Singapore

Bingfeng Miao
National Laboratory of Solid
    State Microstructures
Department of Physics
and
Collaborative Innovation
    Center of Advanced
    Microstructures
Nanjing University
Nanjing, People's Republic
    of China

François Jacques Morvan
Key Laboratory of Magnetic
    Materials and Devices
Ningbo Institute of
    Material Technology and
    Engineering
Chinese Academy of Sciences
Ningbo, People's Republic
    of China

Indra Purnama
Division of Physics and
    Applied Physics
School of Physical and
    Mathematical Sciences
College of Science
Nanyang Techonological
    University
Singapore

Nian Ran
College of Physics and
    Electronic Engineering
Sichuan Normal University
Chengdu, People's Republic
    of China

Sujoy Roy
Advanced Light Source
Lawrence Berkeley National
  Laboratory
Berkeley, California

Andreas Schmid
National Center for Electron
  Microscopy (NCEM),
  Molecular Foundry
Lawrence Berkeley National
  Laboratory
Berkeley, California

Liang Sun
National Laboratory of Solid
  State Microstructures
Department of Physics
and
Collaborative Innovation
  Center of Advanced
  Microstructures
Nanjing University
Nanjing, People's Republic
  of China

Mingliang Tian
High Magnetic Field Laboratory
Chinese Academy of Sciences
Hefei, People's Republic
  of China
and
Collaborative Innovation
  Center of Advanced
  Microstructures
Nanjing University
Nanjing, People's Republic
  of China

Han Wang
Shenyang National
  Laboratory for Materials
  Science
Institute of Metal Research
Chinese Academy of Sciences
Shenyang, People's Republic
  of China

Yizheng Wu
Department of Physics
State Key Laboratory of
  Surface Physics
and
Collaborative Innovation
  Center of Advanced
  Microstructures
Fudan University
Shanghai, People's Republic
  of China

Jing Xia
School of Science and
  Engineering
The Chinese University of
  Hong Kong
Shenzhen, People's Republic
  of China

Weixing Xia
Key Laboratory of Magnetic
  Materials and Devices
Ningbo Institute of
  Materials Technology and
  Engineering
Chinese Academy of Sciences,
Ningbo, People's Republic
  of China

Yongbing Xu
York-Nanjing Joint Center
    for Spintronics and Nano
    Engineering
School of Electronic Science
    and Engineering
Nanjing University
Nanjing, People's Republic
    of China

Long Yang
York-Nanjing Joint Center
    for Spintronics and Nano
    Engineering
School of Electronic Science
    and Engineering
Nanjing University
Nanjing, People's Republic
    of China

Jiadong Zang
Department of Physics and
    Material Science Program
University of New Hampshire
Durham, New Hampshire

Xichao Zhang
School of Science and
    Engineering
The Chinese University of
    Hong Kong
Shenzhen, People's Republic
    of China

Zhidong Zhang
Shenyang National
    Laboratory for Materials
    Science
Institute of Metal Research
Chinese Academy of Sciences
Shenyang, People's Republic
    of China

Guoping Zhao
College of Physics and
    Electronic Engineering
Sichuan Normal University
Chengdu, People's Republic
    of China
and
Key Laboratory of Magnetic
    Materials and Devices
Ningbo Institute of
    Material Technology and
    Engineering
Chinese Academy of Sciences
Ningbo, People's Republic
    of China

Yan Zhou
School of Science and
    Engineering
The Chinese University of
    Hong Kong
Shenzhen, People's Republic
    of China

1. Topology of Magnetic Domains

Yingying Dai, Han Wang, and Zhidong Zhang

Chinese Academy of Sciences, Shenyang, People's Republic of China

Chapter 1

Topology, as a branch of mathematics, investigates properties that are invariant under a continuous transformation. Methods and theorems of topology, especially the homotopy theory, have been used in many fields. Topological soliton is an application of topology in condensed matter physics. In this chapter, we discuss the static and dynamical properties of different types of topological solitons based on topology.

1.1   Introduction

Topology originates from Leonhard Euler's well-known paper, the Seven Bridges of Königsberg (Shields 2012), one of the first academic treatises on topology. In this paper, Euler understood that it was impossible to cross each bridge exactly once. The question depended only on the geometry of the position, rather than on the lengths of the bridges, or on the distance between two bridges. Another interesting example in topology is the hairy ball theorem (Fulton 1995), which states that one cannot comb the hairs continuously and have all the hairs flat on a hairy ball without creating a cowlick. The result does not rely on the shape of the sphere, similar to the Bridges of Königsberg. To discern what properties these problems rely on fosters the development of topology.

Topology as an area of mathematics has developed into a means to research the properties of topological spaces that are invariant under a continuous transformation (Armstrong 1983). For example, a doughnut can be continuously deformed into a coffee cup by creating a dimple and gradually enlarging it, and then shrinking the hole into a handle. Therefore, the doughnut is topologically equivalent to the coffee mug. Topology has been developed in many fields, such as biology (Stadler et al. 2001), computer science (Mislove 1998), and physics. In condensed matter physics, the topological phase factor has been found to be closely related to many important physical phenomena (Zhang 2015), such as Aharonov–Bohm effects (Weinberg 1959), Berry phase (Berry 1984), Josephson effect (Josephson 1962), quantum Hall effects (Klitzing et al. 1980; Tsui et al. 1982; Laughlin 1983), and de Haas–van Alphen effect (de Hass and van Alphen 1930; Landau 1930). Recently, topological solitons in magnets have aroused much attention, such as vortices, merons, bubbles, and skyrmions. Investigation of the static and dynamical properties of these solitons is aimed at improving the magnetic properties of materials.

A soliton as a wave packet exists widely in nature, such as a tidal wave or typhoon. When it moves through a nonlinear dispersive medium, its shape and velocity do not change (Menzel 2011). In mathematics, a soliton is a solution to a nonlinear partial differential equation, such as the Landau–Lifshitz equation and the Yang–Mills equation. Solitons with topological protection cannot decay into the trivial solution, meaning that a topologically protected solution is not equivalent to the trivial one. There are many topological solitons in materials (Menzel 2011), such as screw dislocations in crystals, domain walls in ferromagnets, and other spin textures like vortices, merons, bubbles, and skyrmions. Manipulation of the topological solitons may lead to applications in magnetic storage and spintronics (Parkin et al. 2008; Pfleiderer and Rosch 2010; Jung et al. 2011; Zhang et al. 2015).

This chapter introduces different kinds of topological solitons existing in magnets and is organized as follows: Section 1.2 captures the essential features of topology, such as topological equivalence and invariants, surfaces and knots, winding numbers, and applications of homotopy theory in topological defects. Section 1.3 is devoted to the Gibbs free energy for a magnetic system. Sections 1.4 to 1.8 address different kinds of topological spin textures in magnetism, in particular a domain wall, vortex, meron, bubble, and skyrmion.

1.2   Topology

1.2.1   Topological Equivalence and Invariants

Topology studies the properties of topological spaces which are invariant under a continuous transformation, called topological equivalence. One kind of topological equivalence is homeomorphism. Suppose X and Y to be topological spaces. A function $f: X{\rightarrow}Y$ is called a homeomorphism when it is bijection (i.e., one-to-one and surjective), continuous, and has a continuous inverse function f^{-1} (Armstrong 1983). When such a function exists, X and Y are homeomorphic. As an example, Figure 1.1 shows the homeomorphism of a polyhedron and a sphere. For any point x on the surface of the polyhedron in Figure 1.1a, we can draw a line passing through both the point x and the center of gravity O of the polyhedron, and reach the point $f(x)$ of the sphere with its center of gravity also at O. Doing the same procedure, we can map the polyhedron to the sphere point by point, and the faces of the polyhedron project to curvilinear polygons on the sphere as shown in Figure 1.1b. Similarly, we can project the sphere to the polyhedron point by point as a continuous inverse procedure. Therefore, a polyhedron and a sphere are homeomorphic, and from a topological viewpoint they are the same.

Another example of homeomorphism is shown in Figure 1.2. In topology, these four figures are considered to be the "same space" (Armstrong 1983). A special homeomorphism from space (c) to space (d) is as follows: the points of space (c) are specified by cylindrical

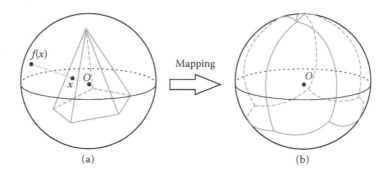

FIGURE 1.1   Continuous mapping from a regular polyhedron to a sphere. (a) The regular polyhedron with the centre of gravity at point O before mapping. (b) After mapping, the faces of the polyhedron becomes curvilinear polygons on the sphere.

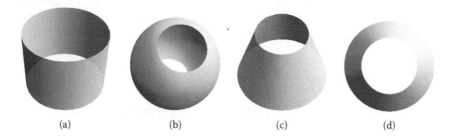

FIGURE 1.2   (a) A cylinder with finite height, excluding the two circles at the ends. (b) A sphere with two holes at the north and south poles. (c) A frustum given by the equation $z = -3\sqrt{x^2 + y^2} + 10$. (d) An annulus specified by $2 < r < 4$.

polar coordinates (r, θ, z) and those of space (d) by plane polar coordinates (r, θ). For space (c), when θ is zero, we can obtain a line $3x + z = 10$. Mapping this line to space (d) where θ is zero, we can get a line $\{ (x, y) \mid 2 < x < 4, y = 0 \}$. For each value of θ varying from 0 to 2π, doing a similar procedure in a continuous manner, we can project space (c) to space (d).

Another type of topological equivalence is called homotopy equivalence. Suppose that $f, g: X \to Y$ are maps. If there is a map $F: X \times I \to Y$ such that $F(x, 0) = f(x)$ and $F(x, 1) = g(x)$ for all points $x \in X$, f is homotopic to g (Armstrong 1983), as shown in Figure 1.3a. The equivalence class of loop f is denoted by $[f]$. The set of homotopy classes of loops in X based at x forms the fundamental group under the multiplication $[f] \circ [g] \equiv [f \circ g]$ and is written as $\pi_1(X, x)$ (Braun 2012). The fundamental group may either be Abelian or non-Abelian. Figure 1.3c shows an example of a space with a non-Abelian fundamental group (Mermin 1979). Though loops f and g are freely homotopic, they are not homotopic at the base point, because one cannot deform the loop f into the loop g while holding

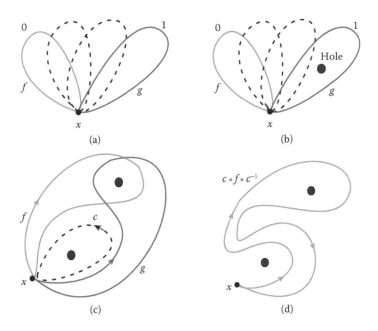

FIGURE 1.3 (a) Two loops f and g are homotopic. (b) f and g are homotopically inequivalent. (c) Loops f and g are not homotopic at the base point x even though they are freely homotopic. (d) A loop $c \circ f \circ c^{-1}$ is homotopic at x to g, where c is the dashed loop in (c).

the base point x. Figure 1.3d shows that g and $c \circ f \circ c^{-1}$ are homotopic. Since f is not homotopic at x to g, $c \circ f$ is not homotopic to $f \circ c$, thus the fundamental group is non-Abelian.

Homotopy theory is used in description and classification of topological defects. A topological defect exists when the mapping from real space S^n to an order-parameter space S^m cannot be shrunk into a single point. In other words, topological defects exist when $\pi_n(S^m) \neq 0$. The nth homotopy group of the n-sphere is isomorphic to the group of integers $\pi_n(S^n) \cong \mathbb{Z}$ for all $n \geq 1$ (Braun 2012). Therefore, topological defects corresponding to $\pi_n(S^n) \neq 0$ can be characterized by the winding number.

A topological invariant, such as a geometric property of a space, the Euler number, or a group constructed from the space, is a property of a topological space which is preserved by a homotopy (Armstrong 1983). That is, if a space X possesses a topological property, every space topologically equivalent to X possesses that property. According to the property, if a topological invariant of two spaces are calculated to be different, we can conclude that the two spaces cannot continuously be deformed into each other. In topology, a common problem is to develop ways to decide whether two topological spaces are topologically equivalent or not.

Chapter 1

1.2.2 Surfaces and Knots

Topology has to deal with the topological equivalence of spaces. Spaces with bounded configurations occurring naturally in Euclidean space are interesting, such as the unit circle, torus, Möbius strip, Klein bottle, and so on. Figure 1.4 shows some interesting examples existing in three-dimensional (3D) space. The former two spaces are orientable, which have consistent 'orientation' over the entire surfaces, while the latter two are nonorientable.

The Möbius strip, as an important two-dimensional (2D) object in topology, has only one side and one edge and therefore arouses much attention. To prove that the Möbius strip has only one side, starting from a central point of the "edges" of the Möbius strip and drawing a line along the strip's surface until you return to the starting point without lifting the pen, you will find the line is continuous, which proves that the strip has only one side. Cutting the Möbius strip along the center line yields an amazing result: a single strip with a 2π twist, rather than two separate strips. The long strip is not a Möbius strip anymore because it has two edges. Interestingly, if you cut around the Möbius strip one-third of the way from the edge, it creates two strips: one is a Möbius strip with one-third of the width and the same length of the original one, and the other is a longer non-Möbius strip with one-third of the width and twice the length of the original strip.

A drastic change such as tearing apart or gluing parts together can destroy the topological equivalence of a space (Sarma et al. 2006). For example, gluing two Möbius strips together along their edges can produce the Klein bottle, but the Möbius strip is not topologically equivalent to the Klein bottle. This procedure cannot be done in ordinary 3D Euclidean space without creating self-intersections but can be represented without crossing itself in four-dimensional (4D) space (Martin et al. 2010).

In topology, apart from the various surfaces, knots are also interesting objects. A knot is a subspace of 3D Euclidean space which is not homeomorphic to the circle (Armstrong 1983). Figure 1.5 shows a few

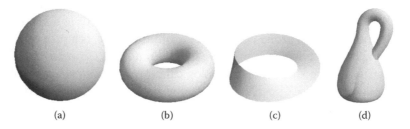

(a)   (b)   (c)   (d)

FIGURE 1.4   Some surfaces in Euclidean space: (a) sphere; (b) torus; (c) Möbius strip; and (d) Klein bottle.

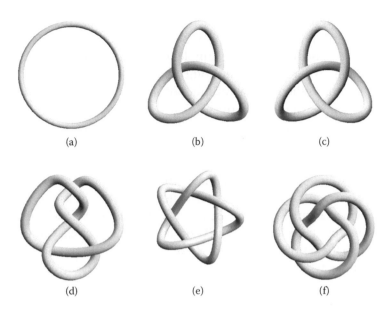

(a)   (b)   (c)

(d)   (e)   (f)

FIGURE 1.5 Different knots in three-dimensional Euclidean space: (a) unknot; (b) trefoil with left-handed chirality; (c) trefoil with right-handed chirality; (d) figure of eight; (e) pentafoil; and (f) torus knot.

examples of knots. The unit circle is called a trivial knot or unknot (Figure 1.5a). We cannot convert any of the knots in Figure 1.5b, d through f into the trivial one, nor any one into another simply by wobbling the string around (Armstrong 1983). To realize the transformation, we must cut the string and make up the knot we need, which means in some sense that these knots are all different.

Two knots k_1 and k_2 are topologically equivalent if there is a homeomorphism f of 3D Euclidean space which satisfies that $f(k_1) = k_2$ (Armstrong 1983). Reflection in a plane as a good homeomorphism can transform a knot into its mirror image, but a chiral knot is an exception which is not equivalent to its mirror image, as shown in Figure 1.5b and c. These two trefoil knots have opposite chiralities, left-handed and right-handed, which are mirror images of each other. But, these two knots are not equivalent to each other, that is, they are not amphichiral. The chirality of a knot is a knot invariant.

1.2.3   Winding Numbers

In mathematics, the winding number of a closed planar curve for a given point is the total number of times the curve travels around the point (Mcintyre and Cairns 1993), which is an integer. The sign of the

Chapter 1

winding number is related to the orientation of the curve, which is positive for the curve traveling around the point counterclockwise and negative for clockwise. As an example, Figure 1.6 shows a closed curve C traveling around the origin point O. Point A is an arbitrary point in the curve, so the winding number of the curve around the point O is calculated by the total number of turning of vector $\overrightarrow{OA}$ around the curve. Therefore, we can get the winding number $n(C, O) = 2$.

Winding numbers play an important role in many fields, such as vector calculus, geometric topology, and physics. In condensed matter physics, topological defects have aroused much attention. Taking planar spins with a field $s(r)$ of unit vector in a 2D plane as an example, as shown in Figure 1.7, suppose $s(r)$ is continuous everywhere in the plane except at the singularity P and consider any circle centered on P, then $s(r)$ is on the circle. Any possible value of $s(r)$ can be specified by an angle, which can be respected as a point on the circular order-parameter space (Mermin 1979), as shown in Figure 1.8. As a result, we can get the winding number of the planar spins around the singularity.

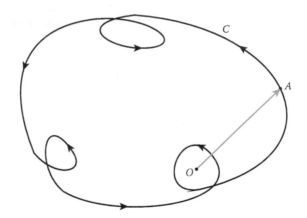

FIGURE 1.6 Winding number of a curve.

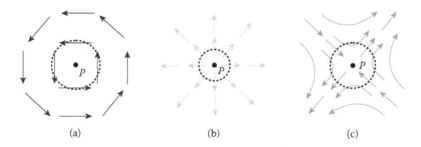

FIGURE 1.7 Planar spins in two-dimensional spaces with the winding (a) $n = +1$; (b) $n = +1$; and (c) $n = -1$.

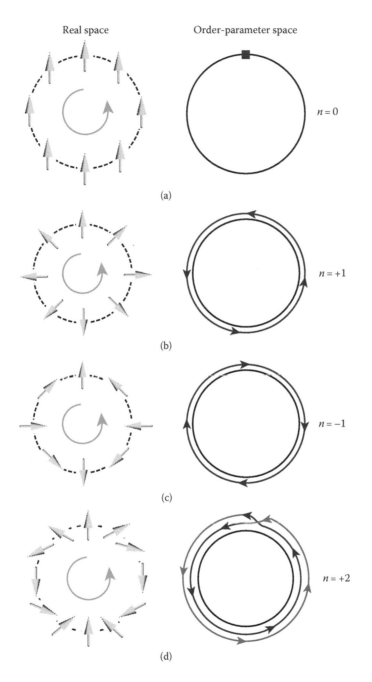

FIGURE 1.8 Spin configurations $s(r)$ on circles (left) and the maps of $s(r)$ onto the order-parameter space (right). (a) The spins are uniform and therefore mapped onto a single point in the order-parameter space. (b–d) The spins are nonuniform with the winding number of +1, −1, and +2, respectively.

The winding number n is just the number of times the mapping wraps the closed curve around the circle.

A winding number is a topological invariant which cannot be changed by a homotopy. That is, a configuration with winding number n can be continuously deformed into another one with the same winding number, but two mappings with different winding numbers cannot be continuously transformed into one other (Mermin 1979; Braun 2012). In Figure 1.7, the winding numbers of the singularities are nonzero, which means that the singularities are topologically stable, and cannot be smoothly removed without affecting the continuity in the far region, nor be obliterated by a fluctuation in the local configuration (Mermin 1979).

A pair of topological defects can be transformed into a configuration with the total net winding number, that is, a defect pair is equivalent to the configuration with the winding number equaling to the sum of the winding number of the pair (Mermin 1979). As a special case, Figure 1.9 shows that a pair of defects with opposite winding numbers is equivalent to a nonsingular configuration, which can be used to annihilate a defect in a bounded region without rearranging the order-parameter field at a large distance. For example, bringing in another stable defect with an opposite winding number from infinity and moving the two defects close to each other can eliminate a stable defect.

If the planar spins are replaced by ordinary 3D spin vectors, the order-parameter space becomes the surface of a 3D sphere instead of a circle (Braun 2012). Then the winding number, also called a topological charge or skyrmion number, of a magnetic topological spin texture is defined by (Rajaraman 1987):

$$S = \frac{1}{4\pi}\iint q\,d^2r,\; q \equiv \boldsymbol{m}\cdot\left(\frac{\partial \boldsymbol{m}}{\partial x}\times\frac{\partial \boldsymbol{m}}{\partial y}\right), \tag{1.1}$$

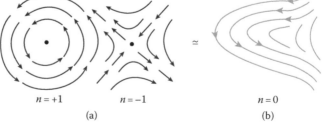

$n=+1$ $n=-1$ $n=0$

(a)   (b)

FIGURE 1.9 (a) Two planar spin defects with the winding number of +1 and −1. (b) Nonsingular configuration as a topological equivalence of (a).

where $m(r)$ is the unit vector of local magnetization and q the topological density. In polar coordinates, $r = (r\cos\varphi, r\sin\varphi)$. Using spherical coordinates and for a spin configuration with radial symmetry, $m(r)$ can be expressed as:

$$m(r) = (\cos\phi(\varphi)\sin\theta(r), \sin\phi(\varphi)\sin\theta(r), \cos\theta(r)). \tag{1.2}$$

Inserting Equation 1.2 into Equation 1.1, S is obtained as (Nagaosa and Tokura 2013):

$$S = \frac{1}{4\pi} \int_0^\infty dr \int_0^{2\pi} d\varphi \frac{d\theta(r)}{dr} \frac{d\phi(\varphi)}{d\varphi} \sin\theta(r) = \frac{1}{4\pi} \cos\theta(r) \Big|_{r=0}^{r=\infty} \phi(\varphi) \Big|_{\varphi=0}^{\varphi=2\pi}. \tag{1.3}$$

The topological charge indicates the number of times that the field configuration covers the 3D sphere of the order-parameter space. The vorticity of the configuration is defined by $m = \phi(\varphi)\big|_{\varphi=0}^{\varphi=2\pi}/2\pi$.

1.3 Magnetic Gibbs Free Energy

Micromagnetics are based on a variational principle derived from thermodynamic principles (Landau and Lifshits 1935; Kronmüller 1966). According to this principle, when the total free energy reaches an absolute or a relative minimum value under the constraint $m^2 = 1$, the vector field of magnetization $m(r) = M(r)/M_s$ is obtained (Hubert and Schäfer 1998). For a magnetic system, its total Gibbs free energy is expressed by (Kronmüller 2007):

$$E_{\text{tot}} = U - \mu_0 \int M \cdot H_{\text{ext}} \, dV, \tag{1.4}$$

where U, M, and H_{ext} denote the internal energy, local magnetization, and an external magnetic field, respectively. If the magnetoelastic energy is not considered, the internal energy includes the exchange, magnetocrystalline, surface anisotropy, asymmetric exchange, and stray field energies. Therefore, the total energy can be written as:

$$E_{\text{tot}} = E_{\text{ex}} + E_{\text{K}} + E_{\text{d}} + E_{\text{H}} + E_{\text{DM}} + E_{\text{s}}, \tag{1.5}$$

Chapter 1

where E_{ex}, E_K, E_d, E_H, E_{DM}, and E_s are the exchange, magnetocrystalline, demagnetization, Zeeman, asymmetric exchange, and surface anisotropy energies, respectively.

1.3.1   Exchange Energy

1.3.1.1   Short-Range Exchange Interactions

In the case of localized electrons with only nearest-neighbor interactions, the Hamiltonian of two nearest neighboring atoms is expressed by the Heisenberg model (Heisenberg 1928):

$$\hat{H}_{ex} = -2J S_1 \cdot S_2, \tag{1.6}$$

where J is the exchange constant between the two spins. For a ferromagnet, J is positive, and the exchange interaction makes the two spins parallel to each other; while for an antiferromagnet, J is negative, and the exchange interaction leads to the two spins antiparallel to each other. For a magnetic system, if only the exchange interactions between the nearest atoms are considered and the exchange constant is assumed the same, the Hamiltonian of the system can be written as follows:

$$\hat{H}_{ex} = -\sum_{ij} J S_i \cdot S_j, \tag{1.7}$$

where the Σ denotes a sum over the nearest neighbors only. In a continuum approximation, the discrete nature of the lattice is ignored; therefore, the exchange-interaction energy can be written by (Blundell 2001):

$$E_{ex} = A \int \left[(\nabla m_x)^2 + (\nabla m_y)^2 + (\nabla m_z)^2 \right] dV, \tag{1.8}$$

where $A = 2J S^2 z/a$, a is the nearest-neighbor distance and z the number of sites in a unit cell. For a simple cubic, $z = 1$; for a body-centered cubic, $z = 2$; for a face-centered cubic, $z = 4$.

1.3.1.2   Long-Range Exchange Interactions

Apart from the direct exchange interaction between nearest-neighbor atoms, an indirect exchange interaction is also important in the metallic ferromagnets. This indirect exchange interaction was originally developed by Zener, Ruderman, Kittel, Kasuya, and Yosida, known as an RKKY interaction (Zener 1951; Ruderman and Kittel 1954; Kasuya 1956; Yosida 1957). The indirect exchange interaction is based on the interaction between localized spins of d or f electrons and the delocalized s electrons. A localized spin moment polarizes the spins of the s electrons and the polarization of those, in turn, couples to a neighboring localized spin moment (Blundell 2001). The surface energy density of the interaction can be described by (Rührig et al. 1991; Miltat and Donahue 2007):

$$e_{coupl} = J_1(1-\boldsymbol{m}_1 \cdot \boldsymbol{m}_2) + J_2[1-(1-\boldsymbol{m}_1 \cdot \boldsymbol{m}_2)^2]. \qquad (1.9)$$

Here $\boldsymbol{m}_1$ and $\boldsymbol{m}_2$ are the magnetization vectors at the interface, and J_1 and J_2 are the bilinear and "biquadratic" coupling constants, respectively.

1.3.1.3   Asymmetric Exchange Interaction

The antisymmetric exchange interaction, also called the Dzyaloshinskii–Moriya interaction (DMI) (Dzyaloshinskii 1958; Moriya 1960), was introduced for low-symmetry crystals. The DMI is described as:

$$E_{DM} = D \int\int \boldsymbol{M} \cdot (\nabla \times \boldsymbol{M}) d^2\boldsymbol{r}, \qquad (1.10)$$

where D is the continuous effective DMI constant, which favors non-uniform magnetic structures. The DMI can also play an important role in surface or interface of ultrathin films, where the inversion symmetry is broken (Fert 1991; Bode 2007). More details about the DMI will be discussed later.

1.3.2   Magnetocrystalline Anisotropy Energy

The magnetocrystalline anisotropy energy depends on the direction of the magnetization related to the orientation of the single crystals. It originates from both the coupling between spin and orbital moments (L–S coupling) and the interaction between ions and the crystal field (Kronmüller 2007). The angular dependence of the magnetocrystalline anisotropy energy is conformed to the symmetry of the crystals. According to this principle, here are some examples of the magnetocrystalline anisotropy energy, which only considers the lowest-order terms of expansions.

1.3.2.1   Cubic Anisotropy

The basic formula for the anisotropy energy density of a cubic crystal is written as (Kronmüller 2007):

$$e_{K_c} = K_0 + K_1 \left(m_x^2 m_y^2 + m_x^2 m_z^2 + m_y^2 m_z^2 \right) + K_2 m_x^2 m_y^2 m_z^2 + ..., \qquad (1.11)$$

Chapter 1

where m_x, m_y, and m_z are the magnetization components along the cubic axes. Mostly, the material constant K_2 and higher-order terms can be ignored, and the sign of K_1 determines the easy magnetic directions (either along the [100] or the [111] directions).

1.3.2.2   Hexagonal Anisotropy

Hexagonal and tetragonal crystals have a uniaxial anisotropy, whose energy density up to fourth-order terms can be expressed by (Kronmüller 2007):

$$e_{K_u} = K_{u1} \sin^2 \theta + K_{u2} \sin^4 \theta, \tag{1.12}$$

where θ denotes the angle between the magnetization and the hexagonal or tetragonal c-axis. A large positive K_{u1} gives rise to an easy axis, while a large negative K_{u1} leads to an easy plane perpendicular to the c-axis.

1.3.2.3   Surface and Interface Anisotropy

Surface anisotropy, introduced by Néel (1953), becomes significant for very thin films and multilayers. For a cubic crystal, surface anisotropy to first approximation is described by (Brown 1940):

$$e_s = K_{s1} \left(1 - m_x^2 n_x^2 - m_y^2 n_y^2 - m_z^2 n_z^2 \right)$$
$$\tag{1.13}$$
$$- 2K_{s2} \left(m_x m_y n_x n_y + m_x m_z n_x n_z + m_y m_z n_y n_z \right),$$

where K_{s1} and K_{s2} are the anisotropy constants, and n is the surface normal. If $K_{s1} = K_{s2}$, Equation 1.13 becomes the isotropic formula:

$$e_s = K_s[1-(m \cdot n)^2]. \tag{1.14}$$

If $K_{s1} > 0$, e_s is minimum for the magnetization perpendicular to the surface. This kind of energy can also exist at interfaces between ferromagnetic and nonmagnetic materials.

1.3.3   Magnetostatic Energies

1.3.3.1   External Field (Zeeman) Energy

The magnetic field energy can be separated into two parts: the external and stray field energies. The first part is the interaction energy between the magnetization vector field and an external field, which can be expressed by:

$$E_H = -\mu_0 \int M \cdot H_{\text{ext}} \, dV. \tag{1.15}$$

When the magnetization is parallel to the external field, the external field energy is the smallest. For a uniform external field, the external field energy depends only on the average magnetization, not on the domain structure or sample shape.

1.3.3.2   Stray Field Energy

The second part of the magnetic field energy is related to the magnetic field generated by the magnetic body itself. According to Maxwell's equation $\nabla \cdot B = 0$, when the external field is zero, the stray field H_d generated by the divergence of the magnetization is defined as follows:

$$\nabla \cdot H_d = -\nabla \cdot M. \tag{1.16}$$

The stray field energy is (Hubert and Schäfer 1998):

$$E_d = \frac{1}{2}\mu_0 \int\limits_{\text{all space}} H_d^2 dV = -\frac{1}{2}\mu_0 \int\limits_{\text{sample}} H_d \cdot M\, dV. \tag{1.17}$$

When $\rho = -\nabla \cdot M$ denotes the volume charge density and $\sigma = M \cdot n$ denotes the surface charge density, where n is the outward directed surface normal, the potential energy of the stray field at position r, given by an integration over r', can be expressed as (Kronmüller 2007):

$$U(r) = \frac{1}{4\pi}\left[\int \frac{\rho(r')}{|r-r'|}dV' + \int \frac{\sigma(r')}{|r-r'|}dS'\right]. \tag{1.18}$$

The stray field can be derived from $H_d(r) = -grad\, U(r)$, which has the form (Miltat and Donahue 2007):

$$H_d(r) = \frac{1}{4\pi}\left[\int \rho(r')\frac{r-r'}{|r-r'|^3}dV' + \int \sigma(r')\frac{r-r'}{|r-r'|^3}dS'\right]. \tag{1.19}$$

The stray field energy can be obtained as (Hubert and Schäfer 1998):

$$E_d = \frac{1}{2}\mu_0\left[\int \rho(r)U(r)dV + \int \sigma(r)U(r)dS\right]. \tag{1.20}$$

The energy terms which have to be taken into account are dependent on the real problem to be solved. Usually, not all of these energy terms are considered. For instance, when treating Bloch wall, the stray and external field energies may be omitted, whereas in the treatment of soft magnetic materials, the magnetocrystalline anisotropy energy may be ignored.

Chapter 1

Competition of these different energy terms results in different topological solitons, such as domain walls, vortices, bubbles, and skyrmions.

1.4   Domain Wall

Domain wall exists in an intermediate region between two magnetic domains, where the magnetization rotates gradually from the direction in one domain to that in the other. There are different kinds of domain walls, such as Bloch walls, Néel walls, and chiral domain walls. Bloch walls are more stable in thick films, while Néel walls exist in thin films.

The most common type of 180° wall is a Bloch wall, in which the spins rotate in a plane perpendicular to the dashed line, as shown in Figure 1.10a. For an infinite sample, the Bloch wall creates no divergence of the magnetization; thus, the total energy includes the contribution from the exchange and the anisotropy energies. For a uniaxial material, the free energy of the domain wall is (Coey 2010):

$$E_{tot} = E_{ex} + E_K = \int \left[A\left(\partial\theta/\partial x\right)^2 + K_u \sin^2\theta \right] dx. \qquad (1.21)$$

If $K_u > 0$, the spins prefer to line up along $\theta = 0$ or $\theta = \pi$. Minimizing Equation 1.21 is equivalent to solving the Euler equation

$$\frac{\partial E_{tot}}{\partial \theta} - \frac{d}{dx}\left(\frac{\partial E_{tot}}{\partial \theta'} \right) = 0,$$

where $\theta' = \partial\theta/\partial x$:

$$\frac{d^2\theta}{dx^2} = \frac{K_u}{A} \sin\theta \cos\theta. \qquad (1.22)$$

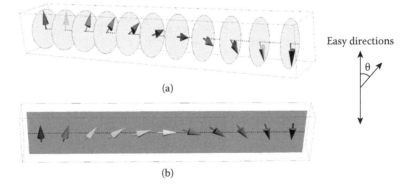

Easy directions

(a)

(b)

FIGURE 1.10   (a) A Bloch wall in which the spins rotate in a plane perpendicular to the dashed line (*x*-direction). (b) A Néel wall in which the spins rotate in the plane parallel to the dashed line.

Solving Equation 1.22 yields the domain equation:

$$x = \sqrt{\frac{A}{K_u}} \ln \tan\left(\frac{\theta}{2}\right). \tag{1.23}$$

The Bloch wall width and the wall energy per unit area are:

$$\delta_{BW} = \pi\sqrt{A/K_u}, \tag{1.24}$$

$$\sigma_{BW} = 4\sqrt{AK_u}. \tag{1.25}$$

In a Néel wall, the spins rotate in the plane parallel to the dashed line, as shown in Figure 1.10b. Consider a planar 180° Néel wall anchored in a uniaxial bulk material, the domain wall width and the wall energy per unit area are written by (Kronmüller 2007):

$$\delta_{NW} = \frac{\pi}{\left(\dfrac{K_u}{A} + \dfrac{\mu_0 M_s^2}{2A}\right)^{1/2}}. \tag{1.26}$$

$$\sigma_{NW} = 4\sqrt{A\left(K_u + \frac{1}{2}\mu_0 M_s^2\right)}. \tag{1.27}$$

Equations 1.25 and 1.27 show that the Néel wall has a larger wall energy than the Bloch wall in bulk materials. For an ultrathin film, however, the energy difference between the two domain walls decreases as the film become thinner (Tarasenko et al. 1998). For an ultrathin film with perpendicular easy axis and inversion symmetry broken along the film normal (z-axis), considering the case that the magnetization only changes along the x-direction, the total micromagnetic energy of the system is (Heide et al. 2008):

$$E[\theta(x)] = \int dx \left[A(\partial\theta/\partial x)^2 + D\,\partial\theta/\partial x + K\sin^2\theta \right], \tag{1.28}$$

where $K = K_u - 1/2\mu_0 M_s^2$ is an effective anisotropy constant, which takes into account the shape anisotropy. Solving the Euler equation $\dfrac{\partial E_{tot}}{\partial\theta} - \dfrac{d}{x}\left(\dfrac{\partial E_{tot}}{\partial\theta'}\right) = 0$ yields:

$$\frac{d^2\theta}{dx^2} = \frac{K}{A}\sin\theta\cos\theta = \frac{\sin\theta\cos\theta}{\Delta^2}, \tag{1.29}$$

where $\Delta = \sqrt{A/K}$ is the Bloch wall width parameter (Hubert and Schäfer 1998). Integrating Equation 1.29 leads to:

$$\left(\frac{d\theta}{dx}\right)^2 = \frac{C + \sin^2\theta}{\Delta^2},\tag{1.30}$$

where C is an integration constant. If D is so small that no cycloid state develops, C is zero and the domain wall width and the wall energy per unit area are (Heide et al. 2008):

$$\delta = \pi\sqrt{A/K},\tag{1.31}$$

$$\sigma = 4\sqrt{A/K} \mp \pi D,\tag{1.32}$$

Interestingly, the DMI introduces chirality to the domain wall, which depends on the sign of D [left (right) rotating walls for positive (negative) D], other than changing the shape of the domain wall when D is small. The critical DMI energy constant $D_c = 4\sqrt{A/K}/\pi$ can be obtained from Equation 1.32. In addition, a cycloid state develops with m rotating in the xz plane (Rohart and Thiaville 2013).

The winding number of a domain wall depends on its spin distribution as shown in Figures 1.11 and 1.12. A traditional domain wall, like a Bloch or Néel wall, has no winding, while a chiral domain wall carries topological charge (Menzel 2011; Chen et al. 2013), as shown in Figure 1.11. In this case, Equation 1.1 simplifies to $S = \dfrac{1}{2\pi}\displaystyle\int \frac{\partial\theta}{\partial x}dx$.

A transverse domain wall contains a pair of defects with opposite winding numbers (Kunz and Reiff 2009), as shown in Figure 1.12b. Kunz and Reiff (2009) have studied the field dependence of injecting a domain wall into nanowires. They found that the position of the topological defect with a positive winding number determined the value of the fields.

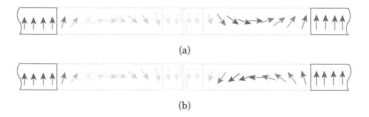

(a)

(b)

FIGURE 1.11   Schematic drawings of domain walls with the winding number equaling to (a) 0 and (b) 1.

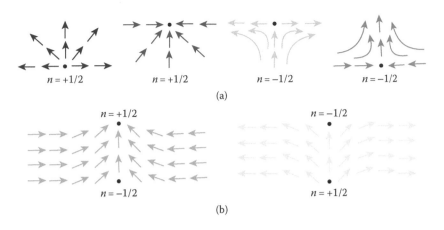

FIGURE 1.12 (a) Sketches of the spin distributions for 2D topological defects with different winding numbers (n). Edge defects have half-integer winding numbers. (b) Transverse domain walls contain a pair of defects with opposite winding numbers, resulting in the net winding number equaling to zero.

1.5   Vortex

Vortex is a topological phenomenon that widely exists in nature, such as a typhoon, nebula, and screw dislocation. Vortex as a topological spin texture also exists in 2D easy-plane magnets. Its spin configuration has a net 2π twist around a particular point or a vortex core. In magnets, there are actually two types of vortices, "in-plane" and "out-of-plane" ones, which depend on the out-of-plane spin component of the stationary vortex. The Hamiltonian of a Heisenberg model for a vortex is defined by (Wysin et al. 1994):

$$\hat{H}_{ex} = -J\sum_{(n,m)}\left(S_n \cdot S_m - \delta S_n^z S_m^z\right). \tag{1.33}$$

where J is the exchange integral, S_n (or S_m) is a 3D spin vector at site n (or m), $0 \le \delta \le 1$, and the sum is over nearest neighbors. When $\delta = 0$, Equation 1.33 corresponds to the isotropic Heisenberg model and a vortex forms with a vortex core; when $\delta = 1$, Equation 1.33 corresponds to the XY model where none of the spins has out-of-plane component, that is, a vortex forms without a core. Sketches of the two kinds of vortices are shown in Figure 1.13. According to Equation 1.3, for an "out-of-plane" vortex with the boundary conditions $m(r \to \infty) = (\cos\phi, \sin\phi, 0)$ and $m(r = 0) = (0, 0, \pm 1)$, the topological charge is calculated to be $\pm 1/2$. It indicates that the spins cover only half of the surface of a 3D unit sphere when mapping an "out-of-plane" vortex to

Chapter 1

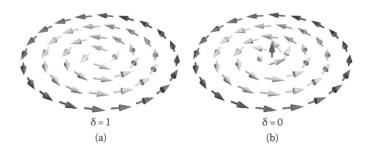

$\delta = 1$ $\delta = 0$

(a)   (b)

FIGURE 1.13 Two types of vortices: (a) "in-plane" vortex and (b) "out-of-plane" vortex.

the order-parameter space. Therefore, a vortex is half of a skyrmion which covers all the surface of the sphere. A vortex can attract an opposite-charged one and repulse a same-charged one, which can be used to reverse the vortex core (Van Waeyenberge et al. 2006; Hertel et al. 2007; Guslienko et al. 2008; Gaididei et al. 2010). When the velocity of a vortex core is higher than a critical velocity, it may create a vortex–antivortex pair. Due to their opposite topological charges, they attract each other and annihilate when they meet; then another vortex with opposite polarity is created.

Magnetic vortex has been studied for many years due to its potential applications in magnetic random access memory (MRAM). Vortex can exist stably in magnetic nanostructures, especially in submicron disk-shaped nanodots, owing to the competition between the exchange and the magnetic dipole–dipole interaction. Uhlíř et al. (2013) proposed using a magnetic vortex to realize multibit storage, that is, one bit has four possible states according to the circulation and polarity of a vortex, and thus can carry more information. To realize the multibit storage, controlling the circulation and polarity independently is of great importance. In the past 20 years, the dynamics of a vortex core has been studied intensively, especially its reversal and resonant excitations, which can be stimulated by an external magnetic field or the spin-transfer torque (Thiaville et al. 2003; Yamada et al. 2007).

The topological density distribution of a vortex is shown in Figure 1.14. The topological density is located near the vortex core, and it is zero where the position is far away from the core, so the distribution is local. In addition, on account of its small core, deformation of the vortex core can be omitted when the vortex is moving. Therefore, most of the time, a vortex can be treated as a particle without a mass

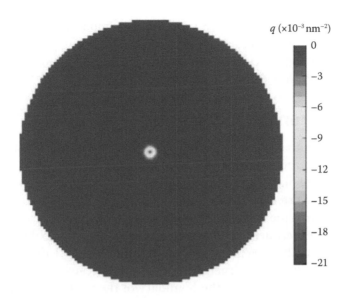

$q\ (\times 10^{-3}\,\mathrm{nm}^{-2})$

0

−3

−6

−9

−12

−15

−18

−21

FIGURE 1.14   Topological density distribution of a static vortex.

when dealing with vortex dynamics. Its motion can be described by the Thiele equation (Thiele 1973):

$$G \times \dot{X} + F - D_d \dot{X} = 0, \tag{1.34}$$

where G is the gyrovector, F a net force on the vortex, X the position of the vortex core, $\dot{X}$ the velocity of the vortex core, and D_d the damping parameter. According to Equation 1.34, the trajectory of a gyrotropic motion of a nonmass vortex is a circle. But for some special cases, we also consider a vortex with a mass. For example, when dealing with the motion of a coupled vortex pair, the mass has to be considered. In this situation, the motion of a vortex is described by the generalized Thiele equation (Cherepov et al. 2012):

$$G_3 \times \ddot{X} - M_{eff} \ddot{X} + G \times \dot{X} + F - D_d \dot{X} = 0, \tag{1.35}$$

where G_3 is a higher-order gyrovector and M_{eff} is the effective mass of the vortex. If the mass is considered, the dynamics of the vortex may show nonlinear behaviors (Ivanov et al. 2010). The nonlinearity is determined only by the potential force in Equation 1.35 but is achieved with the existence of the effective mass and the third-derivative terms.

Chapter 1

The gyrotropic motion of a vortex core is important to the applications of the vortex, because it can not only realize the reversal of the vortex core and control the circulation of the vortex but also can be used as microwave devices (Kiselev et al. 2003).

1.6 Meron

Meron as another type of topological-protect state, which exists only in pairs, was introduced as a classic solution to the Yang–Mills equation (De Alfaro et al. 1976). It has a core at which the spin is up or down (similar to a vortex), and other spins in the plane point radially outward or inward from the core, as shown in Figure 1.15. A meron pair has been discovered experimentally in a NiFe/Cr/NiFe or Co/Rh/NiFe nanodisk (Phatak et al. 2012; Wintz et al. 2013). From Figure 1.15, we can see that a meron has higher energy than a vortex, because a meron carries many magnetic dipoles in the boundary of the nanodisk which claim significant magnetostatic energy. Thus, to stabilize a meron pair in a nanodisk, another kind of energy has to be added to the system apart from the exchange and demagnetization energies, which is offered by the RKKY interaction. Phatak et al. (2012) have discovered that the antiferromagnetic interaction between the two magnetic layers plays an important role in the formation of the meron pair.

The topological charge carried by a meron can be calculated using Equation 1.3. For a meron, the boundary conditions are $m(r \to \infty) = (\cos\phi, \sin\phi, 0)$ and $m(r = 0) = (0, 0, \pm 1)$, so its topological charge is $\pm 1/2$. A meron can attract an opposite-vorticity one and repulse a same-vorticity one, which is exactly the same as a vortex in the classical XY model (Brown and Rho 2010). Interaction between two merons is related to their charges which is attractive for opposite-charged merons and repulsive for same-charged ones.

The topological density distribution of a meron is shown in Figure 1.16. The distribution is located near the core, and zero far away

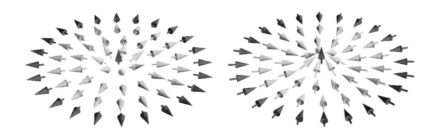

FIGURE 1.15   Two spin textures of merons.

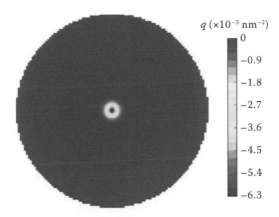

$q\ (\times 10^{-3}\ nm^{-2})$

0

−0.9

−1.8

−2.7

−3.6

−4.5

−5.4

−6.3

FIGURE 1.16   Topological density distribution of a static meron.

from the core, which is similar to the topological density distribution of a vortex (Figure 1.14). Therefore, the dynamics of a meron may have similar behaviors to that of a vortex.

1.7   Bubble

In the 1960s, bubble has been discovered in magnetic films with high perpendicular anisotropy (Malozemoff and Slonczewski 1979; O'dell 1981). Its spin distribution is shown in Figure 1.17. The magnetization in the inner region is antiparallel to that in the outer one, and in both the regions, the magnetization is out of plane. Between the inner and outer regions, there is a narrow domain wall, in which the magnetization rotates gradually from a direction parallel to one of the regions to the opposite direction. The magnetization distribution of a bubble claims large demagnetization energy due to the magnetic dipoles at the plane surface, even though its demagnetization energy is reduced compared to the uniformly out-of-plane magnetic state. Therefore, to stabilize the bubble state, the materials need to have high perpendicular anisotropy to balance the effect of the demagnetization energy and to minimize the total energy.

Figure 1.17 shows two types of bubbles: one has a topological charge of 1, and the other one has a topological charge of 0. The difference is the spin distribution at the domain wall. For the former one, the spins at the domain wall form a circle rotating around the disk center, while for the latter one, there is a pair of vertical Bloch lines (VBLs) at the domain wall (Moutafis et al. 2009). During gyrotropic motion of a bubble, the two kinds of bubbles can be transformed to each other. In a strong field, the bubble with a topological charge of 1 can change to the one with a topological

Chapter 1

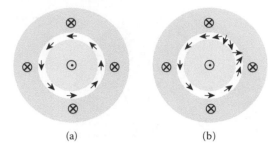

FIGURE 1.17 Sketches of spin textures of bubbles with topological charge of (a) 1 and (b) 0.

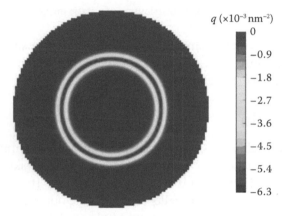

FIGURE 1.18 Topological density distribution of a magnetic bubble with topological charge of 1.

charge of 0; otherwise, under the action of an external magnetic-field gradient, the one with the charge of 0 can transform to the other one.

Topological density distribution of a bubble with topological charge of 1 is shown in Figure 1.18, which shows that the topological density is located at the domain wall with the highest value in the center of the domain wall. In yellow and green regions, the topological density is zero. Moutafis et al. (2009) have investigated the gyrotropic motion of the bubble in a magnetic nanodisk. They used two ways to describe the dynamics of the bubble. One is the mean position (X, Y) of the bubble domain which is defined by:

$$X = \frac{\int x(m_z - 1)dV}{\int (m_z - 1)dV}, \quad Y = \frac{\int y(m_z - 1)dV}{\int (m_z - 1)dV}, \tag{1.36}$$

where m_z is the magnetization component. The other is the guiding center (R_x, R_y) defined by the following equations (Papanicolaou and Tomaras 1991):

$$R_x = \frac{\int xq\,dV}{\int q\,dV}, \quad R_y = \frac{\int yq\,dV}{\int q\,dV}, \quad (1.37)$$

where q is the topological density defined in Equation 1.1. Equation 1.37 denotes the location of the nontrivial topological structure of the bubble.

Under a magnetic-field gradient, the domain wall moves a distance away from the equilibrium position, and then switching off the field, the bubble rotates around its equilibrium, called the gyrotropic motion. The trajectory of (X, Y) is a polygon, while that of (R_x, R_y) is almost a circle (Moutafis et al. 2009). Makhfudz et al. (2012) explained the polygonal trajectory by introducing an effective mass to the Thiele equation. The motion of a bubble is determined by:

$$-M_{\text{eff}}\ddot{X} + G \times \dot{X} - KX = 0, \quad (1.38)$$

where M is the effective mass and K the stiffness coefficient. Equation 1.38 has two circular modes with eigenfrequencies (Makhfudz et al. 2012):

$$\omega_{\pm} = \frac{G}{2M_{\text{eff}}} \pm \sqrt{\left(\frac{G}{2M_{\text{eff}}}\right)^2 + \frac{K}{M_{\text{eff}}}}. \quad (1.39)$$

Motion with the two modes results in the polygonal trajectory. We can see from Equation 1.39 that the eigenfrequencies depend on the effective mass M_{eff}, so changing the mass of the bubble can obtain different eigenfrequencies and thus control the gyrotropic motion of the bubble (Moon et al. 2014).

1.8   Skyrmion

Skyrmions were originally introduced by the British particle physicist Tony Hilton Royle Skyrme to describe localized, particle-like configurations in the field of pion particles (Skyrme 1962). In recent years, they have been directly observed in experiments by means of

Chapter 1

neutron scattering (Mühlbauer et al. 2009), Lorentz transmission electron microscopy (LTEM) (Yu et al. 2010, 2011), and spin-polarized scanning tunneling microscopy (SP-STM) (Heinze et al. 2011). The spin texture of a skyrmion is shown in Figure 1.19. The spin is down at the center and up at the boundary, and it rotates gradually from down to up in the intermediate regions. As we know, the exchange interaction leads to two spins parallel or antiparallel to each other, to form the canted spin texture; other kinds of interactions need to be added into the system, such as the long-ranged magnetic dipolar interaction (Dai et al. 2013), DMI (Yu et al. 2010, 2011), frustrated exchange interaction (Okubo et al. 2012), and four-spin exchange interaction (Heinze et al. 2011).

Skyrmion as a topological field configuration defines a nontrivial surjective mapping from real space to an order-parameter space, which can be obtained by projecting a hedgehog or a combed hedgehog configuration onto the plane (Everschor 2012), as demonstrated in Figure 1.19. Arrows pointing to the north pole correspond to the arrows at the boundary, pointing to the south pole correspond to the arrows in the center, and pointing to the rest of the sphere surface correspond to the intermediate regions. The reverse process indicates that a skyrmion can be mapped to the unit sphere of an order-parameter space and can cover the whole surface once, thus the topological charge is 1 (Everschor 2012). These two configurations are topologically equivalent.

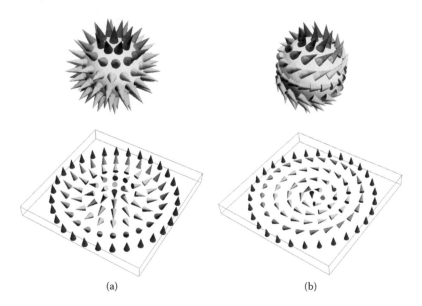

(a)   (b)

FIGURE 1.19   (a) From a hedgehog to a skyrmion configuration. (b) From a combed hedgehog to a chiral skyrmion configuration.

The topological charge carried by a skyrmion can be calculated by means of Equation 1.3. The boundary conditions of a skyrmion are $m(r \to \infty) = (0, 0, 1)$ and $m(r = 0) = (0, 0, -1)$, so its topological charge is calculated to be +1. The vorticity of a skyrmion is +1, while that of an antiskyrmion is −1. Similar to other topological spin textures, a skyrmion can be attractive to an opposite-charged one and repulsive to a same-charged one.

Figure 1.20 shows the corresponding topological density distribution of a static skyrmion in Figure 1.19. It shows that the distribution is nonlocal and has nonzero values in the whole plane, having the highest value in the center and the lowest value at the boundary. The nonlocal topological density distribution distinguishes the skyrmion from other topological spin textures with local distribution, such as vortices, bubbles, and merons. When the skyrmion is moving, the topological density distribution may obtain large deformation; therefore, skyrmion dynamics is expected to exhibit novel properties compared with others, which will be discussed in detail in chapter 10. In this case, the dynamics of the skyrmion can be described by Equation 1.35 with consideration of the effective mass. In other cases, dynamics of the skyrmion may also have similar behaviors to that of the vortex, because they are both topological solitons. For a skyrmion existing in a ferromagnetic background, its topological charge locates at the area of the skyrmion; thus, the distribution is similar to that of a vortex or a meron. Therefore, the deformation of the distribution may be omitted, and the dynamics may be determined by Equation 1.34.

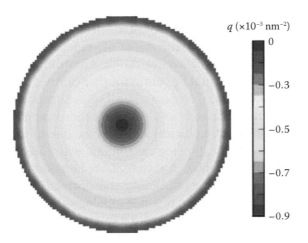

q ($\times 10^{-3}$ nm^{-2})

0

−0.3

−0.5

−0.7

−0.9

FIGURE 1.20   Topological density distribution of a static skyrmion.

Chapter 1

FIGURE 1.21   A skyrmion is composed of a meron pair with the same charge and opposite vorticity.

In addition to the above two skyrmion configurations, there are other types of skyrmions. For example, a closely bound pair of merons with the same charge and opposite vorticity can be viewed as a skyrmion, as shown in Figure 1.21 (Brown and Rho 2010). Two merons with opposite core spins can form one skyrmion (Senthil et al. 2004). Half-skyrmion spin texture can be formed in an antiferromagnetic system doped by a single hole (Brown and Rho 2010).

Note that the mapping of a vortex or a meron from real space to an order-parameter space can cover half of the unit sphere, while that of a skyrmion or a bubble covers the whole surface of the sphere. Therefore, a vortex or a meron can be regarded as half of a skyrmion, and a bubble can be continuously deformed into a skyrmion.

1.9   Summary

Topological properties are important in determining the static and dynamic properties of magnetic domains, especially the topological equivalence and the topological invariants. According to these properties, a trivial solution cannot be continuously deformed into a topological soliton and vice versa. For example, all spins that point in the same direction (ferromagnetic state as a trivial solution) cannot be continuously deformed into a topological spin texture due to the topological protection. Furthermore, topological solitons with the same topological invariant, such as a topological charge, are homotopic to each other, and otherwise cannot be continuously transformed to others. In this sense, a transverse domain wall is not topologically equivalent to any other topological spin textures. A vortex can be continuously deformed into a meron due to the same topological charge. A bubble with a topological charge of 1 can be changed to a skyrmion by a homotopy.

As for dynamics, a topological spin texture can be considered as a charged particle with or without the effective mass because of the topological protection under the continuous deformation. Topological spin textures carrying different topological charges may exhibit different dynamical properties.

Acknowledgments

The work is supported by the National Natural Science Foundation of China (Grant No. 51331006 and 51590883) and the Key Research Program of Chinese Academy of Sciences (Grant No. KJZD-EW-M05-3).

References

Armstrong MA (1983). *Basic Topology*. New York: Springer.
Berry MV (1984). Quantal phase factors accompanying adiabatic changes. *Proceedings of the Royal Society of London A: Mathematical, Physical and Engineering Sciences.* 392:45–57.
Blundell S (2001). *Magnetism in Condensed Matter.* Oxford: Oxford University Press.
Bode M, et al. (2007). Chiral magnetic order at surfaces driven by inversion asymmetry. *Nature.* 447:190–193.
Braun H-B (2012). Topological effects in nanomagnetism: From superparamagnetism to chiral quantum solitons. *Advances in Physics.* 61:1–116.
Brown GE and Rho M (2010). *The Multifaceted Skyrmion.* Hackensack, NJ: World Scientific.
Brown WF (1940). Theory of the approach to magnetic saturation. *Physical Review.* 58:736–743.
Chen G, et al. (2013). Novel chiral magnetic domain wall structure in Fe/Ni/Cu(001) films. *Physical Review Letters.* 110:177–204.
Cherepov SS, et al. (2012). Core-core dynamics in spin vortex pairs. *Physical Review Letters.* 109:097204.
Cocy JMD (2010). *Magnetism and Magnetic Materials.* Cambridge: Cambridge University Press.
Dai YY, et al. (2013). Skyrmion ground state and gyration of skyrmions in magnetic nanodisks without the Dzyaloshinskii-Moriya interaction. *Physical Review B.* 88:054403.
de Hass WJ and van Alphen PM (1930). Note on the dependence of the susceptibility of diamagnetic metals on the field. *Proceedings of Koninklijke Akademie van Wetenschappen Te Amsterdam.* 33:680–682.
De Alfaro V, Fubini S and Furlan G (1976). A new classical solution of the Yang-Mills field equations. *Physics Letters B.* 65:163–166.
Dzyaloshinskii I (1958). A thermodynamic theory of "weak" ferromagnetism of antiferromagnetics. *Journal of Physics and Chemistry of Solids.* 4:241–255.
Everschor K (2012). *Current-induced dynamics of chiral magnetic structures skyrmions, emergent electrodynamics and spin-transfer torques* (PhD diss., Universität zu Köln).
Fert A (1991). Magnetic and transport properties of metallic multilayers. *Materials Science Forum.* 59–60:439–480.

Chapter 1

Fulton W (1995). *Algebraic Topology: A First Course*. New York: Springer-Verlag.

Gaididei Y, Kravchuk VP and Sheka DD (2010). Magnetic vortex dynamics induced by an electrical current. *International Journal of Quantum Chemistry*. 110:83–97.

Guslienko KY, Lee K-S and Kim S-K (2008). Dynamic origin of vortex core switching in soft magnetic nanodots. *Physical Review Letters*. 100:027203.

Heide M, Bihlmayer G and Blügel S (2008). Dzyaloshinskii-Moriya interaction accounting for the orientation of magnetic domains in ultrathin films: Fe/W(110). *Physical Review B*. 78:140403.

Heinze S, et al. (2011). Spontaneous atomic-scale magnetic skyrmion lattice in two dimensions. *Nature Physics*. 7:713–718.

Heisenberg W (1928). Zur theorie des ferromagnetismus. *Zeitschrift fur Physik*. 49:619–636.

Hertel R, Gliga S, Fähnle M and Schneider CM (2007). Ultrafast nanomagnetic toggle switching of vortex cores. *Physical Review Letters*. 98:117201.

Hubert A and Schäfer R (1998). *Magnetic Domains: The Analysis of Magnetic Microstructures*. New York: Springer.

Ivanov BA, et al. (2010). Non-Newtonian dynamics of the fast motion of a magnetic vortex. *JETP Letters*. 91:178–182.

Josephson BD (1962). Possible new effects in superconductive tunnelling. *Physics Letters*. 1:251–253.

Jung H, et al. (2011). Tunable negligible-loss energy transfer between dipolar-coupled magnetic disks by stimulated vortex gyration. *Scientific Reports*. 1:59.

Kasuya T (1956). A theory of metallic ferro- and antiferromagnetism on Zener's model. *Progress of Theoretical Physics*. 16:45–57.

Kiselev SI, et al. (2003). Microwave oscillations of a nanomagnet driven by a spin-polarized current. *Nature*. 425:380–383.

Klitzing Kv, Dorda G and Pepper M (1980). New method for high-accuracy determination of the fine-structure constant based on quantized hall resistance. *Physical Review Letters*. 45:494–497.

Kronmüller H (1966). Magnetisierungskurve der ferromagnetika. In: *Moderne Probleme der Metallphysik* (Seeger A, ed.), pp. 24–156. New York: Springer.

Kronmüller H (2007). General micromagnetic theory. In: *Handbook of Magnetism and Advanced Magnetic Materials* (Kronmüller H and Parkin S, eds.), pp. 1–30. Chichester, UK: Wiley.

Kunz A and Reiff SC (2009). Dependence of domain wall structure for low field injection into magnetic nanowires. *Applied Physics Letters*. 94:192504.

Landau L (1930). Diamagnetismus der Metalle. *Zeitschrift fur Physik*. 64:629–637.

Landau L and Lifshits E (1935). On the theory of the dispersion of magnetic permeability in ferromagnetic bodies. *Physikalische Zeitschrift der Sowjetunion*. 8:153–169.

Laughlin RB (1983). Anomalous quantum hall effect: An incompressible quantum fluid with fractionally charged excitations. *Physical Review Letters*. 50:1395–1398.

Makhfudz I, Krüger B and Tchernyshyov O (2012). Inertia and chiral edge modes of a skyrmion magnetic bubble. *Physical Review Letters*. 109:217201.

Malozemoff AP and Slonczewski JC (1979). *Magnetic Domain Walls in Bubble Materials*. New York: Academic.

Martin S, Thompson A, Coutsias EA and Watson J-P (2010). Topology of cyclo-octane energy landscape. *The Journal of Chemical Physics*. 132:234115

Mcintyre M and Cairns G (1993). A new formula for winding number. *Geometriae Dedicata*. 46:149–159.

Menzel M (2011). *Non-collinear magnetic ground states observed in iron nanostructures on iridium surfaces* (PhD diss., Universität Hamburg).

Mermin ND (1979). The topological theory of defects in ordered media. *Reviews of Modern Physics*. 51:591–648.

Miltat JE and Donahue MJ (2007). Numerical micromagnetics: Finite difference methods. In: *Handbook of Magnetism and Advanced Magnetic Materials* (Kronmüller H and Parkin S, eds.). Chichester, UK: Wiley.

Mislove MW (1998). Topology, domain theory and theoretical computer science. *Topology and Its Applications*. 89:3–59.

Moon K-W, Chun BS, Kim W, Qiu ZQ and Hwang C (2014). Control of skyrmion magnetic bubble gyration. *Physical Review B*. 89:064413.

Moriya T (1960). Anisotropic superexchange interaction and weak ferromagnetism. *Physical Review*. 120:91–98.

Moutafis C, Komineas S and Bland JAC (2009). Dynamics and switching processes for magnetic bubbles in nanoelements. *Physical Review B*. 79:224429.

Mühlbauer S, et al. (2009). Skyrmion lattice in a chiral magnet. *Science*. 323:915–919.

Nagaosa N and Tokura Y (2013). Topological properties and dynamics of magnetic skyrmions. *Nature Nanotechnology*. 8:899–911.

Néel L (1953). L'anisotropie superficielle des substances ferromagnétiques. *Journal de Physique et le Radium*. 237:1468–1470.

O'dell TH (1981). *Ferromagnetodynamics: The Dynamics of Magnetic Bubbles, Domains, and Domain Walls*. New York: Wiley.

Okubo T, Chung S and Kawamura H (2012). Multiple-q states and the skyrmion lattice of the triangular-lattice Heisenberg antiferromagnet under magnetic fields. *Physical Review Letters*. 108:017206.

Papanicolaou N and Tomaras TN (1991). Dynamics of magnetic vortices. *Nuclear Physics B*. 360:425–462.

Parkin SSP, Hayashi M and Thomas L (2008). Magnetic domain-wall racetrack memory. *Science*. 320:190–194.

Pfleiderer C and Rosch A (2010). Condensed-matter physics: Single skyrmions spotted. *Nature*. 465:880–881.

Phatak C, Petford-Long AK and Heinonen O (2012). Direct observation of unconventional topological spin structure in coupled magnetic discs. *Physical Review Letters*. 108:067205.

Rajaraman R (1987). *Solitons and Instantons*. Amsterdam: North-Holland.

Rohart S and Thiaville A (2013). Skyrmion confinement in ultrathin film nanostructures in the presence of Dzyaloshinskii-Moriya interaction. *Physical Review B*. 88:184422.

Rührig M, et al. (1991). Domain observations on Fe-Cr-Fe layered structures. Evidence for a biquadratic coupling effect. *Physica Status Solidi (a)*. 125:635–656.

Ruderman MA and Kittel C (1954). Indirect exchange coupling of nuclear magnetic moments by conduction electrons. *Physical Review*. 96:99–102.

Sarma SD, Freedman M and Nayak C (2006). Topological quantum computation. *Physics Today*. 59:32–38.

Senthil T, Vishwanath A, Balents L, Sachdev S and Fisher MPA (2004). Deconfined quantum critical points. *Science*. 303:1490–1494.

Shields R (2012). Cultural topology: The seven bridges of Königsburg, 1736. *Theory, Culture & Society*. 29:43–57.

Skyrme THR (1962). A unified field theory of mesons and baryons. *Nuclear Physics*. 31:556–569.

Stadler BMR, Stadler PF, Wagner GP and Fontana W (2001). The topology of the possible: Formal spaces underlying patterns of evolutionary change. *Journal of Theoretical Biology*. 213:241–274.

Chapter 1

Tarasenko SV, Stankiewicz A, Tarasenko VV and Ferré J (1998). Bloch wall dynamics in ultrathin ferromagnetic films. *Journal of Magnetism and Magnetic Materials.* 189:19–24.

Thiaville A, García JM, Dittrich R, Miltat J and Schrefl T (2003). Micromagnetic study of Bloch-point-mediated vortex core reversal. *Physical Review B.* 67:094410.

Thiele AA (1973). Steady-state motion of magnetic domains. *Physical Review Letters.* 30:230–233.

Tsui DC, Stormer HL and Gossard AC (1982). Two-dimensional magnetotransport in the extreme quantum limit. *Physical Review Letters.* 48:1559–1562.

Uhlíř V, et al. (2013). Dynamic switching of the spin circulation in tapered magnetic nanodisks. *Nature Nanotechnology.* 8:341–346.

Van Waeyenberge B, et al. (2006). Magnetic vortex core reversal by excitation with short bursts of an alternating field. *Nature.* 444:461–464.

Weinberg S (1959). Interference effects in leptonic decays. *Physical Review.* 115:481–484.

Wintz S, et al. (2013). Topology and origin of effective spin meron pairs in ferromagnetic multilayer elements. *Physical Review Letters.* 110:177201.

Wysin GM, Mertens FG, Volkel AR and Bishop AR (1994). Mass and momentum for vortices in two-dimensional easy-plane magnets. In: *Nonlinear Coherent Structures in Physics and Biology* (Spatschek KH and Mertens FG, eds.), New York: Plenum.

Yamada K, et al. (2007). Electrical switching of the vortex core in a magnetic disk. *Nature Materials.* 6:270–273.

Yosida K (1957). Magnetic properties of Cu-Mn alloys. *Physical Review.* 106:893–898.

Yu XZ, et al. (2010). Real-space observation of a two-dimensional skyrmion crystal. *Nature.* 465:901–904.

Yu XZ, et al. (2011). Near room-temperature formation of a skyrmion crystal in thin-films of the helimagnet FeGe. *Nature Materials.* 10:106–109.

Zener C (1951). Interaction between the d shells in the transition metals. *Physical Review.* 81:440–444.

Zhang X, et al. (2015). Skyrmion-skyrmion and skyrmion-edge repulsions in skyrmion-based racetrack memory. *Scientific Reports.* 5:7643.

Zhang ZD (2015). Magnetic structures, magnetic domains and topological magnetic textures of magnetic materials. *Acta Physica Sinica.* 64:067503.

2. Experimental Observation of Magnetic Skyrmions

Hubin Luo and Weixing Xia
Chinese Academy of Sciences, Ningbo, People's Republic of China

Haifeng Du
Chinese Academy of Sciences, Hefei, People's Republic of China

J. Ping Liu
University of Texas at Arlington, Arlington, Texas

Chapter 2

The magnetic interactions in a magnet, combined with the external magnetic field, result in different types of magnetic domains. Among them, the magnetic skyrmion is the newly found magnetic configuration that is promising in various applications. Knowing why the magnetic skyrmions occur and how they can be manipulated is critical to the applications, which requires state-of-the-art observation techniques. This chapter tries first to establish the connections between different types of magnetic domains by referring to their formation backgrounds and spatial configurations, and then introduces the observation techniques that have been used to study them. More attention is paid to the Lorentz transmission electron microscopy (TEM) due to its ability to image magnetic skyrmions in real space. The theoretical origin of the phase stability of magnetic skyrmions is then introduced. Finally, the experimental observations of magnetic skyrmions using both neutron scattering and Lorentz TEM are reviewed.

2.1   Introduction

Magnetic skyrmions are vortex-like quasiparticles in which the magnetic moments rotate coherently and chirally across the center. Thus, a magnetic skyrmion is a special case of magnetic domains. Experimental observation of magnetic domains is based on the magnetic polarization in the materials by using different techniques such as the magneto-optical Kerr effect, magnetic force microscopy (MFM), neutron scattering, and Lorentz TEM. The methods based on optical and magnetic force mechanisms find their applications in the observation of many conventional domains. However, the chiral feature of magnetic skyrmions requires sophisticated techniques to reveal the detailed arrangements of the magnetic moments within. To date, the most frequently adopted technique to identify the magnetic skyrmions is Lorentz TEM.

Most of the above-mentioned methods provide images of various magnetic domains. Besides the direct "seeing" of the images in real space, magnetic skyrmions can be also observed by neutron diffraction and soft X-ray diffraction (see Chapter 3). The neutron scattering can give a skyrmion lattice in the reciprocal space. In 2009, the first evidence of a magnetic skyrmion lattice in MnSi was obtained by Mühlbauer et al. using small-angle neutron scattering.[1] A real-space image of the magnetic skyrmions was first observed using Lorentz TEM in 2010 by Yu et al.[2] Since then, the study of magnetic skyrmions has drawn tremendous attention, especially in the skyrmion stability[3–8] and their manipulation.[9–12] In view of the importance of magnetic skyrmion observations, we will first introduce briefly the widely investigated conventional magnetic domain structures and the methods to observe them. We will then focus on the techniques and their applications to the identification of magnetic skyrmions. Based on these, we will go further from a theoretical point of view to explain the origin and stability of the skyrmion phases. At last, the experimental observations of magnetic skyrmions in different magnetic materials using neutron scattering and Lorentz TEM are reviewed.

2.2 Magnetic Domains

Generally, the existence of magnetic domains is due to the competition of related magnetic energies, including magnetostatic energy. To reach the energetically preferred demagnetized state, bulk or thin film ferromagnetic materials with a certain size usually contain magnetic domains so that the overall magnetic energy can be reduced to the minimum. Since a demagnetization process is closely related to the shape of the sample and also the internal microstructures, determination of the process is a formidable task in most cases. In addition, the interplay of the demagnetization effect and, e.g., external field, exchange interaction, and spin–orbit coupling, may cause abundant microscopic mechanisms of the formation of magnetic domains. Usually, the occurrence of magnetic domains is characterized by the formation of domain walls that fall into different types.

The well-known domain walls are Bloch wall and Néel wall. In the Bloch wall, the magnetic moments gradually rotate in the plane parallel to the wall, whereas in the Néel wall, the magnetic moments rotate in the plane normal to the wall, as shown in Figure 2.1a. The latter usually occurs in magnetic thin films because the magnetic moments tend to lie in the film plane to lower the magnetostatic energy. It is worth mentioning that the angle between the magnetic moments in the two adjacent domains is not unique. It can be 180° or less than 180° (e.g., 90° domain wall), especially in soft magnetic materials. However, in magnetic materials with a high uniaxial magnetocrystalline anisotropy, the 90° domain wall is not preferred due to its high energy and only 180° domain walls are allowed.

The common feature of the above-mentioned domain walls is that they have net magnetic charge. In some magnetic nanostructures, e.g., films or nanowires, there also exist magnetic charge-free domain walls. A typical example is the magnetic vortex, as shown in Figure 2.1b, in which the stray field of the domain wall is absent. The vortex core is, in a sense, a domain but has only one magnetic moment. This kind of

Chapter 2

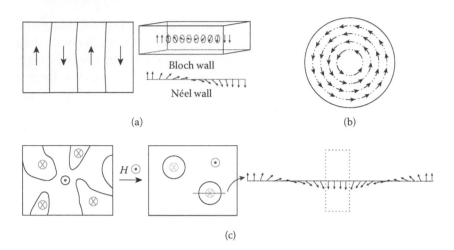

FIGURE 2.1 Different domain structures: (a) conventional domains with Bloch walls or Néel walls; (b) the magnetic vortex; and (c) the magnetic bubble structure evolved from conventional domains.

magnetic charge–free domain walls is usually found in magnetic films somewhat thicker than that with a conventional Néel wall.

If an external field is applied to a demagnetized magnetic material to favor one domain magnetization, e.g., the case in Figure 2.1c, the size of the domains will change. A possible result is that the preferred domain is enlarged while the inversely magnetized areas shrink to form cylindrical domains. They are the so-called magnetic bubbles. Supposing the domain walls are all Néel-type, the moment directions across the domain wall of the bubble may rotate clockwise or counterclockwise. The size of the bubble is not constant but can be changed by an external field. It should be noted that inside the bubble domain, the moments do not rotate. If the bubble further shrinks, as an extreme case, only domain walls exist and the moments continuously rotate across the domain walls (cf. Figure 2.1c). Then, it becomes the chiral bubble-like skyrmion.

2.3 Basic Concepts of Experimental Techniques for Magnetic Domain Observations

The magnetic polarization can interact with an electric or magnetic field, as well as certain matter and electromagnetic waves like neutron and X-ray beams. These kinds of interactions form the fundamental mechanisms of different techniques for magnetic domain imaging or diffraction. In this section, an introduction is given to several frequently used techniques.

2.3.1   Magneto-Optical Kerr Effect

Magneto-optical effects originate from the interaction between the light and the magnetic field in a magnetic material, which include the Faraday effect and Kerr effect. The former is restricted to transparent materials since it involves refractive behavior, while the latter takes advantage of the reflected light in the condition that the surface of the material is smooth enough, and thus is widely used. Considering a linearly polarized incident light, the electric field interacts with the magnetization and finally results in a Lorentz force acting on the vibrational motion of the light: $-M \times E$. This may cause a rotation of the vibration plane of the electric field which can be detected in the reflected polarized light. The typical geometric relations of the incident beam and magnetization are schematically shown in Figure 2.2.

It is clear that, for the case in Figure 2.2a, the rotation angle α increases when θ decreases, and reaches the maximum at perpendicular incidence. On the contrary, for the case in Figure 2.2b, α decreases when θ decreases and the magneto-optical rotation vanishes at a perpendicular incidence. Although the Lorentz force in the case in Figure 2.2c is not zero, its vector lies in the vibration plane of the electric field of polarized light, and thus no magneto-optical rotation will be detected. It is easy to find that, if we rotate E for 90° around the incident direction, the magneto-optical effects can still be detected for the cases in Figure 2.2a and b but cannot be detected for that in Figure 2.2c. In addition, the magneto-optical rotation angle is proportional to the magnetization if other conditions are kept unchanged, which can

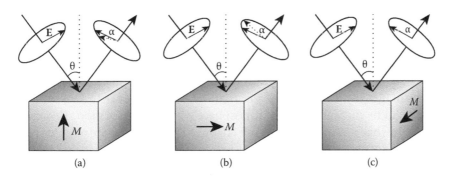

(a)   (b)   (c)

FIGURE 2.2 The schematic pictures of the Kerr effect that may result in rotation of the vibration plane of the polarized light depending on the orientation of the magnetization. The magnetization is (a) normal to the surface plane, (b) parallel to both the surface plane and vibration planes, and (c) parallel to the surface plane but normal to the vibration plane.

Chapter 2

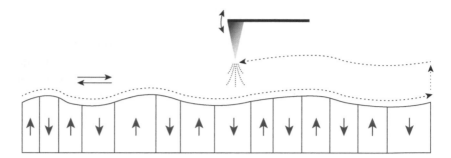

FIGURE 2.3 Two scanning runs are required in the magnetic force microscopy. The first records the surface profile and the second records the magnetic domains.

be described as $\alpha = kM$. The detected magneto-optical rotation thus differs from domain to domain, which gives the information of the patterns of magnetic domains.

2.3.2 Magnetic Force Microscopy

The MFM takes advantage of both the short-range interatomic and long-range magnetostatic interactions between the sharp probe and the magnetic material surface. A schematic picture of this technique is shown in Figure 2.3. There are two stages for probing the magnetic domains. In the first scanning run, the probe intermittently "knocks" the surface and measures the strong repulsive force due to the short-range interatomic interaction between the probe tip and the surface. This scanning run records the landscape of the surface, which is actually the atomic force microscopy. In the second scanning run, the magnetic probe is lifted to a certain height to avoid the short-range repulsive force and moves according to the obtained surface profile. The long-range magnetostatic interaction due to the existence of the stray field from the domains will cause variations of the vibration phase and magnitude of the probe. By analyzing these variations, the domain pattern can be reconstructed. The advantages of this technique are that the sample is easy to be prepared and can be measured in the atmosphere with high resolution. However, the drawback is that the scanning is slow. MFM is not only useful to measure conventional domain, recent investigation shows that it can also be used to measure the magnetic skyrmions.[13]

2.4   Neutron Diffraction

Neutron diffraction is a powerful tool to study the magnetic structures of magnetic materials. Neutrons have no electric charge but carry spins, which results in a weak magnetic moment of about 0.001 μ_B.[14] The neutron scattering originates from two sources. One is the scattering due to the collision with nuclei, and the other is caused by magnetic dipole interactions between neutrons and electrons. Within the reciprocal space, the former gives the information resembling that from conventional X-ray diffractions, whereas the latter provides the scattering information of the magnetic structure.

As shown in Figure 2.4a, the neutron diffraction happens when the Bragg equation is satisfied, i.e., $2d \sin \theta = n\lambda$, in which λ is the neutron wave length and $n\lambda$ is an integral multiple of wave length representing the difference in the path length of waves. To make the equation in the form with a single wave length, we rewrite it as:

$$2(d/n)\sin\theta = 2d_{HKL} \sin\theta = \lambda. \tag{2.1}$$

The (HKL) may not be real atomic planes. Supposing that $\mathbf{S}$ and $\mathbf{S}'$ are the unit vectors for the incident and scattered waves, respectively, it is easy to obtain the relation

$$|\mathbf{S}' - \mathbf{S}| = 2\sin\theta = \lambda / d_{HKL} \quad \text{or} \quad \frac{2\pi}{\lambda}|\mathbf{S}' - \mathbf{S}| = \frac{2\pi}{d_{HKL}}. \tag{2.2}$$

Let $2\pi\mathbf{S}/\lambda$ be $\mathbf{k}$ and $2\pi\mathbf{S}'/\lambda$ be $\mathbf{k}'$, representing the wave vectors of incident and scattered neutrons, respectively. Note that $2\pi/d_{HKL}$ is the norm of a reciprocal vector, $\mathbf{G} = H\mathbf{a}^* + K\mathbf{b}^* + L\mathbf{c}^*$, which corresponds

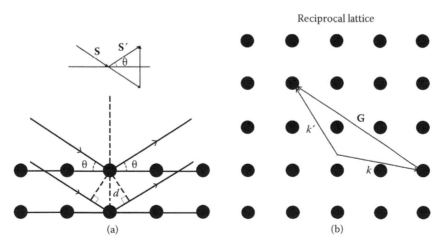

FIGURE 2.4   The requirement of neutron diffraction presented in two ways (in real and reciprocal spaces): (a) $2d \sin\theta = n\lambda$ and (b) $\mathbf{k}' - \mathbf{k} = \mathbf{G}$.

Chapter 2

to the (*HKL*) planes (**G** is normal to these planes). Hence, we arrive at a simple general form of the requirement for the diffraction, $\mathbf{k'} - \mathbf{k} = \mathbf{G}$, which is schematically shown in Figure 2.4b.

The diffraction originating in neutron scattering by nuclei carries the information of crystal structure. In contrast, the diffraction due to the neutron scattering by magnetic moments reflects the magnetic structure of the sample. For a ferromagnetic material, the diffractions due to the magnetic scattering coincide with those from nucleus scattering, and the magnetic structure is also the same as the crystal structure. However, for antiferromagnetic or ferrimagnetic structures, the diffractions due to magnetic scattering cause additional peaks because in these cases magnetic superlattices form based on the crystal structures. To separate the magnetic diffractions from the nucleus diffractions, we can obtain two diffraction patterns of the material (below and above the Curie/Néel temperature). The difference between them is the diffraction information of the magnetic structure. If polarized neutrons are used, one can also measure the spin-flip scattering to obtain the magnetic information of the materials.[14,15]

For the magnetic skyrmion lattice considered in this chapter, the magnetic unit cell is usually much larger than the crystal unit cell, because skyrmions are chirally formed in the mesoscopic scale. Taking the two-dimensional (2D) hexagonal skyrmion lattice as an example, the corresponding reciprocal lattice is also hexagonal, as depicted in Figure 2.5a. If the incident neutron beam is normal to the lattice plane, then the requirement of magnetic diffractions can be at least approximately satisfied, as drawn schematically in Figure 2.5b. According to the diffraction mechanism, observation of skyrmions using neutron scattering is feasible only when they form a lattice. This limitation does not exist if a real-space observation technique is adopted.

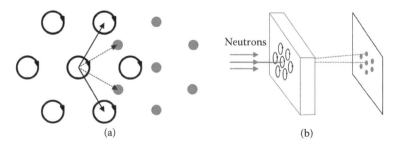

FIGURE 2.5   A hexagonal skyrmion lattice and its reciprocal lattice (solid symbols) (a) and the schematic picture for a neutron scattering setup for the observation of a skyrmion lattice (b).

2.5   Lorentz Transmission Electron Microscopy

For the characterization of topological spin structures—magnetic skyrmions—Lorentz TEM is an effective method. The skyrmion phase was predicted long time ago in chiral magnet[1] and was differentiated with helical and conical spin textures by using the neutron scattering method. Like X-ray diffraction, neutron scattering provides the reciprocal information of skyrmions. However, a real-space characterization is important to understand the "real" spin configuration, the formation condition, and the physical characters of magnetic skyrmions. Since the first observation of skyrmion textures in real space by Yu et al. using Lorentz TEM, more aspects of skyrmions in different materials and dimensions as well as their potential applications have been investigated.[2,4,9,16–21] In most materials, the formation of a skyrmion needs two basic experimental conditions, external magnetic field and low temperature, which can be easily satisfied by Lorentz TEM. In this chapter, the principle of Lorentz TEM and the experimental method to observe skyrmions are introduced.

2.5.1   Lorentz TEM Observation of Magnetic Domains

2.5.1.1   180° and 90° Domain Walls

When an electron passes through an area with a magnetic field, it is influenced by the Lorentz force obeying the right-hand rule. The magnetic domain of a ferromagnetic material can be considered as a kind of magnetic field acting on electrons; therefore, Lorentz microscopy can be used to observe the domain structure. The Lorentz force by a magnetic domain can be expressed by $F = -e(\vartheta \times M)$, where e is the elementary electrical charge, ϑ is the electron velocity, and M is the magnetization vector. Because the direction of electron motion is along the TEM column, i.e., perpendicular to the specimen plane, only the horizontal component of magnetization can be detected by Lorentz TEM. Through a more detailed calculation,[22] the reflection angle, θ, of electron due to Lorentz force caused by a magnetic domain with thickness t and horizontal magnetization M_s, is, $\theta = \dfrac{et}{m\vartheta} M_s$, where m is the electron mass at velocity ϑ. For an iron film with a thickness $t = 50$ nm and saturation magnetization $M_s = 2.1$ T, the deflection angle is 0.1 mrad for 100 keV electron microscope, which is far smaller than Bragg angles of crystalline diffraction.

Figure 2.6 schematically shows the deflection of an electron beam when it passes through domain walls. W1 and W2 are two 180° domain walls, and the directions of magnetization in both sides of walls are

Chapter 2

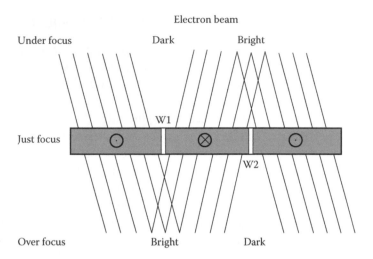

FIGURE 2.6 Schematic of electron deflection and Lorentz TEM imaging.

perpendicular to paper. The figure does not show the actual position of the specimen in TEM, it represents the image position in an imaging system. At over-focus state, the positions of domain walls emerge as dark lines due to the beam divergence or bright lines due to convergence. While at under-focus state, the dark lines become bright lines, and vice versa.

Figure 2.7 shows the closure domain structure in a cross section of perpendicular $Y_3Fe_5O_{12}$(YIG) thin film.[23] Two typical domain walls, an 180° wall and a 90° wall, are clearly observed in (b) and (c), respectively, which are confirmed by the lines of magnetic flux (d) obtained by electron holography technology.[24] The closure domain structure decreases the demagnetizing energy in comparison with the stripe domain structure with 180° walls only. Low perpendicular magnetic anisotropy is necessary for forming closure domains. Figure 2.8 shows the changes in domain structure observed by in situ Lorentz TEM. With increased magnetic field applied in TEM in a horizontal direction, the domain with a magnetization direction the same as a magnetic field enlarges and, finally, the magnetization is saturated in the whole specimen, whereas the period of the closure domains keeps constant before the final saturation.

2.5.1.2 Magnetic Vortex

If the magnetization directions between two adjacent domains have no large angle, the difference in Lorentz forces by two domains is not large enough to form contrast in Lorentz image and the domain walls cannot

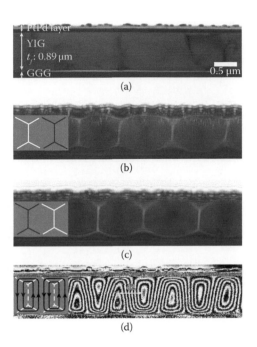

FIGURE 2.7 (a) Bright-field image, Lorentz micrographs for (b) over-focus and (c) under-focus condition, and (d) holography image indicating the lines of magnetic flux. On the left-hand sides of (b) and (c), schematic diagrams are added to show domain walls with white and black lines; the corresponding lines of magnetic flux are depicted in (d).

be observed; on the other hand, in the case of the circular direction of magnetization in a magnetic vortex, the electron beam converges to a bright spot (or diverges to a dark spot) at the center of the vortex in Lorentz images.[25] Figure 2.9 shows the vortex structure in a fabricated disk of amorphous FeSiB soft magnetic material by a focused ion-beam technique. Figure 2.9a is the bright-field image at just-focus state. In Lorentz images Figure 2.9b and c, dark and bright spots are observed, respectively. Figure 2.9d shows the magnetic flux lines obtained by electron holography, indicating that only the core area has the magnetization perpendicular to the disk, which is schematically shown in Figure 2.9e.

2.5.1.3 Magnetic Bubbles

In TEM, the image of the specimen is formed and amplified by a magnetic lens. Generally, there exists a fixed magnetic field of several Teslas from an object lens with a direction perpendicular to the specimen plane, which saturates the magnetization of the specimen. This kind of

Chapter 2

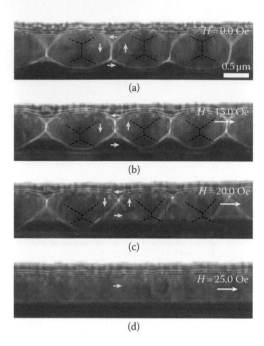

FIGURE 2.8 In situ observation of the magnetization process in Lorentz imaging when applying a horizontal magnetic field, indicated by white arrows; (a) is for zero magnetic field; (b), (c), and (d) show domain walls for 15.0, 20.0, and 25.0, respectively. White lines indicate domain walls in under-focus condition. Black lines are used to emphasize domain walls in black contrasts. Arrows in the middle indicate the magnetization direction of domains.

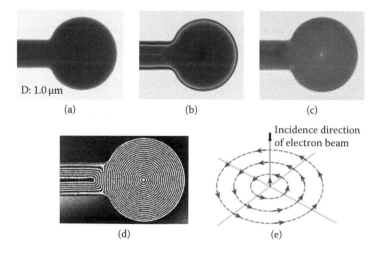

FIGURE 2.9 (a) Just-focus, (b) over-focus, and (c) under-focus Lorentz images of a magnetic disk. (d) Reconstructed phase image obtained by electron holography. (e) Schematic of spin structure of magnetic vortex.

saturation should be avoided in order to observe the domain structure at a demagnetized state. In most cases of in situ observation of magnetic domain dynamics under a magnetic field, the perpendicular field of object lens is made approximately zero, and a horizontal field parallel to the specimen plane is used in order to avoid the influence of a demagnetization field. On the other hand, in the observation of a magnetic bubble and skyrmion, an alterable perpendicular field is required. Therefore, the TEM should work in Lorentz mode when observing magnetic bubbles and skyrmions. In this mode, operation is done by first switching off the objective lens and using an additional Lorentz lens to keep the TEM at a high magnification, and then applying on the objective lens a variable voltage to produce a magnetic field on the specimen, which is generally much lower than that of a standard observation condition. The low-temperature condition for stable bubbles and skyrmions can be easily realized by a cooling TEM holder with liquid nitrogen or helium.

A magnetic bubble is a stable magnetization region in a magnetic thin film with its direction perpendicular to the film plane. Magnetic bubbles were originally found in some single-crystal garnet materials, of which the magnetic anisotropy is perpendicular to the film plane and the anisotropic energy is larger than the demagnetization energy $2\pi M_s^2$. The formation, motion, and annihilation of magnetic bubbles can be observed by a magneto-Kerr microscope or Lorentz TEM. Generally, some snaky stripe domain structures are observed at the demagnetized state with half magnetization orienting upwards and half downwards. With an increasing magnetic field perpendicular to the film plane, the region with a magnetization direction parallel to the field direction enlarges and the region with opposite direction shrinks to magnetic bubbles. There exist magnetic bubble walls in the edge of bubbles, where the magnetization direction in the center of the bubble wall lies in the film plane, which makes the electrons deflect and form contrast in Lorentz TEM. So far, a few types of ferrimagnetic iron-garnets $R_3(Fe, M)_5O_{12}$ (M: Al, Ga, In) and spin-canted orthoferrites $RFeO_3$ (R: rare earth element) have been reported to produce magnetic bubbles. Recently, magnetic bubbles were found in a ferromagnetic multiorbital Mott insulator $La_{7/8}Sr_{1/8}MnO_3$.[26] In the 1/8 Sr compound, the ferromagnetic metallic phase transits to the ferromagnetic insulator phase at 150 K. Figure 2.10 shows the dynamic process of a magnetic bubble of $La_{7/8}Sr_{1/8}MnO_3$ in a ferromagnetic insulator phase at 100 K. With a continuously changing magnetic field, the width of the snaky domain gradually changes, and nanoscale elliptical magnetic

Chapter 2

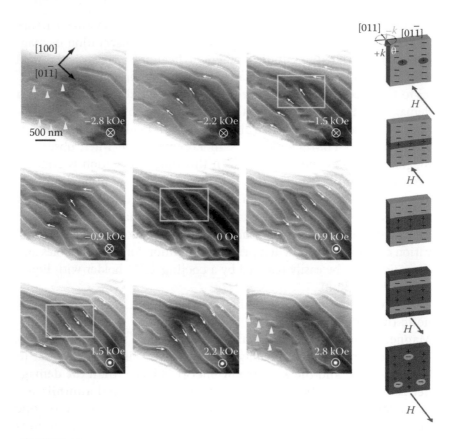

FIGURE 2.10 Lorentz images of $La_{7/8}Sr_{1/8}MnO_3$ at 100 K for magnetic fields changing from −2.8 to 2.8 kOe. The domains indicated by arrows are deteriorated by the fields, resulting in nanoscale elliptical magnetic bubbles shown by triangles. The inset shows a schematic of the formation of magnetic bubbles. (From Nagai, T., et al., *Appl. Phys. Lett.*, 101, 162401, 2012. With permission.)

bubbles are observed with fields of −2.8 and 2.8 kOe. The inset in Figure 2.10 schematically shows the magnetization directions in bubbles and their formation and annihilation.

2.5.2   Transport of Intensity Equation

In order to determine the direction of magnetization and hence the spin structure of a ferromagnetic sample by using Lorentz TEM images, a method named transport-of-intensity equation (TIE) is generally adopted, which is briefly introduced as follows. When a plane wave of electron passes through a TEM specimen, the propagation direction is varied by the interaction between electron wave and specimen, and the image intensity is produced. The changes in image intensity include both the amplitude and phase information of the transmitted electron wave. TIE can be used to detect the phase distribution from the intensity distribution as follows:[27]

$$\frac{2\pi}{\lambda}\frac{\partial I(xyz)}{\partial z} = \nabla_{xy}\left[I(xyz)\nabla_{xy}\phi(xyz)\right], \tag{2.3}$$

where $I(xyz)$ and $\phi(xyz)$ are the intensity and phase distributions of transmitted electron wave, respectively. The phase shift of electron wave corresponds to the integration of electrical potential and magnetic vector potential along the propagating path, which is mainly contributed from the specimen, and the contribution from vacuum area can generally be neglected. $\nabla_{xy}\phi(xyz)$ stands for the differentiation of the phase of an electron beam in a specimen plane, which represents the sum of electrical potential and magnetization of the specimen. The electrical potential corresponds to the inner potential of the specimen which is related to the specimen composition and thickness. In the case of uniform composition and thickness, a uniform electrical potential is simply superimposed on the magnetic potential and hence does not affect the evaluation of the magnetic information. While for magnetic potential, only the component parallel to the electron wave or z-component contributes to the phase shift of electron beam. Therefore, the magnetization distribution in specimen plane can be obtained by a differentiation of phase distribution of an electron beam with respect to z

In order to get $\dfrac{\partial I(xyz)}{\partial z}$ from the experimental data, $\dfrac{\partial I(xyz)}{\partial z}$ is approximately expressed as

$$\frac{\partial I}{\partial z} \approx \frac{I(x,y,z_0+\Delta z)-I(x,y,z_0-\Delta z)}{2\Delta z} \tag{2.4}$$

where z_0 is the focus distance of the objective lens, $2\Delta z$ the distance between the over-focus and under-focus planes, and $2\Delta z \ll z_0$, as schematically shown in Figure 2.11. $I(x,y,z_0-\Delta z), I(x,y,z_0)$ and $I(x,y,z_0+\Delta z)$ are intensity distributions of Lorentz images at under-focus, just-focus and over-focus conditions, respectively.

Figure 2.12a shows one example of Lorentz images of magnetic skyrmion lattices in FeGe under a magnetic field of 0.1 T and at 260 K.[4] Bright dots and dark dots are observed in the left part and right part, which are due to the convergence and divergence of

Chapter 2

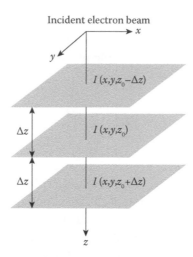

FIGURE 2.11   Schematic of Lorentz images at under-focus, just-focus, and over-focus conditions. In order to get a good analysis of the phase distribution, Lorentz images should be taken at appropriate defocus distance, which depends on the magnetic property of the sample and TEM experimental condition.

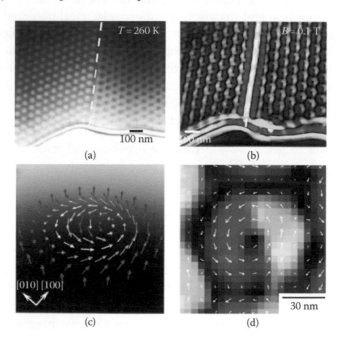

FIGURE 2.12   (a) Under-focus Lorentz image and (b) lateral magnetization distribution of FeGe under magnetic field 0.1 T and at 260 K. (From Yu, X.Z., et al., *Nat. Mater.*, 10, 106–109, 2011. With permission.) Schematic spin structure (c) and lateral magnetization distribution (d) of a skyrmion. (From Yu, X., et al., *Nature*, 465, 901–904, 2010. With permission.) (b) and (d) are obtained by TIE method.

electron beam, respectively. As shown in Figure 2.12c, a skyrmion has a spin structure with the center and edge part parallel to an electron beam and swirling up from inside to outside.[2] The lateral or in-plane component of magnetization acts on an electron beam to form the bright or dark contrast in Lorentz images, as clearly shown in Figure 2.12d which is obtained by TIE analysis. The bright and dark contrasts in Figure 2.12a are due to the opposite circular directions of skyrmions.

2.6   Phase Diagram of Skyrmion Formation

Generally, the formation of conventional domains, magnetic vortex and bubbles can be attributed to the competition between exchange and magnetostatic energies (including Zeeman energy). However, in most cases, this analysis does not suit the formation of magnetic skyrmions. Here we discuss the formation mechanisms of magnetic skyrmions.

2.6.1   Magnetic Interactions

In a spin system, there are different magnetic interactions due to, for example, magnetic exchange, spin–orbit coupling, and magnetostatic effect. The magnetic exchange tends to form ferromagnetic or antiferromagnetic alignments, while the spin–orbit coupling introduces favored orientation of the moments in the crystal (magnetic anisotropy). In comparison, the magnetostatic effect is less "microscopic," since it prefers the magnet to be demagnetized, leading to formation magnetic domains.

Particularly, for a magnetic material with spin–orbit coupling and a broken inversion symmetry, there exists Dzyaloshinskii–Moriya (DM) interaction[28,29] between spins. In this regard, the Hamiltonian to describe the magnetic system can be written as

$$H = -\sum_{i,j} J_{ij} \boldsymbol{M}_i \cdot \boldsymbol{M}_j - \sum_i K\left(M_i^z\right)^2 - \sum_i BM_i^z + E_d - \sum_{i,j} \boldsymbol{D}_{ij} \cdot (\boldsymbol{M}_i \times \boldsymbol{M}_j).$$

(2.5)

On the right side, the first four terms represent exchange, magnetocrystalline, Zeeman, and demagnetization energies, respectively. The last term is the DM interaction which is of most importance here. Moriya discussed in detail the direction of vector $\boldsymbol{D}_{ij}$ according to the symmetry of the crystal.[29] If this term is nonzero, it tends to align $\boldsymbol{S}_i$ vertically to $\boldsymbol{S}_j$. The coexistence of these interactions results in the abundant magnetic pictures that have been found in magnetic materials.

Chapter 2

2.6.2 Stability of Magnetic Skyrmions

As mentioned above, the exchange and DM interactions compete with each other and the spins will deviate from a collinear alignment. Regardless of the other terms in the Hamiltonian, the final result is that the spins are arranged in a helical order with an approximate wave vector of the spiral being $k \approx D/J$.[30] The helical magnetic structure is chiral which can be "right-handed" or "left-handed" in either a cycloidal- or transverse-spiral way, depending on the direction of the vector $\boldsymbol{D}_{ij}$. Due to the symmetry of the crystal, multiple wave vectors of the spin spiral are allowed and are degenerate in energy. Under an external field, the energy of ferromagnetic state ($k = 0$) is lowered. The superposition of the spirals characterized by these wave vectors and the ferromagnetic state may form skyrmions (Figure 2.13), as was found in MnSi, $(Fe_{1-x}Co_x)Si$, and other chiral magnets.[1,2,4,7]

The first evidence of the skyrmion was provided by Pfleiderer and his group in the MnSi chiral magnet, which was identified as the A-phase occurring adjacent to the Curie temperature when applying a certain external magnetic field to the material.[1] Following experiments showed that the skyrmion can be both stable and metastable.[2-4,7] In Figure 2.14, we draw a schematic magnetic phase diagram for compounds with stabilized skyrmions. Theoretically, the stability of the skyrmion can be discussed from several aspects.

For the bulk chiral magnets at a low temperature, the helical phase is initially stabilized. With an external magnetic field applied, the helix continuously transforms from proper-screw to conical along the external field and finally to field-polarized (Figure 2.14). When the temperature increases, multiple spirals can be excited due to the entropy. For instance, the observed hexagonal skyrmion lattice is consistent with the superposition of three spirals at an angle of 120° perpendicular to the external field. It can be expected that the skyrmion phase may be

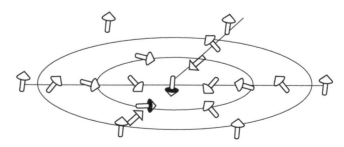

FIGURE 2.13 A schematic picture of the skyrmion. The spins align in a spiral order along the diameter.

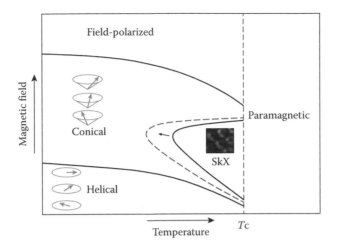

FIGURE 2.14 The schematic magnetic phase diagram for compounds with stabilized skyrmion (SkX). The vertical dashed line separates the polarized phases and the paramagnetic phase on the atomic scale. The arrow indicates that the phase region of skyrmion may be stretched toward a lower temperature to cover larger B–T space, as discussed in the text.

stabilized in a larger temperature range if the conical phase is successfully suppressed. This is achieved by lowering one dimension of the bulk materials, i.e., by fabricating thin films. When the thickness of thin film is smaller than the helical wave length, the direction of the helical wave vector is confined within the plane of the film, and the skyrmion phase was found to be stabilized even at a temperature of 5 K.[2] In addition, the observation in FeGe further demonstrated the thickness-dependent stability of the skyrmion phase.[4]

It should be stressed that the deposited thin film itself has the nature of broken inversion symmetry at the interface. Actually, the skyrmion phase has already been found in the ultra-thin Fe and PdFe bilayer film on Ir surface.[3,12] The first-principles calculations combined with spin dynamics calculations have also predicted that, by depositing Fe on heavy transition metals, the strong spin–orbit coupling provided by the matrix will result in a strong DM interaction and stabilize the skyrmion phase at low temperatures (Figure 2.15).[31] By tuning the exchange interactions using an additional Pd overlayer with different stacking, the period and also the stability of spin spiral (and thus the skyrmion) can be effectively tailored. Moreover, it is interesting to find that the exchange interactions in Pd/Fe/Ir(111) favor the formation of skyrmions when the stacking of a Pd overlayer is hcp, which indicates the possibility to design the skyrmion materials by use of finely tailored magnetic exchange and DM interaction.

Chapter 2

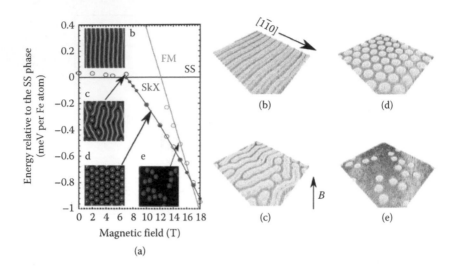

FIGURE 2.15 The low temperature phase diagram (a) and spin textures (b–e) of Pd/Fe/Ir(111) with face-centered cubic stacking of the Pd overlayer. There are two sets of (b–e). The (b–e) outside the phase diagram are the 3D views of the counterparts inside. The energy of spin spiral (SS) has been chosen as the reference. (From Dupé, B., et al., *Nat Commun.*, 5, 4030, 2014. With permission.)

Table 2.1   The Available Curie Temperatures of the Widely Investigated Chiral Magnets in Which the Skyrmion Phase Has Been Observed

Materials	Curie Temperature (K)	Reference
MnSi	29.5	1
$Fe_{1-x}Co_xSi$	<60	2, 38, 47
FeGe	278, 280	4, 5
Cu_2OSeO_3	58	7

In an application sense, the skyrmion phase should be stable adjacent to the room temperature for data storage. As can be seen from Figure 2.14, the upper limit of the stability region cannot exceed the Curie temperature. It is thus important that the exchange interaction should be sufficiently strong to maintain the stability of the ordered magnetic states. However, as mentioned above, the variation in exchange interaction meanwhile has a significant effect on the formation of skyrmions, which should be treated seriously. In Table 2.1, we list the available Curie temperatures of the reported chiral magnetic materials with the skyrmion phase. All these widely investigated materials have Curie temperatures typically lower than 300 K.

The separation of the magnetic atoms by nonmagnetic ones seems to be the reason. In this regard, fabricating Fe or Co layers on the heavy metal surfaces is an alternative approach since the Fe and Co have high Curie temperatures. However, this is still challenging because the DM interaction that is required to stabilize the skyrmion is produced adjacent to the interface which limits the thickness of the magnetic layer. The exchange interaction in an ultrathin ferromagnetic film is also weak due to the lack of magnetic coordinates, as can be deduced from the experiment in Pd/Fe/Ir.[12] Despite this, some heartening experimental results have been reported recently.[32–34]

Apart from the DM interaction due to the breaking of centrosymmetry of the magnetic crystals, the demagnetization field (or dipolar interaction) is also found to cause the skyrmion phase. Using micromagnetic simulation, Dai et al. concluded that the skyrmion can exist in Co/Ru/Co nanodisks even without taking into account the DM interaction.[35] The details can be found in Chapter 10 of this book. In this case, no requirement is needed for an ultrathin film and thus the skyrmion phase is stable even at high temperature. The disadvantage is that the resultant skyrmions cannot move and its existence is confined to nanodisk since the demagnetization field depends sensitively on the specimen shape. Another astonishing result of the magnetic dipolar interaction is that it may generate so-called *biskyrmions* in thin plates, which look like a bound state of two skyrmions with opposite spin helicities and the two skyrmions share the spins in the overlapped region.[36] The thin plate possesses conventional stripe-like domains with the magnetic moments lying within to minimize the magnetostatic energy. The moderate perpendicular magnetic anisotropy guarantees the formation of 180° Bloch-type domain walls. Under the external field, the evolution of a vortex-like spin texture from the conventional domains is mediated by the magnetic anisotropy, which cooperatively stabilize the *biskyrmion* state with inherited spin helicities. However, the dynamics of the *biskyrmion* state still needs more investigation regarding especially that its spin texture is not topologically protected in a strict sense.

2.7   Neutron Scattering Findings of Magnetic Skyrmions

The first evidence of the magnetic skyrmion was given by Mühlbauer et al. using neutron scattering,[1] which showed the appearance of 2D skyrmion lattice in MnSi perpendicular to the applied weak external

Chapter 2

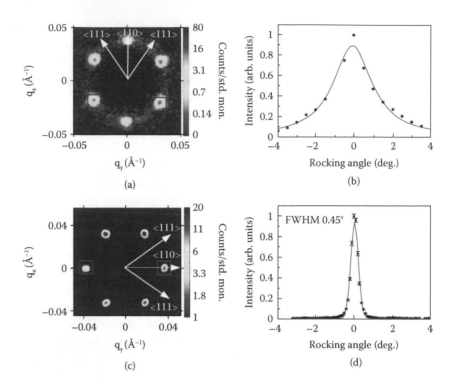

FIGURE 2.16 Typical neutron scattering data of skyrmion lattice in MnSi. (a) The intensity pattern in a thick cylindrical sample with $H = 0.16$ T. (b) The rocking scan result for the data of (a). (c) The intensity pattern in thin platelet with $H = 0.2$ T. (d) The rocking scan result for the data of (c). (From Adams, T., et al., *Phys. Rev. Lett.*, 107, 217206, 2011. With permission.)

magnetic field, regardless of the field direction relative to the crystal lattice. The skyrmion phase is stable at the border between the helimagnetic and paramagnetic states near T_C (29.5 K).[1] The typical small-angle neutron scattering intensities of the skyrmion lattice of MnSi are given in Figure 2.16, which shows the dependence of the skyrmion lattice on the sample shape and the applied external field. A more systematic study using polarized neutron scattering showed that the skyrmions in MnSi are completely left-handed and dynamically disordered.[37] Later, Münzer et al. reported the skyrmion phase in $Fe_{0.8}Co_{0.2}Si$ using small-angle neutron scattering and attributed its wide temperature range stability to the site disorder of the Fe and Co atoms.[38] Similar phase stability of magnetic skyrmions was also found in the B20 sibling material Cu_2OSeO_3 by using neutron scattering.[6,39]

The skyrmion phase can possibly be stabilized as the ground state according to the investigation of MnGe by Kanazawa et al.[40] Under zero field, the neutron scattering shows a Debye-ring-like pattern due to the random orientation of the helical wave vectors. The pattern transits to intense peaks when applying an external field, indicating the appearance of skyrmion lattice. Interestingly, the skyrmion lattice can be preserved even when the external field is removed, showing at least a metastable behavior. Recently, Gilbert et al. further demonstrated that the ground-state skyrmion lattice at room temperature can be artificially realized in hybrid nanostructures such as the asymmetric Co nanodot arrays over the underlayer Co/Pd thin film with a perpendicular magnetic anisotropy, which was directly identified by the neutron reflectometry.[32]

2.8 Lorentz TEM Images of Magnetic Skyrmions

Just after the discovery of a magnetic skyrmion by neutron diffraction,[1] real-space observation gave direct images of skyrmions in helical magnet $Fe_{0.5}Co_{0.5}Si$ in 2010 by using Lorentz TEM, as reported by the group led by Tokura et al.[2] This group has also observed a helical ground state in the same material, $Fe_{0.5}Co_{0.5}Si$, in 2006.[41] Compared with other image techniques of domain wall, the prominent feature in Lorentz TEM lies in its high resolution. As the size of skyrmions is usually on the order of 3–100 nm,[42] Lorentz TEM provides a natural tool to image such small-size vortex spin textures. However, with conventional Lorentz TEM, the magnetic field cannot be added perpendicular to the thin specimen plane so that only a helical state is observed. With the advance of TEM techniques, the objective lens current can be adjusted continuously to control the magnetic fields applied to the specimen along the z-axis. Accordingly, the group successfully obtained the first real-space images of the magnetic skyrmions.[2]

Figure 2.17 shows the initial magnetic contrast of spin textures in $Fe_{0.5}Co_{0.5}Si$ by using Lorentz TEM.[2] At low field, a helical state appears, which is followed by the skyrmion lattice phase at the elevated field. At the highest field, the skyrmion state transfers into the field-driven ferromagnetic state via a mediated state. More importantly, compared with the tiny temperature–magnetic field (T–B) window of skyrmions in bulk materials, the constructed phase diagram shows significantly increased T–B region in the 2D

Chapter 2

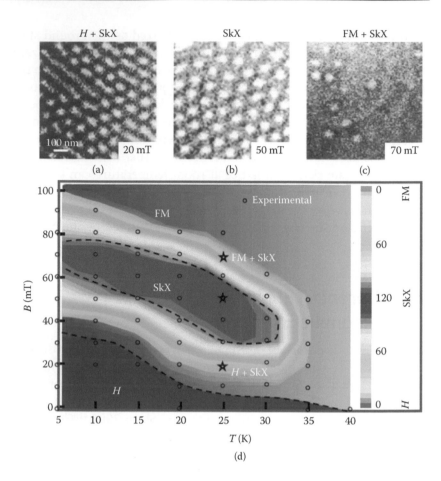

FIGURE 2.17 Phase diagram and spin textures in a thin film of $Fe_{0.5}Co_{0.5}Si$. (a–c) intact magnetic contrast of spin textures in the sample. H, SkX, and FM represent helical, skyrmions, and field-driven ferromagnetic states, respectively. Stars in (d) indicate (T, B) conditions for the images shown in (a–c). (From Mühlbauer, S., et al., *Science*, 323, 915–919, 2009. With permission.)

film, indicating a high stability of 2D helimagnets. This discovery is further confirmed in the 2D FeGe samples with a varied thickness.[4] In this case, the thinner the crystal plate is, the wider the SkX plane region can be found in the plane of vertically applied magnetic field versus temperature. A highly stable skyrmion state has also been observed in 2D or quasi-2D MnSi materials.[43] The extended skyrmion state in 2D helimagnets has been explained by the spatial confinement effects.[44] When the thickness of the film is below a threshold, a type of 3D skyrmions, characterized by a superposition of conical modulations along the skyrmion axis and double-twist rotation in the perpendicular plane, is thermodynamically stable in a broad *T–B* range with *B* normal to the film plane. This mechanism makes sense only if the film thickness is comparable to or less than the skyrmion lattice constant.

After obtaining the real-space imaging of SkX in thin samples including metal, semiconductors, and insulators by using the special TEM technique (Lorentz Mode), TEM is further used to detect the three-dimensional (3D) structure of skyrmion lattices by using the electron holography.[45] Electron holography technique using the wave nature of electrons to directly visualize a quantized magnetic flux possesses nanometer resolution and reliable ability to separate the magnetic signal from unwanted contributions from local variations around the specimen. This technique directly detects the phase shift of the electron waves due to the electromagnetic field. Figure 2.18 shows the first magnetic flux flow of skyrmions in the $Fe_{0.5}Co_{0.5}Si$ sample at the temperature of 10.6 K. By further detecting the dependence of the phase shifts on the sample thickness, it is concluded that skyrmions show the 3D spin structure with all the same magnetic arrangements along the magnetic field direction.

Asides from the static magnetic properties of skyrmions in 2D helimagnets, the current-driven dynamics of skyrmions has also been performed by Lorentz TEM. To do so, a microdevice composed of the helimagnet FeGe is fabricated to integrate into the in situ Lorentz TEM (Figure 2.19) and demonstrate the motion of skyrmions driven

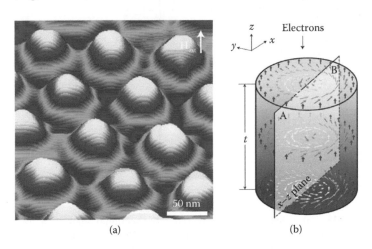

(a)   (b)

FIGURE 2.18   (a) Surface plots of the phase image in a thin film of $Fe_{0.5}Co_{0.5}Si$ and (b) schematic illustration showing the three-dimensional spin structure of a skyrmion. (From Park, H.S., et al., *Nat. Nanotechnol.*, 9, 337–342, 2014. With permission.)

Chapter 2

by electrical currents.[9] They discovered that the nanoscale spin vortex "skyrmion" can be driven by the electric current near room temperature. The required current density for moving skyrmions is about five orders of magnitude lower than that for moving ferromagnetic domains (Figure 2.19).

The similar device is also fabricated for a bilayered manganite with centrosymmetric structure, in which the DM interaction is absent. A *biskyrmion* state, defined by a molecular form of two bound skyrmions, is found in this material below 60 K.[36] The *biskyrmion* state is derived from the conventional domains with a stripe-like pattern. The opposite spin helicities are determined by bare Lorentz TEM images with under- and over-focus (Figure 2.20). It has been demonstrated that the critical current density to drive the *biskyrmions* is on the order of 10^8 A/m^2, which is about four orders of magnitude lower than that for the conventional ferromagnetic domain walls (10^{12} A/m^2).[36] Recent investigation by Wang et al. using Lorentz TEM showed that the *biskyrmion* state can be stabilized in a wide temperature and magnetic field range in MnNiGa with a centrosymmetric Ni$_2$In-type structure, as shown in Figure 2.21.[46]

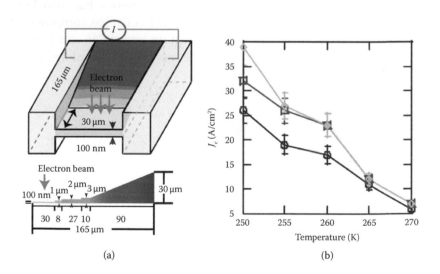

(a)   (b)

FIGURE 2.19 (a) Schematic diagram and cross-sectional view of a microdevice with a trapezoidal FeGe plate that is composed of a 100-nm-thick thinner terrace for electron-beam transmission and another trapezoidal thicker part for supporting the thinner part. (b) The temperature dependence of the critical current density, J_C, as a function of the temperature for the 2D sample. The diamonds, squares and circles represent the data for the different areas. (From Yu, X.Z., et al., *Nat. Commun.*, 3, 988, 2012. With permission.)

The interesting point is that this is a metallic compound which facilitates the manipulation of the *biskyrmions* by using electric current. The alloying effect on the behavior of *biskyrmions* also warrants further investigation.

In summary, TEM has played a key role in identifying complex spin textures in helimagnets including the spin structures and the dynamics. Future demonstration of skyrmion-based devices will still require more applications of the advanced TEM techniques.

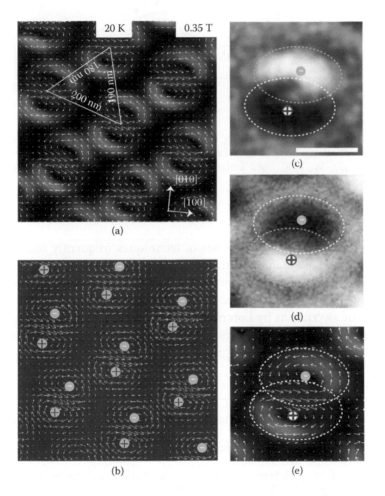

FIGURE 2.20   The magnetic configurations of the *biskyrmions* in bilayered manganite. (a, b) The in-plane magnetic components in the *biskyrmion* lattice. The over- (c) and under-focus (d) Lorentz TEM images for the *biskyrmion*. (e) A magnified picture for the spin texture of *biskyrmion*. The "plus" and "minus" signs indicate clockwise and counter-clockwise spin helicities, respectively. (From Yu, X.Z., et al., *Nat. Commun.*, 5, 3198, 2014. With permission.)

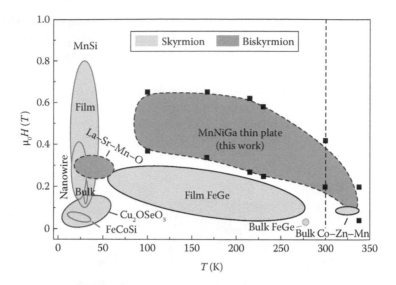

FIGURE 2.21 The comparison of the T–H phase diagram of the skyrmion lattice in different materials with the *biskyrmion* lattice found in the Ni_2 In-type MnNiGa. (From Wang, W., et al., *Adv. Mater*, 28, 6887, 2016, DOI: 10.1002/adma.201600889. With permission.)

2.9 Conclusion

This chapter introduced a series of techniques frequently used in the observation of magnetic microstructures, including conventional domains and magnetic skyrmions. For each technique, the basic concept and experimental procedure were outlined. The observation of the magnetic skyrmions by Lorentz transmission electron microscopy was discussed in detail, in combination with the theoretical and experimental phase stability of magnetic skyrmions.

References

1. S. Mühlbauer, et al. Skyrmion lattice in a chiral magnet. *Science* 2009(323): 915–919.
2. X. Yu, et al. Real-space observation of a two-dimensional skyrmion crystal. *Nature* 2010(465): 901–904.
3. S. Heinze, et al. Spontaneous atomic-scale magnetic skyrmion lattice in two dimensions. *Nat. Phys.* 2011(7): 713–718.
4. X. Z. Yu, et al. Near room-temperature formation of a skyrmion crystal in thin-films of the helimagnet FeGe. *Nat. Mater.* 2011(10): 106–109.
5. S. X. Huang and C. L. Chien. Extended skyrmion phase in epitaxial FeGe(111) thin Films. *Phys. Rev. Lett.* 2012(108): 267201.
6. S. Seki, et al. Formation and rotation of skyrmion crystal in the chiral-lattice insulator Cu_2OSeO_3. *Phys. Rev. B* 2012(85): 220406.

7. S. Seki, X. Z. Yu, S. Ishiwata and Y. Tokura. Observation of skyrmions in a multiferroic material. *Science* 2012(336): 198–201.
8. J. H. Yang, et al. Strong Dzyaloshinskii-Moriya interaction and origin of ferroelectricity in Cu_2OSeO_3. *Phys. Rev. Lett.* 2012(109): 5.
9. X. Z. Yu, et al. Skyrmion flow near room temperature in an ultralow current density. *Nat. Commun.* 2012(3): 988.
10. A. Fert, V. Cros and J. Sampaio. Skyrmions on the track. *Nat. Nano technol.* 2013(8): 152–156.
11. W. Jiang, et al. Blowing magnetic skyrmion bubbles. *Science* 2015(349): 283–286.
12. N. Romming, et al. Writing and deleting single magnetic skyrmions. *Science* 2013(341): 636–639.
13. P. Milde, et al. Unwinding of a skyrmion lattice by magnetic monopoles. *Science* 2013(340): 1076–1080.
14. G. L. Squires. *Introduction to the theory of thermal neutron scattering.* Cambridge, UK: Cambridge University Press, 2012.
15. E. Ressouche. Polarized neutron diffraction. *École thématique de la Société Française de la Neutronique* 2014(13): 02002.
16. B. M. Tanygin. Symmetry theory of the flexomagnetoelectric interaction in the magnetic vortices and skyrmions. *Physica B-Condensed Matter* 2012(407): 868–872.
17. A. Tonomura, et al. Real-space observation of skyrmion lattice in helimagnet MnSi thin samples. *Nano Lett.* 2012(12): 1673–1677.
18. X. Yu, et al. Magnetic stripes and skyrmions with helicity reversals. *PNAS* 2012(109): 8856–8860.
19. A. Bauer, M. Garst and C. Pfleiderer. Specific Heat of the Skyrmion Lattice Phase and field-induced tricritical point in MnSi. *Phys. Rev. Lett.* 2013(110): 177207.
20. M. Janoschek, et al. Fluctuation-induced first-order phase transition in Dzyaloshinskii-Moriya helimagnets. *Phys. Rev. B* 2013(87): 134407.
21. E. Moskvin, et al. Complex chiral modulations in FeGe close to magnetic ordering. *Phys. Rev. Lett.* 2013(110): 077207.
22. D. Shindo and T. Oikawa. *Analytical electron microscopy for materials science.* Springer Science & Business Media, 2013.
23. W. Xia, et al. Investigation of magnetic structure and magnetization process of yttrium iron garnet film by Lorentz microscopy and electron holography. *J. Appl. Phys.* 2010(108): 123919.
24. A. Tonomura. *Electron holography.* Springer, 1999.
25. W. Xia, et al. Magnetization distribution of magnetic vortex of amorphous FeSiB investigated by electron holography and computer simulation, *J. Electron Microsc.* 2012(61): 71–76.
26. T. Nagai, et al. Formation of nanoscale magnetic bubbles in ferromagnetic insulating manganite La7/8Sr1/8MnO3. *Appl. Phys. Lett.* 2012(101): 162401.
27. K. Ishizuka and B. Allman. Phase measurement of atomic resolution image using transport of intensity equation. *J. Electron Microsc.* 2005(54): 191–197.
28. I. Dzyaloshinskii. A thermodynamic theory of "weak" ferromagnetism of anti-ferromagnetics. *J. Phys. Chem. Solids* 1958(4): 241–255.
29. T. Moriya. New mechanism of anisotropic superexchange interaction. *Phys. Rev. Lett.* 1960(4): 228–230.
30. S. V. Grigoriev, et al. Magnetic structure of MnSi under an applied field probed by polarized small-angle neutron scattering. *Phys. Rev. B* 2006(74): 214414.
31. B. Dupé, M. Hoffmann, C. Paillard and S. Heinze. Tailoring magnetic skyrmions in ultra-thin transition metal films. *Nat. Commun.* 2014(5): 4030.

Chapter 2

32. D. A. Gilbert, et al. Realization of ground-state artificial skyrmion lattices at room temperature. *Nat. Commun.* 2015(6): 8462.

33. O. Boulle, et al. Room-temperature chiral magnetic skyrmions in ultrathin magnetic nanostructures. *Nat. Nanotechnol.* 2016(11): 449–454.

34. S. Woo, et al. Observation of room-temperature magnetic skyrmions and their current-driven dynamics in ultrathin metallic ferromagnets. *Nat. Mater.* (15): 501–506.

35. Y. Y. Dai, et al. Skyrmion ground state and gyration of skyrmions in magnetic nanodisks without the Dzyaloshinskii-Moriya interaction. *Phys. Rev. B* 2013(88): 054403.

36. X. Z. Yu, et al. Biskyrmion states and their current-driven motion in a layered manganite. *Nat. Commun.* 2014(5): 3198.

37. C. Pappas, et al. Chiral paramagnetic skyrmion-like Phase in MnSi. *Phys. Rev. Lett.* 2009(102): 197202.

38. W. Münzer, et al. Skyrmion lattice in the doped semiconductor$Fe_{1-x}Co_xSi$. *Phys. Rev. B* 2010(81): 041203.

39. T. Adams, et al. Long-wavelength helimagnetic order and skyrmion lattice phase in Cu_2OSeO_3. *Phys. Rev. Lett.* 2012(108): 237204.

40. N. Kanazawa, et al. Possible skyrmion-lattice ground state in the B20 chiral-lattice magnet MnGe as seen via small-angle neutron scattering. *Phys. Rev. B* 2012(86): 134425.

41. M. Uchida, Y. Onose, Y. Matsui and Y. Tokura. Real-space observation of helical spin order. *Science* 2006(311): 359–361.

42. K. Shibata, et al. Towards control of the size and helicity of skyrmions in helimagnetic alloys by spin-orbit coupling. *Nat. Nano technol.* 2013(8): 723–728.

43. X. Yu, et al. Observation of the magnetic skyrmion lattice in a MnSi nanowire by Lorentz TEM. *Nano Lett.* 2013(13): 3755–3759.

44. F. N. Rybakov, A. B. Borisov and A. N. Bogdanov. Three-dimensional skyrmion states in thin films of cubic helimagnets. *Phys. Rev. B* 2013(87): 094424.

45. H. S. Park, et al. Observation of the magnetic flux and three-dimensional structure of skyrmion lattices by electron holography. *Nat. Nanotechnol.* 2014(9): 337–342.

46. W. Wang, et al. A centrosymmetric hexagonal magnet with superstable biskyrmion magnetic nanodomains in a wide temperature range of 100–340 K. *Adv. Mater.* 2016(28): 6883.

47. S. V. Grigoriev, et al. Magnetic structure of Fe1-xCoxSi in a magnetic field studied via small-angle polarized neutron diffraction. *Phys. Rev. B* 2007(76): 224424.

48. T. Adams, et al. Long-range crystalline nature of the skyrmion lattice in MnSi. *Phys. Rev. Lett.* 2011(107): 217206.

3. Resonant X-Ray Scattering Studies on Skyrmions

Sujoy Roy, Matthew Langner, and James Lee

Lawrence Berkeley National Laboratory, Berkeley, California

Topological states in condensed matter are current research topics of tremendous interest due to both their unique physics and their potential in device applications. Skyrmions are topologically twisted spin textures that often form dense hexagonal lattices. A novel Hall effect caused by skyrmion spin-texture topology allows these lattices to be moved coherently over macroscopic distances with very low current densities. These features make magnetic skyrmions appealing for low-power memory and information processing applications. In this chapter, we review neutron and resonant soft X-ray scattering work on skyrmions. Resonant soft X-ray scattering, in particular, uniquely provides element-specific and orbital-sensitive electronic and magnetic structure information. This sensitivity of X-rays now makes it possible to perform electronic spectroscopy on skyrmions.

3.1 Introduction

In classical physics, a system can be described by rigid masses that are connected by springs and governed by an equation of motion with appropriate boundary conditions. In field theories, the description of the

Chapter 3

system happens through solving the appropriate wave equation, and excitations are described by a wavefunction. British physicist Tony Skyrme found a field theory of interacting pions that has a solution with topologically protected excitations in a stable field configuration [1]. These particle-like topological entities are called skyrmions. Although originally predicted in the context of nuclear physics, skyrmions find relevance in diverse topics in condensed matter physics such as a two-dimensional (2D) quantum Hall system [2,3], liquid crystals [4–6], multiferroics [7,8], ferroelectrics [9,10], and even Bose condensate [11].

The magnetic skyrmion has a "hedgehog" spin texture which when projected into a 2D plane has a knot-like structure (Figure 3.1) [12–15]. The knot cannot be untied continuously, which means a skyrmion cannot be morphed into a ferromagnet continuously by a deformation force. It is therefore a topologically stable particle and classified according to the winding number. An intuitive way to think about the winding number is the number of times the spins wrap around the great circle of a sphere. The winding number is an integer and, on a unit sphere in spin space, is given by:

$$W = \frac{1}{4\pi} \int \widehat{M} \cdot (\partial_x M \times \partial_y M) \cdot dx \, dy. \tag{3.1}$$

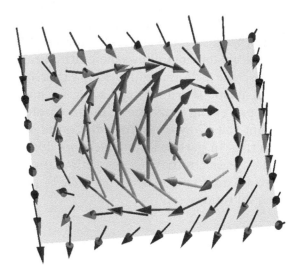

FIGURE 3.1   Schematic of a skyrmion spin texture. The spins at infinity are pointed out of the plane, whereas the spin at the origin is pointed vertically downward. The magnetization is continuously varying as a function of distance from origin.

The skyrmion texture is nonsingular and finite everywhere. Thus, spins continuously evolve in space such that the direction of spin at the origin is opposite to the spins at infinity. This spin texture is quite different from a magnetic vortex because, unlike vortices, all the spins at infinity in a skyrmion point in the same direction. The texture is also different than chiral spin structures because W acts as an axial vector. Skyrmions are different from spin density waves because density waves have no winding numbers and have a single momentum transfer vector q associated with it. The skyrmion on the other hand is a $3q$ object.

In magnetic systems, the concept of topological quasiparticles was described by Belavin and Polyakov [16]. They argued that in 2D if fluctuations are considered, then the correlation length does not diverge at T = 0 K; rather, quasiparticles quite similar to skyrmions form. Bogdanov and Yablonskiï showed that for the crystal symmetry of C_{nv} class, a magnetic vortex state arises in magnetic fields [17]. The important point here is that the metastable state was a result of an inhomogeneous interaction energy term that has the symmetry of the Dzyaloshinskii–Moriya (DM) interaction term. The theoretical treatment was made more general by explicitly incorporating chiral exchange interactions [18]. In a different context, skyrmions in ferromagnetic quantum Hall systems are predicted to form with the right Zeeman and Coulomb energy combinations [2].

The breakthrough experiment confirming the existence of magnetic skyrmions came from the work of Mühlbauer et al., who found a hexagonal scattering pattern in neutron scattering while a magnetic field was applied to the helical magnet MnSi [13]. This was followed by the real-space observation of a skyrmion lattice using Lorentz transmission electron microscopy (LTEM) [19]. The neutron scattering study was able to identify the skyrmion lattice, but the LTEM work showed the internal structure of an individual skyrmion. For the first time, it was possible to see the spin swirls with a center spin pointing downwards and the spins at the periphery pointing out of plane. Yu et al. used a thinned single crystal of $Fe_{0.5}Co_{0.5}Si$ and applied a magnetic field perpendicular to the film. They observed the formation of a helical spin structure that gradually morphed into a skyrmion crystal with a lattice spacing of 90 nm, as the applied magnetic field was increased. Scanning tunneling microscopy was used to show that a single skyrmion could be manipulated using small current pulses in an ultrathin film of Fe grown on Ir(111) [20]. In addition to helimagnets, skyrmions also appear in dipolar interaction-based magnets [21,22]. It turns out that both magnetic bubble domains and skyrmions can be treated within the same group theory. A bubble without a twist, which can be called

Chapter 3

a "trivial bubble," has a winding number of zero, while a bubble with winding number unity is a skyrmion [23]. The spin textures of bubbles and stripe domains turn out to be much richer than those of spin helices and skyrmions in helical magnets. Since skyrmions are also obtained by tailoring the strength of dipolar interactions, skyrmions can be synthesized in polycrystalline or amorphous thin films of technological relevance. Indeed, room temperature skyrmions have been found in sputtered ultrathin Pt/Co/MgO nanostructures at room temperature and zero external magnetic field [24].

From the above experimental results, it is clear that various competing interactions can give rise to a skyrmion texture. We know that symmetric exchange interaction gives rise to a long-range order. However, the presence of antisymmetric exchange interaction, such as the relativistic DM interaction of the form D ($\mathbf{S}_i \times \mathbf{S}_j$), where D is the Dzyaloshinskii vector, gives rise to spin canting. Helical magnets with DM interaction show spiral spin structures. In a cubic crystal, the superposition of the spirals along the three axes gives rise to a skyrmion texture. Skyrmions can also result from a competition between symmetric exchange and dipolar interaction. In perpendicular magnetic systems that have been extensively studied, the dipolar interaction wants to make the spins parallel to the film plane while the anisotropy energy wants the spins to be out of the film plane. The resultant domain structure is that of a stripe domain with Bloch domain walls, and under appropriate field conditions the stripe domains can morph into skyrmions.

In the remainder of this chapter, we will review the neutron scattering studies of MnSi and resonant X-ray scattering studies on multiferroic Cu_2OSeO_3 and see how they provide insights into the mystery of the skyrmion state.

3.2 Skyrmions in MnSi

The transition metal compound MnSi is a typical example of a weakly magnetic d-electron compound. MnSi is cubic in structure (space group $P2_13$) with lattice parameter of 0.46 nm. The Mn and the Si atoms are displaced in opposite [111] directions from the nominal face-centered cubic position. There are four Mn/Si atoms in a unit cell. The cubic structure lacks inversion symmetry, and the chemical arrangement does not have a spiral structure.

MnSi exhibits a rich phase diagram as determined by different experimental techniques (Figure 3.2) [24–27]. MnSi has an ordering temperature of 29 K with an ordered moment of 0.4 μ_B per Mn atom. At zero external field, MnSi has a helical spin structure with a pitch

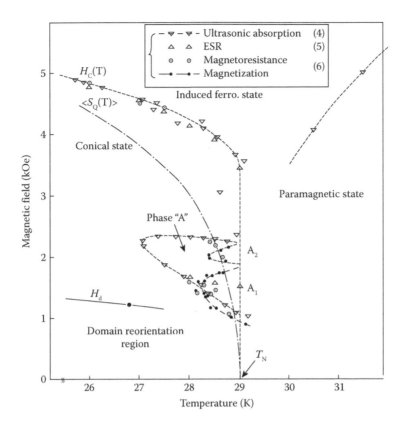

FIGURE 3.2   Phase diagram of MnSi. Particularly interesting is the phase "A," which was identified as the skyrmion phase. (After Ishikawa, Y., and Arai, M., *J. Phys. Soc., Jpn.*, 53, 2726, 1984. With permission.)

of 18 nm and wave-vector **q** along the [111] direction. The wave-vector orients itself along the applied field direction although the periodicity remains same. The magnetization saturates at a field of ≥ 6 kOe and the saturation moment is less than 1.4 μ_B per Mn atom as obtained from the Curie–Weiss region. Interestingly, the moment in the paramagnetic region is larger than the ordered moment below the transition temperature. Theoretical work [27] shows that the ferromagnetic spiral in MnSi is caused due to a weak spin–orbit mediated DM interaction, which arises due to the nonsymmetric arrangement of the Mn atoms in the unit cell. The sign of the Dzyaloshinskii vector sets the handedness of the helix. (Without the DM interaction, both right- and left-handed spirals will occur with the same probability.)

Measurement of resistivity as a function of temperature at applied pressure reveals that for applied pressure $p_C \leq 14.6$ kbar, a kink appears indicative of magnetic ordering (Figure 3.3) [28]. With increase

Chapter 3

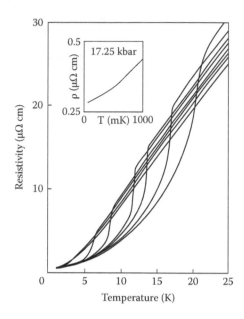

FIGURE 3.3 Resistivity as a function of temperature at applied pressures of 5.55, 8.35, 10.40, 11.40, 12.90, 13.55, 14.30, and 15.50 kbar in MnSi. Below 15 kbar, a shoulder develops in the resistivity curves. No such shoulder exists at higher applied hydrostatic pressures. (From Pfleiderer, C., et al., *Phys. Rev., B*, 55, 8330, 1997. With permission.)

in pressure, the transition temperature decreases and appears to move with a bigger slope toward zero temperature as the pressure approaches 14.6 kbar. The resistivity has a T^2 dependence for $p \leq p_C$ as expected for a Fermi liquid. At p_C, however, the resistivity deviates from T^2 behavior, indicative of a pressure-induced quantum phase transition. Although bulk measurements indicate that magnetic ordering vanishes at $p_C = 14.6$ kbar, neutron scattering showed a significant magnetic intensity even at pressure higher than p_C [29]. In the non-Fermi liquid phase, the intensity was found to be broad within a radius of $q = 0.043$ Å^{-1} rather than concentrated near the Bragg peak and was highest along the [110] direction. Thus, not only the spiral continues to exist well above the ordering temperature, but also the intensity sphere would suggest that the direction of the spirals is no longer fixed.

The unusual transport properties together with a complex phase diagram make MnSi a candidate for new phases. Specifically, the three main phases in MnSi are: a zero-field helical phase; a conical phase resulting from an unwinding of the spin helix for $T \ll T_C$ and fields around 1 kOe; and a ferromagnetic state above 5 kOe. The interesting phase is the "A" phase that lies ≈ 1 K below T_C. Neutron scattering performed as a function of field shows that the behavior of the intensity at 28 K is quite different than the conical state, indicating that the "A" phase is a new phase that penetrates the conical phase [30].

The small angle neutron scattering (SANS) work of Mühlbauer et al. revealed the nature of this "A" phase (Figure 3.4) [13]. Their experiment was carried out in an atypical SANS geometry. Usually, neutron scattering

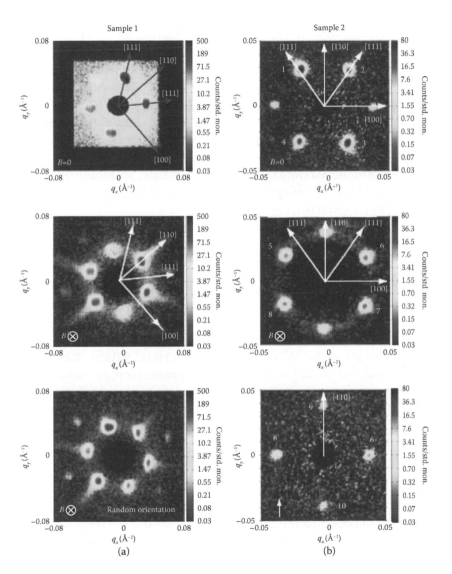

FIGURE 3.4 Small angle neutron scattering intensities from MnSi. (a) Six diffraction spots due to spin helices along the crystallographic axes. (b) The new spots that appear when a magnetic field is applied. Note that the spots rotate and appear along the [110] direction. (After Mühlbauer, S., et al., *Science*, 323, 915, 2009. With permission.)

Chapter 3

studies of MnSi were done with the magnetic field applied perpendicular to the beam propagation direction. Mühlbauer et al. applied a magnetic field parallel to the beam direction. Since the skyrmions develop in a plane perpendicular to the applied field, this means in the SANS setup the skyrmions will form in a plane perpendicular to the beam propagation direction, increasing the sensitivity of the neutron beam to the skyrmions.

In agreement with previous studies, Mühlbauer et al. found the helical Bragg spots along the [111] direction. However, in the "A" phase they discovered magnetic intensities along the [110] directions as well, and the scattering exhibited a sixfold symmetry. The sixfold symmetric intensity pattern persisted irrespective of the magnetic field orientation with respect to the atomic lattice. This means that the plane perpendicular to the magnetic field has a spin texture with hexagonal symmetry, which can be constructed by three **q** vectors.

Having found a sixfold symmetric spin texture, Mühlbauer et al. then addressed what type of spin structure this could be. The free energy of the system is given by:

$$F = \int d^3r (aM^2 + J(\nabla M)^2 + U(M^2)^2 + 2DM \cdot (\nabla \times M) - B \cdot M), \quad (3.2)$$

where the first two terms represent the usual square of magnetization and its variation. The fourth term is the DM interaction term, and the last term indicates the interaction of magnetization with an applied magnetic field. The term with the fourth power of M stabilizes the spin structure at the "A" phase. In atomic crystal lattices, their stability depends on the lattice having no net momentum, i.e., whether the **q**-vectors of their charge density waves add up to zero. In this case, the stability of the spin crystal would depend on the three **q**-vectors corresponding to the three helices adding up to zero:

$$\sum_{i=1}^{3} q^i = 0. \quad (3.3)$$

The zero net momentum condition is satisfied if the three vectors lie in a plane and have a 120° angle between them, giving rise to a magnetization field:

$$M = M_f + \sum_{i=1}^{3} M_s (n_1^i \cos(q^i \cdot (r + \delta r^i)) + n_2^i \sin(q^i \cdot (r + \delta r^i))). \quad (3.4)$$

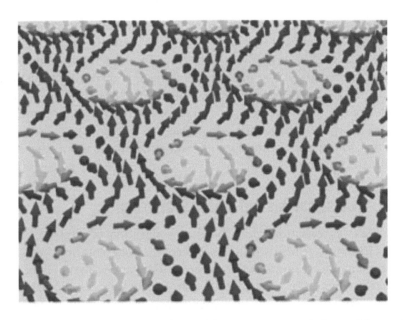

FIGURE 3.5   A skyrmion texture obtained from Equation 3.4. (After Mühlbauer, S., et al., *Science*, 323, 915, 2009. With permission.)

For particular values of δr, a spin crystal can be obtained that resembles a skyrmion crystal (Figure 3.5). It was shown that the skyrmion phase is more stable when fluctuations are taken into account. (Due to the nearness of T_C, fluctuations could be significant.) The conical state is otherwise a stable phase for other parts of the phase diagram.

3.3   Resonant X-Ray Scattering

X-ray scattering is a sensitive probe of electronic order. Traditionally, the short wavelength of X-rays has been used to measure charge density variations of materials, such as the crystal structure of solids or the roughness of interfaces and surfaces. X-ray energies are also comparable to the binding energies of core-level atomic orbital states. This enables X-rays to measure orbital and magnetic ordering. Orbital and magnetic states usually have miniscule scattering cross-sections. These can be enhanced by orders of magnitude, however, by tuning X-ray energies to core-level atomic-binding energies. This excites resonances associated with coherent electronic dipole transitions. The equivalent of crystal structure and interfacial roughness can, thereby, be measured for orbital or magnetic order in materials.

Chapter 3

To explain how X-ray scattering reveals magnetic ordering, which is relevant to studies of magnetic skyrmions, we will begin by discussing the classical basis of X-ray scattering, emphasizing the role of the susceptibility tensor. The susceptibility tensor is quantum mechanical in nature. Thus, we will discuss the microscopic basis of X-ray scattering and derive an expression for the susceptibility relevant to resonant elastic X-ray scattering. This will conceptually reveal why X-rays are sensitive to the spatial variations of magnetic states as well as the vector nature of the magnetic ordering.

3.3.1   Classical Basis of X-Ray Scattering

The Maxwell equations describe the propagation of electromagnetic fields in matter if nearly all the scattering is coherent, i.e., there is no difference in wavelength between the incident and scattered light. This is a safe assumption since, in the soft X-ray range, elastic scattering cross-sections of electrons in elements such as iron are many orders of magnitude greater than the inelastic scattering cross-sections [31].

The scattering is dependent on the microscopic properties of the system, which do not enter into the Maxwell equations except through the macroscopic quantities, such as the electric polarization $\bar{P}$, and the magnetization $\overline{M}$. We may use X-ray scattering to characterize $\bar{P}$ and $\overline{M}$ and, thereby, provide constraints on microscopic calculations. Considering the case in which there are no free currents and no free charges, and using the usual constitutive equations [32,33]

$$(\nabla^2 + \omega^2 \varepsilon_0 \mu_0)\overline{D} = -\nabla \times \nabla \times \bar{P} - i\omega\varepsilon_0\mu_0 \nabla \times \overline{M}, \tag{3.5}$$

where ε_0 and μ_0 are, respectively, the permittivity and permeability of free space. The fields are assumed to be harmonic.

$\bar{P}$ and $\overline{M}$ describe the response of the material to external fields. If the incident beam is not too intense, the response is linear:

$$\bar{P} = \ddot{\chi}_e \cdot \overline{D}_0 \quad \text{and} \quad \overline{M} = \bar{\chi}_m \cdot \overline{H}_0 = \bar{\chi}_m \cdot (\bar{k}_0 \times \overline{D}_0)/\sqrt{\varepsilon_0\mu_0}. \tag{3.6}$$

In these expressions, $\overline{D}_0$ is the incident field, $\bar{k}_0$ is the direction of the incident wave vector, and the incident field is nearly monochromatic. $\ddot{\chi}_e$ and $\bar{\chi}_m$ are spatially inhomogeneous, i.e., functions of position, which is why there is scattering to begin with. It should also be noted that the dependence on $\overline{D}_0$ amounts to the use of the first-order Born approximation [33].

In the far-field limit, the portion of the scattered field at the detector with polarization $\hat{\epsilon}$ is given by:

$$\hat{\epsilon}^* \cdot \vec{D} = \vec{D}_0 + \frac{k^2 e^{-ikr}}{4\pi r} \int e^{-i\vec{q}\cdot\vec{r}'} \left(\hat{\epsilon}^* \cdot \vec{\chi}_e \cdot \hat{\epsilon}_0 + (\vec{k}' \times \hat{\epsilon}^*) \cdot \vec{\chi}_m \cdot (\vec{k}_0 \times \hat{\epsilon}_0) \right) d\vec{r}', \quad (3.7)$$

where $\vec{q} = \vec{k}_0 - \vec{k}'$ is the momentum transfer, and $\hat{\epsilon}_0$ and $\hat{\epsilon}$ are the incident and scattered polarization vectors, respectively. The first term in the far-field solution is attributed to "charge scattering" and the second term with "magnetic scattering." While such a formal separation can be made and the scattering can be expressed as a function of two separate tensors, a single, effective susceptibility tensor that accounts for the effects of both the polarization and magnetization can be created [34]. The far-field scattering solution is then:

$$\hat{\epsilon}^* \cdot \vec{D}_{sc} = D_0 \frac{k^2 e^{-ikr}}{4\pi r} \int e^{-i\vec{q}\cdot\vec{r}'} (\hat{\epsilon}^* \cdot \vec{\chi}_{sc} \cdot \hat{\epsilon}_0) d\vec{r}', \tag{3.8}$$

from which it is possible to isolate the classical scattering amplitude:

$$A_{sc,cl}(\vec{k}_0, \hat{\epsilon}_0; \vec{k}', \hat{\epsilon}) = \frac{k^2}{4\pi} \int e^{-i\vec{q}\cdot\vec{r}'} (\hat{\epsilon}^* \cdot \vec{x}_{sc} \cdot \hat{\epsilon}_0) d\vec{r}', \tag{3.9}$$

where

$$\hat{\epsilon}^* \cdot \vec{\chi}_{sc,H} \cdot \hat{\epsilon}_0 = -\frac{r_0 \lambda^2}{\pi V_{cell}} \sum_i^{\langle cell \rangle} f_i^{ab}(\vec{q}_H, \omega) e^{i\vec{q}_H \cdot \vec{r}}. \tag{3.10}$$

The scattering of light is, then, dependent on the properties of the generalized susceptibility, as well as the polarization of the incident X-rays.

Note that the effective susceptibility can be decomposed into a scalar component, a traceless symmetric component, and an antisymmetric component, as any second-rank tensor can. The scalar and symmetric contributions express properties that are even under a time-reversal operation, while the antisymmetric contribution expresses properties that are odd under a time-reversal operation.

Chapter 3

3.3.2   Bridge to Quantum Mechanical Results

The quantum mechanical electromagnetic Hamiltonian is:

$$H_{int,i} = \sum_i \frac{e}{m} \overline{A}(\bar{x}_i) \cdot \bar{p}_i + \frac{e^2}{2m} \overline{A}(\bar{x}_i)^2, \qquad (3.11)$$

where m is the rest mass of the electron, $\overline{A}$ is the magnetic vector potential of the radiation field and e is the magnitude of the electron charge.

Using Fermi's Golden Rule, including terms up to second order in the magnetic vector potential, as well as the dipole approximation ($e^{\bar{k}\cdot\bar{r}} \approx 1$), the scattering amplitude for elastic scattering is [35]:

$$A_{sc,q}(\bar{k}, \hat{\epsilon}_0; \bar{k}', \hat{\epsilon}) =$$

$$-r_0 \sum_i^{\langle cell \rangle} \sum_{a_i} \left(\hat{\epsilon}^* \cdot \hat{\epsilon}_0 \left\langle a_i \,|\, e^{i\bar{q}\cdot\bar{r}_i} \,|\, a_i \right\rangle - \sum_{I} \frac{1}{m} \frac{\left\langle a_i |\, \hat{\epsilon}^* \cdot \bar{p}_i \,| I \right\rangle \left\langle I |\, \hat{\epsilon}_0 \cdot \bar{p}_i \,| a_i \right\rangle}{E_I - E_{a_i} - \hbar\omega - i\gamma_I} \right).$$

$$(3.12)$$

In Equation 3.12, the labels a_i are the ground states of all the electrons of the ith atom in the unit cell; I are the intermediate states over which the second-order perturbation terms are summed; E_{a_i} and E_I are energies of the electron ground states and intermediate states, respectively; and γ_I is the inverse lifetime of the intermediate state, typically hundreds of MeV. The first term describes Thomson scattering, whereas the second term describes the resonant scattering of X-rays and is a tensor quantity dependent on the incident and scattered polarization states.

A relationship can be established between the classical scattering amplitude in Equation 3.10 and the quantum mechanical scattering amplitude shown in Equation 3.12:

$$\frac{1}{V_{cell}} f_i^{ab}(\omega) e^{i\bar{q}_H \cdot \bar{r}}$$

$$= \sum_{a_i} \left(\hat{\epsilon}^* \cdot \hat{\epsilon}_0 \left\langle a_i \,|\, e^{i\bar{q}\cdot\bar{r}_i} \,|\, a_i \right\rangle - \sum_{I} \frac{1}{m} \frac{\left\langle a_i |\, \hat{\epsilon}^* \cdot \bar{p}_i \,| I \right\rangle \left\langle I |\, \hat{\epsilon}_0 \cdot \bar{p}_i \,| a_i \right\rangle}{E_I - E_{a_i} - \hbar\omega - i\gamma_I} \right). \qquad (3.13)$$

The above expression is often written in the form

$$f_i^{ab}(\omega) = Z + f_{1,i}^{ab}(\omega) - i f_{2,i}^{ab}(\omega), \tag{3.14}$$

where Z is the atomic number, $f_{1,i}^{ab}$ is the real part of the resonant term, and $f_{2,i}^{ab}(\omega)$ is the imaginary part.

The states in the sum over a_i that are most important to the resonant scattering process are the inner-shell (core) states of the atoms, e.g., 1s or 2p states. When the incident X-ray beam energy $\hbar\omega$ is tuned to the core electron binding energies, $E_I - E_{a_i}$, virtual transitions of electrons from the core level to unoccupied states in the valence band (the intermediate states I) can be excited. Due to the limited lifetime of the excited state, an electron from the valence band state drops into the vacant core state, emitting a photon. The resonant denominator in Equation 3.13 characterizing this coherent dipole transition process enhances the scattering amplitude by orders of magnitude. This makes resonant X-ray scattering a practical probe of electronic order.

3.4   Application to Magnetic Materials

The matrix elements of the resonant scattering process, in Equation 3.13, linearly depend on the momentum operator, which is odd under parity. In 3d transition metals such as Mn, Fe, and Co, the energy differences between their 2p core level orbitals and 3d valence states are hundreds of electron volts. "Soft" X-ray energies lie in the range of E~100–2000 eV and are, thus, well suited to exciting resonances that increase the scattering amplitudes of the 3d valence states. These scattering amplitudes are affected by the local spin polarization. Information about how the spin polarization varies in space is contained in the scattering tensor in Equation 3.10 and reflected in the far-field scattering from the material. There are many examples of such work [36–39].

A general expression about the magnetic scattering cross section can be obtained from Equation 3.13:

$$f_n^{if}(\omega) = f_{0n}(\omega)\left(\vec{\epsilon}^{f*} \cdot \vec{\epsilon}^{i}\right) - i f_{1n}(\omega)\left(\vec{\epsilon}^{f*} \times \vec{\epsilon}^{i}\right) \cdot \hat{m}_n$$
$$+ f_{2n}(\omega)\left(\vec{\epsilon}^{f*} \cdot \hat{m}_n\right)\left(\vec{\epsilon}^{i} \cdot \hat{m}_n\right), \tag{3.15}$$

where $\hat{m}_n$ is the unit magnetization vector of the nth atomic site in the unit cell; the labels i and f denote the incident and outgoing X-ray polarization;

Chapter 3

and the terms $f_{0n}(\omega)$, $f_{1n}(\omega)$, and $f_{2n}(\omega)$ are, respectively, the monopole, dipole, and quadrupole parts of the scattering amplitude.

The dipolar term in Equation 3.15 is the reason why a polarization analysis of the X-ray scattering can reveal the direction of the ordered magnetic moment (the quadrupole term is usually neglected since it is much smaller than the other terms [37]). The cross-product of the incident and scattered X-ray polarization vectors is sensitive to the direction of the local magnetic ordered moment. Changing the polarization state of the incident beam and changing the scattering geometry can control the direction of the dipole term cross-product.

As a concrete example, examine the scattering geometry in Figure 3.6. If the incident polarization is $\varepsilon_\perp$ and the measured polarization of the scattered light is $\varepsilon'_\parallel$, then the overlap of the magnetic moment with the outgoing X-ray's normalized wavevector $\mathbf{k}'$ is measured. On the other hand, if the incident polarization is $\varepsilon_\parallel$ and the measured polarization of the scattered light is $\varepsilon'_\perp$, then it is the overlap of the magnetic moment with the incident wavevector $\mathbf{k}$ that is measured. In terms of resonant magnetic X-ray scattering, these two cases are the relevant ones, since the magnetic scattering process necessarily causes a rotation between the incident and scattered X-ray polarizations, from $\parallel$ to $\perp$ or vice versa.

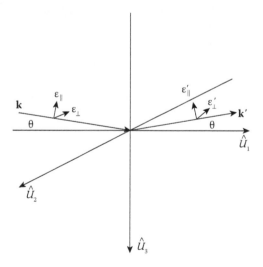

FIGURE 3.6 A typical X-ray scattering geometry. The normalized wavevectors of the incident and scattered X-rays are $\mathbf{k}$ and $\mathbf{k}'$. The components of the polarization vectors, ε and ε', that are parallel with and perpendicular to the scattering plane defined by $\mathbf{k}$ and $\mathbf{k}'$ are denoted by $\parallel$ and $\perp$, respectively. The angle θ is the incident and scattered angles with respect to the sample surface. $\hat{U}_i$ is a unit basis vector. (From van der Laan, G., C. R. *Physique*, 9, 570, 2008. With permission.).

In the following, we will discuss how the resonant X-ray scattering can be used to study skyrmion spin textures.

3.5 Skyrmions in Cu_2OSeO_3

Magnetoelectric materials are those whose electric and magnetic order parameters are coupled. That is, it is possible to induce a magnetization by applying an electric field and, conversely, electric polarization can be induced by an applied magnetic field. Generally, these materials are magnetic but not polar and have a modest magnetoelectric coupling coefficient. Discovery of a skyrmion phase in magnetoelectric Cu_2OSeO_3 material is remarkable as it opens up the possibility of manipulating the skyrmions with an electric field [7].

Cu_2OSeO_3 has significant magnetoelectric coupling but does not have a spontaneous lattice distortion. It has a cubic B20 structure, the same as MnSi alloys. Cu_2OSeO_3 is ferromagnetic with a Curie temperature of 60 K. The two ferrimagnetic sublattices are formed by spins on Cu atoms with different oxygen bonding geometries: specifically, Cu sites that are connected to square pyramidal or trigonal bipyramidal oxygen ligands [7,40]. The Cu atoms have a formal valency of +2 and form a network of distorted tetrahedra. The Cu coordination polyhedrons deviate quite a bit from ideal square pyramidal and trigonal bipyramidal, respectively. Helical magnetic phases arise from a combination of symmetric spin–exchange interactions, such as those that lead to ferromagnetism, and antisymmetric exchange resulting from a DM interaction.

Neutron powder diffraction [40] results show that the magnetic space group has to be lower than a crystal space group since a cubic magnetic structure cannot explain the ferrimagnetic ordering. A two-sublattice model each with Cu1 and Cu2 site was found to fit the data. Rietveld refinement gave the moment of Cu1 and Cu2 to be 0.35 μ_B and −0.35 μ_B, respectively. The total moment is 0.61 μ_B per Cu atom.

The insulating nature of Cu_2OSeO_3 makes the mechanism of conductivity quite different from MnSi. Particularly interesting will be the application of an electric field. Because of strong magnetoelectric coupling, the application of a magnetic field generates spontaneous electric polarization along the [111] direction [7,8]. Recent experiments have shown that the skyrmions can be manipulated using small electric fields [41,42]. It is important to note that, like MnSi, the skyrmion phase in Cu_2OSeO_3 can be stabilized only in a small region below the critical temperature.

Resonant X-ray scattering experiments on Cu_2OSeO_3 were performed at the Advanced Light Source, Lawrence Berkeley National Laboratory [43]. Figure 3.7 shows the experimental geometry as well

Chapter 3

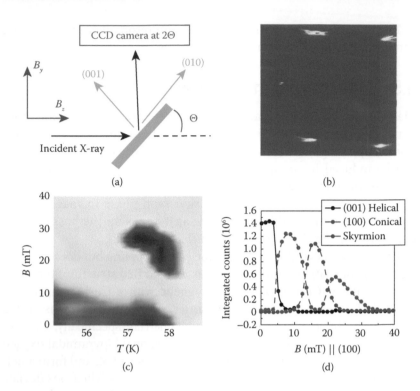

FIGURE 3.7   (a) Experimental geometry to observe skyrmions in Cu_2OSeO_3. The (001) was aligned so as to capture the (001) Bragg peak. The magnetic field direction is fixed in the laboratory reference frame. (b) The five distinct spots around the lattice Bragg peak are due to the skyrmions; and the sixth peak is off camera. For clarity, the (001) lattice peak has been subtracted out. (c) Phase diagram of the skyrmion (red) and helical (blue) peaks as a function of temperature and vertical applied field. (d) Intensity of conical and skyrmion peaks as a function of field along (100) at 57.5 K. (From Langner, M.C., *Phys. Rev. Lett.*, 112, 167202, 2014. With permission.)

as the scattering data. The energy of the incident beam was tuned to the Cu L_3 edge and aligned the (001) lattice Bragg peak in the detector. In Figure 3.7b, the camera images of the diffraction peaks from the skyrmion phase at $T = 57.5$ K, with a field of 20 mT applied in the vertical direction is shown. The vertical field direction corresponds approximately to the [011] lattice direction in the diffraction geometry. The magnetic peaks appear as satellites at (001) $\pm\tau$ around the (001) lattice Bragg peak, where τ represents the q-vectors for the skyrmion ordering. Due to large intensity at the lattice peak, the weak scattering due to the skyrmions is difficult to detect. A beam block was placed to eliminate the (001) lattice peak. A twofold symmetric satellite peaks around (001) $\pm\tau$, with $\tau = (0.103, 0, -0.04)$ nm^{-1}, arising due to the helical phase at $T = 40$ K, and zero applied field was observed.

At a temperature of 57.5 K and an applied magnetic field of 20 mT, we observed Bragg peaks due to the formation of a sixfold symmetric skyrmion lattice (Figure 3.7b). The skyrmion structure forms in a plane perpendicular to the applied field direction. The diffraction spots look distorted due to the projection of the skyrmions onto the (001) plane. The projection takes place because of the angle between the applied field and [001] direction. This geometrical effect causes a reduction in the surface $\mathbf{q}$-vector perpendicular to the X-ray direction and leads to a compression of the sixfold symmetry in the camera image. After correction of the geometrical effects, the calculated q-vectors for both the helical and skyrmion phase correspond to a periodicity of 59.5 ± 5.0 nm, which is consistent with published literature [7].

The phase diagram of the skyrmion phase was determined by the scattering measurements with a vertical applied field. Figure 3.7d shows the intensities of the helical, conical, and skyrmion peaks for a field applied along the [100] lattice direction, with the sample initially in the helical state aligned along (001). Above the 5 mT depinning threshold, the conical phase aligns along (100), resulting in a peak at $(|\tau|, 0, 1)$. At higher fields, the skyrmion phase appears as a peak at $(0, 0, 1 + |\tau|)$, which results from scattering by the skyrmion columns aligned along (100). Further increase in the applied field re-establishes the conical phase. The peaks from the skyrmion phase were observed for fields applied along all available orientations of the applied field, i.e., for fields applied either along the x, y, or z directions of the laboratory coordinate axes, thereby indicating the robustness of the phase.

Resonant scattering experiments have also been reported in $Fe_{0.5}Co_{0.5}Si$ [44]. The skyrmion lattice is 70 nm, which means magnetic peaks at the Fe L_3 edge appear very near the (000) Bragg peak. Performing reflection at those small angles becomes a challenge. To avoid this technical problem, a scattering experiment in transmission geometry is desirable. However, since the soft X-ray penetration depths are short, a single crystal was thinned to about 200 nm, and Yamasaki et al. performed resonant X-ray scattering in small angle geometry (see Figure 3.8). Simultaneous X-ray magnetic circular dichroism measurements were done. From the XMCD data, it was concluded that the Fe orbital magnetic moment is not totally quenched. Whether orbital moment in the material has a role in establishing the skyrmion state remains to be seen. Further, resonant X-ray scattering at the Fe L_3 edge reveals a strong scattering signal due to the helical peaks (Figure 3.9d). These Bragg peaks appear at the $\pm Q$ values corresponding to the skyrmion spacing in the real space.

Chapter 3

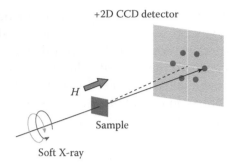

FIGURE 3.8 X-ray experimental setup used to measure helical state in $Fe_{0.5}Co_{0.5}Si$. (From Yamasaki, Y., et al., *J. Phys. Conf. Ser.*, 425, 132012, 2013. With permission.)

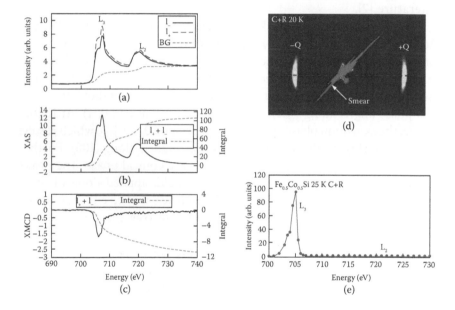

FIGURE 3.9 X-ray absorption and magnetic circular dichroism data (a–c) and resonant X-ray scattering (d, e) due to the helical spin structure in $Fe_{0.5}Co_{0.5}Si$. Both the experiments were done at the Fe L_3 edge. The peaks at the ±Q are the helical Bragg peaks. The direct beam saturates the detector at the center. (From Yamasaki, Y., et al., *J. Phys. Conf. Ser.*, 425, 132012, 2013. With permission.)

The importance of resonant X-ray scattering from the skyrmions is that it now becomes possible to study the specific element that is responsible for the skyrmion formation. Particularly interesting will be chemical contrast-specific measurements toward understanding the mechanisms to control and reconfigure the skyrmion states to create new topological phases. Further, since the X-rays interact with both spins and orbitals, it now becomes possible to study the orbital and spin coupling and its related effects on skyrmion in the Cu_2OSeO_3. Dynamical X-ray measurements will allow for study of the electric-field-induced changes to the skyrmion lattice, relevant to spintronic applications, and time-domain measurements of the forced annihilation of skyrmions and corresponding creation of monopoles, e.g., through laser excitation.

References

1. T H R Skyrme. A unified theory of mesons and baryons. *Nucl. Phys.* **31**, 556 (1962).
2. S L Sondhi, A Karlhede, S A Kivelson and E H Rezayi. Skyrmions and the crossover from the integer to fractional quantum Hall effect at small Zeeman energies. *Phys. Rev. B* **47**, 16419 (1993).
3. D Lilliehook, K Lejnell, A Karlhede and S L Sondhi. Quantum Hall Skyrmions with higher topological charge. *Phys. Rev. B* **56**, 6805 (1997).
4. A N Bogdanov, U K Rößler and A A Shestakov. Skyrmions in nematic liquid crystals. *Phys. Rev. E* **67**, 016602 (2003).
5. A O Leonov, I E Dragunov, U K Rößler and A N Bogdanov. Theory of skyrmion states in liquid crystals. *Phys. Rev. E* **90**, 042502 (2014).
6. P J Ackerman, R P Trivedi, B Senyuk, J van-de Lagemaat and I I Smalyukh. Two-dimensional skyrmions and other solitonic structures in confinement-frustrated chiral nematics. *Phys. Rev E* **90**, 012505 (2014).
7. S Seki, X Z Yu, S Ishiwata and Y Tokura. Observation of skyrmions in a multi-ferroic material. *Science* **336**, 198 (2012).
8. S Seki, et al. Formation and rotation of skyrmion crystal in the chiral-lattice insulator Cu_2OSeO_3. *Phys. Rev. B* **85**, 220406(R) (2012).
9. A K Yadav, et al. Observation of polar vortices in oxide superlattices. *Nature.* **530**, 198 (2016).
10. Y Nahas, et al. Discovery of stable skyrmionic state in ferroelectric nanocomposites. *Nat. Commun.* **6**, 8542 (2015) doi: 10.1038/ncomms9542.
11. T L Ho. Spinor Bose condensates in optical traps. *Phys. Rev. Lett.* **81**, 742 (1998).
12. K Everschor-Sitte and M Sitte. Real-space Berry phases: Skyrmion soccer (invited). *J. Appl. Phys.* **115**, 172602 (2014).
13. S Mühlbauer, et al. Skyrmion lattice in a chiral magnet. *Science* **323**, 915 (2009).
14. B Binz and A Vishwanath. Chirality induced anomalous-Hall effect in helical spin crystals. *Physica B* **403**, 1336 (2008).
15. N Nagaosa and Y Tokura. Toplogical properties and dynamics of magnetic skyrmions. *Nat. Nanotechnol.* **8**, 899 (2013).
16. A A Belavin and A M Polyakov. Metastable states of two-dimensional isotropic ferromagnets. *Pis'ma Zh. Eksp. Teor. Fiz* **22**, 503 (1975).
17. A N Bogdanov and D A Yablonskiĭ. Thermodynamically stable "vortices" in magnetically ordered crystals. The mixed state of magnets. *Zh. Eksp. Teor. Fiz* **95**, 178 (1989).
18. A N Bogdanov and U K Rößler. Chiral symmetry breaking in magnetic thin films and multilayers. *Phys. Rev. Lett.* **87**, 037203 (2001).
19. X Z Yu, et al. Real-space observation of a two-dimensional skyrmion crystal. *Nature* **465**, 901 (2010).

Chapter 3

20. N Romming, et al. Writing and deleting single magnetic skyrmions. *Science* **341**, 636 (2013).

21. X Yu, et al. Magnetic stripes and skyrmions with helicity reversals. *PNAS* **109**, 8856 (2012).

22. H Ewaza. Giant skyrmions stabilized by dipole-dipole interactions in thin ferromagnetic films. *Phys. Rev. Lett.* **105**, 197202 (2010).

23. H-B Braun. Topological effects in nanomagnetism: From superparamagnetism to chiral quantum solitons. *Adv. Phys.* **61**, 1 (2012).

24. O Boulle, et al. Room-temperature chiral magnetic skyrmions in ultrathin magnetic nanostructures. *Nat. Nanotechnol.* **11**, 449 (2016).

25. G Shirane, et al. Spiral magnetic correlation in cubic MnSi. *Phys. Rev. B* **28**, 6251 (1983).

26. Y Ishikawa, G Shirane, J A Tarvin and M Kohgi. Magnetic excitations in the weak itinerant ferromagnet MnSi. *Phys. Rev. B* **16** 4956 (1977).

27. O Nakanishi, A Yanase, A Hasegawa and M Kataoka. The origin of the helical spin density wave in MnSi. *Solid State Comm.* **35**, 995 (1980).

28. C Pfleiderer, G J McMullan, S R Julian and G G Lonzarich. Magnetic quantum phase transition in MnSi under hydrostatic pressure. *Phys. Rev. B* **55**, 8330 (1997).

29. C Pfleiderer, et al. Partial order in the non-Fermi-liquid phase of MnSi. *Nature* **427**, 227 (2004).

30. Y Ishikawa and M Arai. Magnetic phase diagram of MnSi near critical temperature studied by neutron small angle scattering. *J. Phys. Soc.*, **53**, 2726 (1984).

31. J Stöhr. *NEXAFS Spectroscopy* (*Springer Series in Surface Sciences*, Vol. 25), (Springer-Verlag, Heidelberg, 1992).

32. D J Griffiths. *Introduction to Electrodynamics, Third Edition* (Addison Wesley, 1999).

33. J D Jackson. *Classical Electrodynamics, Third Edition* (Wiley, 1999).

34. P S Pershan, Magneto-optical effects, *J. Appl. Phys.* **38**, 1483 (1967).

35. M Blume. Magnetic effects in anomalous dispersion, in *Resonant Anomalous X-ray Scattering: Theory and Applications*, G Materlik, C J Sparks, K Fischer, eds., (Elsevier, 1994).

36. G van der Laan. Soft X-ray resonant magnetic scattering of magnetic nanostructures. *C. R. Physique* **9**, 570 (2008).

37. O Hellwig, G P Denbeaux, J B Kortright and E E Fullerton. X-ray studies of aligned magnetic stripe domains in perpendicular multilayers. *Physica B* **336**, 136 (2003).

38. J B Kortright and Sang-Koog Kim. Resonant magneto-optical properties of Fe near its $2p$ levels: Measurement and applications. *Phys. Rev. B* **62**, 12216 (2000).

39. J B Kortright, et al. Soft-x-ray small-angle scattering as a sensitive probe of magnetic and charge heterogeneity. *Phys. Rev. B* **64**, 092401 (2001).

40. J-W G Bos, C V Colin and T T M Palstra. Magnetoelectric coupling in the cubic ferrimagnet Cu_2OSeO_3. *Phys. Rev. B* **78**, 094416, (2008).

41. J S White, et al. Electric field control of the skyrmion lattice in Cu_2OSeO_3. *J. Phys. Condens. Matter* **24**, 432201 (2012).

42. F Jonietz, et al. Spin transfer torques in MnSi at ultralow current densities. *Science* **330**, 1648 (2010).

43. M C Langner, et al. Coupled skyrmion sublattices in Cu_2OSeO_3. *Phys. Rev. Lett.* **112**, 167202 (2014).

44. Y Yamasaki, et al. Diffractometer for small angle resonant soft x-ray scattering under magnetic field. *J. Phys. Conf. Ser.* **425**, 132012 (2013).

4. Artificial Two-Dimensional Magnetic Skyrmions

Liang Sun, Bingfeng Miao, and Haifeng Ding

Nanjing University, Nanjing, People's Republic of China

Chapter 4

With topological charge in real space, skyrmion crystal bears ample interesting phenomena and has great potential for applications. Experimentally, what has impeded its property exploration is the rare material selection as well as the narrow temperature and magnetic field phase diagram. Utilizing dipolar–dipolar interaction, artificial skyrmion crystal was proposed and experimentally realized recently. The artificial skyrmion can completely bypass the need for Dzyaloshinskii–Moriya interaction (DMI), thus significantly enriching the material selections for skyrmions. In addition, an artificial skyrmion crystal has a robust working regime, including room temperature and above, and even without magnetic field. It was found that the created artificial skyrmion crystal exhibits similar static and dynamic behavior to that of skyrmion crystal with DMI. The finding of artificial skyrmions opens a new door to the exploration of the unique fundamental properties of skyrmions. The reported advance marks the strong potential for practical applications such as high-density storage devices.

In this chapter, we review the recent progress on artificial skyrmion crystals, both theoretically and experimentally. This chapter is organized as follows: Section 4.1 gives a brief introduction to the motivation and research background. Section 4.2 discusses the proposal of Bloch-type and Néel-type artificial magnetic skyrmions without DMI, as well as the potentially formed skyrmion state in ferroelectric nanocomposites. Section 4.3 presents several experimental realizations of artificial skyrmion crystal with different techniques. Finally, in Section 4.4, a micromagnetic study of the dynamic behavior of two-dimensional (2D) artificial skyrmion crystal is given.

4.1 Introduction

The skyrmion is a concept in particle physics, created by Tony Skyrme in 1961 (Skyrme 1961, 1962). It represents a topologically stable field configuration with particle-like properties that can move as a soliton. In the magnetic skyrmion, the spins point in all directions and wrap into a sphere, forming a topologically stable spin structure. The skyrmion in a magnetic solid carries a topological charge and a Berry phase in real space. It is anticipated to produce unconventional spin electronic phenomena, such as the topological Hall Effect (Onose et al. 2005; Lee et al. 2009; Neubauer et al. 2009; Huang and Chien 2012), and to exhibit spectacular dynamic properties (Petrova and Tchernyshyov 2011; Mochizuki 2012; Onose et al. 2012). The skyrmion crystal also attracts interdisciplinary interest as an analog of similar lattice structures in nuclear physics (Skyrme 1962; Klebanov 1985), quantum Hall systems (Sondhi et al. 1993; Brey et al. 1995), and liquid crystals (Bogdanov et al. 2003). Skyrmion is fundamentally interesting regarding not only its topological features but also its great potential for applications. Technologically, the skyrmion crystal may be exploited as a new class of spintronic materials due to its unusual response to the electric charge current and spin current (Jonietz et al. 2010; Yu et al. 2012).

Typically, the formation of skyrmion crystal requires the presence of DMI and was found to be stable only within a narrow temperature–magnetic field region (Rossler et al. 2006; Muhlbauer et al. 2009; Munzer et al. 2010), which impedes its physical exploration and application severely. Although the situation improves in thin films (Yu et al. 2011; Huang and Chien 2012) and DMI-induced room temperature skyrmion crystal was also reported very recently (Chen et al. 2015; Jiang et al. 2015; Tokunaga et al. 2015), its reliance on the DMI strongly limits the material selection of skyrmion crystals. To overcome/bypass these obstacles, Sun et al. and Dai et al. proposed an approach to create artificial skyrmion crystal with ordinary magnetic material through the dipolar–dipolar interaction (Dai et al. 2013; Sun et al. 2013). By embedding a magnetic vortex state into an out-of-plane magnetized environment, the proposed artificial skyrmion crystals have been realized experimentally at room temperature (Li et al. 2014; Miao et al. 2014; Gilbert et al. 2015). Though it is not demonstrated, it can be anticipated that the skyrmion crystal can further extend its phase diagram to even higher temperatures. Without the need of DMI, artificial skyrmion crystals greatly expand the pool of prospective materials in the future skyrmion-based application devices.

Chapter 4

4.2   Theoretical Prediction of Artificial 2D Skyrmion Lattices

4.2.1   The Relation between Skyrmion, Vortex, and Antivortex

Magnetic vortex and antivortex are the common objects in magnetism. For a vortex or an antivortex, the skyrmion number can be simplified as half of the product of its winding number and the polarity. The factor of "half" originates from the fact that the quantum number of a spin is 1/2. Since a magnetic moment is a vector, the winding number is defined differently with that in mathematics. For a magnetic vortex or antivortex, the local magnetic moment rotates around its center. At the center, the magnetic singularity occurs. The local magnetic moments point to the out-of-plane direction and form a core. Depending on which direction it points to, the core has either +1 or −1 polarity. The core also serves as a pole for the surrounding magnetic moments to wind around. Within one period, the magnetization rotates 360° for a vortex or an antivortex. Intuitively, the winding number can be obtained through the curvature of the surrounding magnetic moments. If they curve toward the center, the winding number is +1. Conversely, if they curve away from the center, the winding number is −1. Therefore, the winding number is always +1 for a magnetic vortex, independent of its circulation (clockwise [CW] or counter-clockwise [CCW]). Similarly, the winding number is −1 for an antivortex, also independent in its circulation (Chien et al. 2007). For clarity, the winding numbers, polarities, and skyrmion numbers for the magnetic vortex and antivortex are summarized in Figure 4.1. From Figure 4.1, we can learn that a vortex with a positive polarity has a skyrmion number $Q = 1/2$, with spin swirls around the upper hemisphere (Chien et al. 2007; Tretiakov and Tchernyshyov 2007). Meanwhile, a central negatively magnetized antivortex also has a skyrmion number $Q = 1/2$, with the local moments pointing to all possible directions in the lower hemisphere (Chien et al. 2007; Tretiakov and Tchernyshyov 2007). Thus, the skyrmion can be considered as the pair, a vortex and an antivortex, with opposite polarities (Figure 4.2a). We note that the antivortex core is not located at the skyrmion center; instead, it locates at the periphery where the local magnetic moments orient perpendicularly. For the first glimpse, the skyrmion in the right panel of Figure 4.2a could also be visualized as a vortex at the center with surrounding perpendicular magnetization. Therefore, if a vortex can be implanted into a perpendicular film, a skyrmion might also be

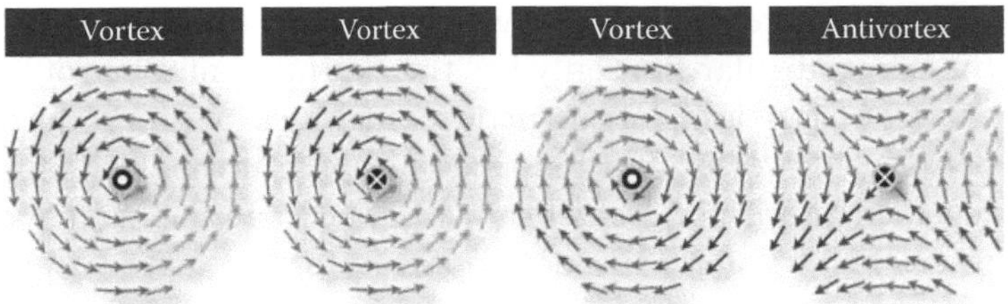

	Vortex	Vortex	Vortex	Antivortex
Circulation	CCW	CCW	CW	
p	1	−1	1	−1
n	1	1	1	−1
Q	1/2	−1/2	1/2	1/2

FIGURE 4.1 The skyrmion number of vortices and antivortices. A vortex or an antivortex with winding number n and core polarity p has a half integer skyrmion number $Q = np/2$. The winding number, or vorticity, is +1 for a vortex, regardless of the clockwise or counterclockwise circulation, and −1 for an antivortex; more generally, it is the change in the angle of the local magnetization, integrated over a loop around the vortex core and divided by 2π. (Adapted from Chien, C.L., et al., *Phys. Today.*, 60, 40–45, 2007. With permission.)

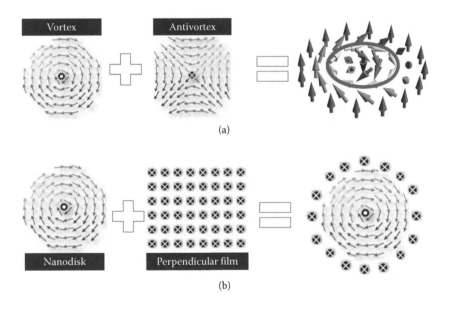

(a)

(b)

FIGURE 4.2 (a) The skyrmion can be considered as the pair of a vortex and an antivortex with opposite polarities. (b) A skyrmion may form through implanting a vortex into a perpendicular film.

Chapter 4

formed (Figure 4.2b). The question arises as to its viability given that the perpendicular film only has the skyrmion number of 0. However, there is a sharp jump of magnetic moment orientation at the connection between the vortex and perpendicular film if they are simply welded together. This is not energetically favorable, and the magnetic exchange interaction will force the local moments to rotate at the connection and an antivortex may form. In such situations, a skyrmion may be realized with the combination of a vortex and perpendicular film. Fortunately, both basic elements, the vortex and the perpendicular film, are common objects and are readily available as modern magnetic materials.

4.2.2 Assembly of 2D Artificial Skyrmion Crystals from Vortex and Antivortex

Figure 4.3 represents the proposed pathway for creating the 2D skyrmion crystal. First, a film with perpendicular anisotropy is prepared. The second step is to fabricate ordered arrays of nano-sized magnetic disks on top of the film. This can be achieved utilizing advanced patterning techniques, such as e-beam lithography and focused ion beam approaches.

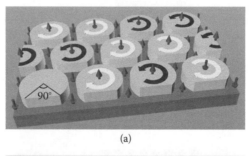

(a)

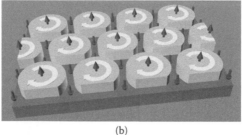

(b)

FIGURE 4.3 Proposed pathway for creating the 2D skyrmion crystal. (a) Ordered arrays of magnetic submicron disks are prepared on top of a film with perpendicular anisotropy. The arrows represent the magnetization orientation of the local moments. (b) Skyrmion lattice creation with the field treatment mentioned in the text. (From Sun, L., et al., *Phys. Rev. Lett.*, 110, 167201, 2013. With permission.)

With proper selection of the aspect ratio between the diameter and thickness, the disks can be magnetically configured into vortex states according to the known magnetic phase diagram (Cowburn et al. 1999; Ding et al. 2005). Due to the random distribution, the vortices would naturally have different circulation and polarity as shown in Figure 4.3a. The skyrmion lattice requires all the vortices to bear the same circulation and polarity (Yu et al. 2010). Therefore, the third step is to format all the vortices into the same configuration with the magnetization of the vortex cores oppositely aligned with that of the surrounding disk, as shown in Figure 4.3b. In such cases, a skyrmion crystal can be created. As demonstrated below, this can be achieved by the combination of slightly modifying the shape of the disks to be asymmetrical and specific magnetic field treatments. In this study, we use edge-cut circular disks with a cutting angle of 90° (see left disk, front row of Figure 4.3a), similar to that used by Dumas et al. (2009). All disks are cut along the same direction.

In the following, the feasibility of the proposal will be discussed utilizing micromagnetic simulations. Co is chosen as the disk material as it has a vortex state in similar geometrical confinement conditions (Ding et al. 2005). For the perpendicular film, CoPt film is used because it has a high perpendicular anisotropy and is conductive starting from a thickness of a few nanometer (Moritz et al. 2008). The material parameters used in calculations are exchange constant: $A^{Co} = 2.5 \times 10^{-11}$ J/m; saturation magnetization: $M_S^{Co} = 1.4 \times 10^6$ A/m for Co (Munzer et al. 2010), and $A^{CoPt} = 1.5 \times 10^{-11}$ J/m and $M_S^{CoPt} = 5.0 \times 10^5$ A/m for CoPt (Maret et al. 1997; Eyrich et al. 2012). A uniaxial anisotropy perpendicular to the film with constant $K_1^{CoPt} = 4.0 \times 10^5$ J/m³ is included for CoPt (Maret et al. 1997). Since sputtered Co films are typically polycrystalline, the calculations were performed assuming zero anisotropy. Additional calculations have also been made by assuming the single crystalline magnetic anisotropy value for bulk Co. The main results remain the same except that the size of the vortex cores has been changed slightly. An interlayer exchange constant between Co and CoPt of 1.9×10^{-11} J/m is also used (Donahue and Porter 1999). The disks with the diameter D and the thickness t_d are aligned into hexagonal arrays. The separation between centers of disks is S. In the calculations, we used 2D periodical boundary conditions within the plane and a grid size of $2 \times 2 \times 1$ nm³, which is smaller in length scale than the exchange length of Co (≈ 11 nm) and CoPt (≈ 5 nm).

As a starting point, we choose disks with $D = 120$ nm and $t_d = 18$ nm, the CoPt film with thickness $t_f = 8$ nm and the separation $S = 150$ nm, which is close in size to the skyrmion crystal geometry reported in Yu (2010). With the initial random spin configuration, the results show that the disks are partially in vortex states and partially in C-states. This is due to the dimensions of the disks being within the bi-stable region even though the vortex state has lower energy (Ding et al. 2005). The polarity and circulation of the vortices are randomly distributed, similar to that shown in Figure 4.3a. To format all the disks into uniform vortices, we proceed with a magnetic field treatment similar to that reported by Pang et al. (2012). We first apply a field of 800 mT perpendicular to the disks to bring them into the vortex state with the same polarity. Then we add an in-plane field pulse of 400 mT along the cutting edge to configure them into the same circulation. After releasing the perpendicular field, we find that all the disks are indeed formatted into uniform vortices of the same polarity and circulation.

During the formatting of the vortex, we find that the surrounding areas of the disks are also magnetized in the same direction of the vortex core due to the applied strong field. A skyrmion crystal, however, requires the magnetization orientation of the vortex core and the surrounding area aligned oppositely. Therefore, we further apply a magnetic field with the opposite direction. The vortex core and the surrounding area of the disk may have different switching fields. By choosing a suitable reversing field, we can switch only one of them and the vortex core and surrounding area of the disk are aligned oppositely. Figure 4.4 represents the calculated zero-field configuration after applying a field of −400 mT. We find

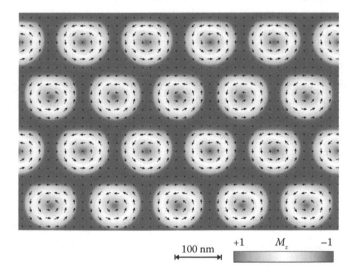

100 nm   +1 M_z −1

FIGURE 4.4   Top view of magnetic configuration of the CoPt layer in a calculated artificial skyrmion crystal at zero field. (Co disk size: $D = 120$ nm and $t_d = 18$ nm; CoPt film thickness $t_f = 8$ nm.) The arrows represent the magnetization orientation of the local moments. (From Sun, L., et al., *Phys. Rev. Lett.*, 110, 167201, 2013. With permission.)

that the CoPt film surrounding the disk can be uniformly magnetized oppositely with the vortex core and without significantly influencing the vortex configuration. Thus, a skyrmion crystal is created.

4.2.3 Exploration of Topological Properties of the Artificial Skyrmion Crystal

To explore the topological properties of the artificial skyrmion crystal, the local skyrmion density $\phi = \dfrac{1}{4\pi}\bar{n} \cdot \left(\dfrac{\partial \bar{n}}{\partial x} \times \dfrac{\partial \bar{n}}{\partial y} \right)$ has been computed following the description given in Muhlbauer et al. (2009), where $\bar{n}$ defines the direction of the local moment. If ϕ integrates to 1 or −1 in a unit cell, a topologically stable knot exists in the magnetic structure. As shown in Figure 4.5, the skyrmion density is finite and oscillates as a function of the position, similar to that shown in Muhlbauer et al. (2009). Moreover,

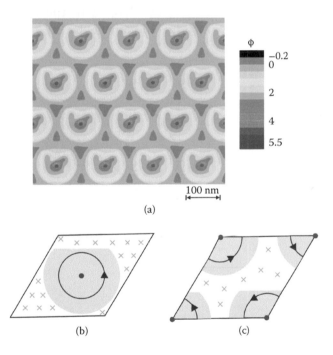

(a)

(b)   (c)

FIGURE 4.5 (a) Calculated local skyrmion density per unit cell for the artificial skyrmion crystal shown in Figure 4.4. (b) Schematic picture of a vortex with core magnetization pointing up (the light gray area). (c) Schematic picture of a distorted antivortex with core magnetization pointing down (outside the light gray area). The black arrows represent the in-plane magnetization orientation. The red dots and blue crosses indicate that the local magnetization orientation, pointing up and down, respectively. (From Sun, L., et al., *Phys. Rev. Lett.*, 110, 167201, 2013. With permission.)

Chapter 4

the skyrmion number per 2D unit cell is quantized and adds up to +1, proving the topological nature of the created artificial skyrmion crystal. This can also be understood within the qualitative picture mentioned above. When a unit cell is split by a boundary where the magnetization is fully in plane (black circle, Figure 4.5b), it can be considered as a combination of a vortex with the core pointing up (green area, Figure 4.5b) and a distorted antivortex whose core magnetization points down (outside green area in Figure 4.5c). We note that the distorted antivortex is not a typical antivortex which has a cross configuration (Hertel and Schneider 2006). But one can readily derive that it has a winding number of −1 and the polarity of −1, yielding a skyrmion number of 1/2. Together with the contribution of the vortex with opposite polarity, a unit cell has a skyrmion number of +1. The fact that a vortex and an antivortex with antiparallel core polarizations have equal skyrmion numbers adding to a total of +1 or −1 was previously pointed out by Tretiakov and Tchernyshyov (2007).

Figure 4.6 presents the hysteresis loop with the magnetic field applied perpendicularly. At large enough fields, the magnetization

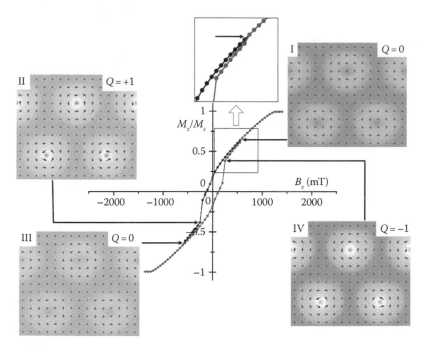

FIGURE 4.6 The hysteresis loop with the magnetic field applied perpendicularly. The calculated hysteresis loop was obtained using parameters defined in the text. The insets show the top view of magnetic configuration of the Co/Pt layer in a calculated artificial skyrmion crystal at different stages.

is saturated along the field direction. In reducing the field slightly from the saturation state, the vortices and antivortices form with their cores align along the same direction (States I and III). Since both the vortex and antivortex have the same polarity, the skyrmion number of the pair equals to 0. After reversing the field to the opposite direction, there are two switching fields (see the inserted amplified view of the calculated hysteresis in Figure 4.6), one corresponds to the switching field of the vortex core and the other corresponds to that of the antivortex core. In between these two switching fields, the vortex and the antivortex have opposite polarity (States II and IV). Therefore, the pair of them have the skyrmion number of either +1 or −1, forming the skyrmion states. It has been pointed out that the switching between different topological states is not smooth and needs to be accompanied by spin-wave excitations (Van Waeyenberge et al. 2006; Hertel et al. 2007; Tretiakov and Tchernyshyov 2007). Figure 4.7 exhibits the in-plane component magnetization along

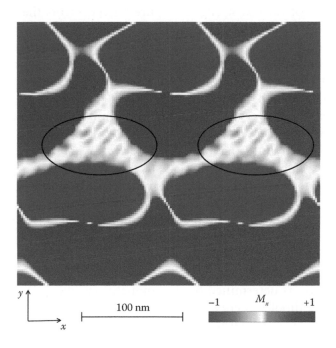

y

x

100 nm

-1 M_x $+1$

FIGURE 4.7 Top view of magnetic configuration during the transition from a non-skyrmion state to a skyrmion state. The spin waves are spotted in the black ellipses. The red and blue colors indicate local moments pointing to the right and left, respectively. To better visualize the spin waves, the contrast was significantly enhanced. (From Miao, B.F., et al., *Phys. Rev., B*, 90, 174411, 2014. With permission.)

Chapter 4

the x-axis during the switching between different states, while the red/blue color denotes the local magnetization pointing to right and left, respectively. One can find spin-wave excitations, as highlighted in the black ellipses, reflecting that the transition is not smooth. The finding of the spin-wave excitation manifests the transition that occurs between different topological states.

4.2.4   The Phase Diagram of Artificial Skyrmion Crystal

In the following, we continue to discuss the stability of the artificial skyrmion crystal. In the calculations, the material parameters of Co and CoPt at room temperature are used. Given the strong magnetic anisotropy of CoPt and the film thickness, it would be expected that the system has a Curie temperature T_C close to its bulk value. Therefore, the magnetic configuration could survive from low temperature to close to T_C, which is well above room temperature. To explore its stability in field, we also performed calculations in magnetic fields of various strengths. Figure 4.8 represents the calculated phase diagram and the field-dependent vortex core diameter, d, which is defined as the full width of the half value of the magnetization in the perpendicular direction (inserted line profile in Figure 4.8). We find that the stability of the skyrmion crystal depends on the direction of the applied field. When the field is applied along the vortex core direction, the crystal can be stabilized until the film magnetization is switched. When the field is applied oppositely, the skyrmion crystal is stable until the flipping of the vortex polarity. In between, the system remains in a skyrmion state. In this particular geometry, we find that the skyrmion crystal can be stable in a wide field range, from about −580 to +360 mT. We note that the stability also depends on the sizes of the disks and their separation, as well as the film thickness. We also find that the vortex core expands when the field is parallel to the magnetization of the vortex core and shrinks when the field is applied oppositely. This can be understood as follows. The vortex core size is determined by the competition among the exchange energy, the dipolar energy, and the perpendicular anisotropy energy. The perpendicular external field can be considered as a unidirectional anisotropy. Depending on its orientation, the field can enhance or decrease the effective anisotropy and result in a change in the vortex core size accordingly. We note that we did not observe the dip zone which is commonly found surrounding a usual vortex core in a soft disk (Ha et al. 2003). It could be related to the disks surrounded by the

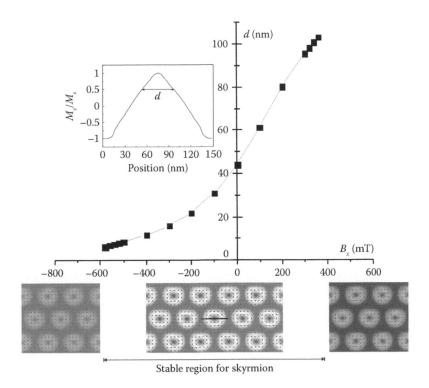

FIGURE 4.8 Calculated phase diagram and the core diameter of vortex in skyrmion configuration as a function of perpendicular field. The detailed dimension is the same as in Figure 4.4. The inset shows the typical magnetization line profile across a vortex center as marked at the bottom. (From Sun, L., et al., *Phys. Rev. Lett.*, 110, 167201, 2013. With permission.)

area with the perpendicular magnetization aligned oppositely to the vortex core, in our case.

The stability of the skyrmion is also explored as the function of both the film and disk thickness and it is found that it can be stabilized in a broad region as shown in Figure 4.9. For $D = 120$ nm and $S = 150$ nm, it exists in most of the combinations of film and disk thickness when $t_d > 20$ nm (Figure 4.9a). Below that, the disks and the area under them are in C-states. Interestingly, we find that the switching of vortex cores and the disks' surrounding area are thickness dependent. When $t_f > 10$ nm, the polarities of the vortices are switched before the disk surrounding area, resulting in a skyrmion lattice with opposite skyrmion density per unit cell. Figure 4.9b represents the phase diagram for $D = 90$ nm and $S = 100$ nm. The stable region is slightly smaller. The disks have the tendency to form C-states

Chapter 4

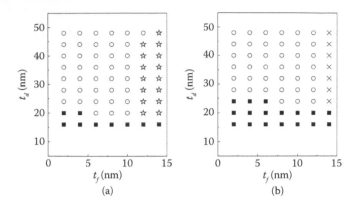

FIGURE 4.9 Calculated phase diagram of the skyrmion stability for systems with (a) $D = 120$ nm and $S = 150$ nm, and (b) $D = 90$ nm and $S = 100$ nm. Black squares represent disks in C-states. The open circles and stars show the systems in skyrmion states with +1 and −1 skyrmion density per unit cell, respectively. The cross represents the systems in states with partially skyrmion and partially topologically normal state (the vortex core and the magnetization of the disk surrounding area are pointing to the same direction). (From Sun, L., et al., *Phys. Rev. Lett.*, 110, 167201, 2013. With permission.)

with decreasing size. The stability is also examined for larger dimensions, e.g., $D = 800$ nm and $S = 900$ nm, and the skyrmion crystal was also obtained. Additional calculations are made for the combination of Co disks and FePt film; the skyrmion state can also be found with proper geometrical configuration.

4.2.5 Theoretical Prediction of Néel–Type Skyrmion Crystal

As discussed in previous chapters, there are two types of skyrmions, Bloch-type and Néel-type. This chapter primarily discusses the Bloch-type skyrmion. The Néel-type artificial skyrmion was also theoretically proposed in Co/Ru/Co nanodisks (Dai et al. 2013). The approach does not require the presence of the DMI, either. By tuning the thickness of an inserted Ru layer, skyrmion with opposite skyrmion numbers can form both within upper and lower Co layers due to the interlayer antiferromagnetic interaction. The authors also demonstrate that the skyrmion can remain stable in the applied magnetic field along the +z direction even up to 0.44 T. More details can be found in Chapter 1. In addition, Xie and Sang (2014) investigated the triple-layer CoPt/Co/CoPt structure and found that the skyrmion-like state can be formed not only in the CoPt layers but also in the middle Co layer.

4.2.6   Proposed Skyrmion State in Ferroelectric Nanocomposites

While skyrmions were amply investigated in magnets due to the presence of chiral interactions, ferroelectrics was somewhat hindered by the absence of intrinsic chiral interactions that are known to stabilize such configurations in noncentrosymmetric magnetic systems. However, it is now unequivocal that such interactions are not a prerequisite for obtaining skyrmion topological patterns (Dai et al. 2013; Nagaosa and Tokura 2013; Sun et al. 2013; Miao et al. 2014). Recently, Nahas et al. (2015) theoretically predicted the skyrmion configuration of polarization via the use of a first-principles-based technique. This work widens the scope and capability of future skyrmion-based applications.

A ferroelectric nanocomposite was considered, consisting of a cylindrical $BaTiO_3$ (BTO) nanowire with a radius of 2.7 nm embedded in a $SrTiO_3$ (STO) matrix in the search for electrical skyrmion. The investigated ferroelectric structure is schematized in Figure 4.10a.

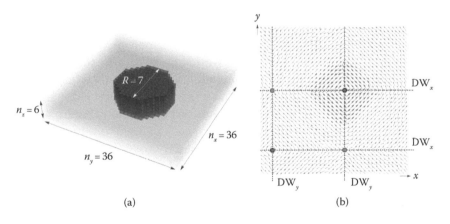

(a)   (b)

FIGURE 4.10   (a) Schematic representation of the periodic supercell used for calculating the ferroelectric skyrmion. The structure consists of a cylindrical $BaTiO_3$ (BTO) nanowire with a radius R of seven lattice constant units (2.7 nm) embedded in a $SrTiO_3$ (STO) matrix with lateral sides along the [100] and [010] directions of $n_x = n_y = 36$ lattice constant units, and a length $n_z = 6$ along the [001] pseudo-cubic direction. (b) Cross-sectional dipolar configuration of the $(V_{xy}|FE_z)$ state characterized by a vortex pattern in the z-planes co-occurring with an electrical polarization along the [001] direction. Arrows correspond to the x and y components of the electric dipoles in an arbitrary (001)-plane of the shifted periodic supercell. Dark and light gray circles specify the location of vortices and antivortices, respectively, all occurring at the intersection of DW_x and DW_y domain walls (dashed lines) separating different configurations of the x and y components of polarization. (From Nahas, Y., et al., *Nat. commun.*, 6, 8542, 2015. With permission.)

Chapter 4

This nanocomposite is mimicked by a $36 \times 36 \times 6$ supercell that is periodic along the x-, y-, and z-axis (which lies along the pseudocubic [100], [010], and [001] directions, respectively). The creation of an electrical skyrmion is presently achieved by a numerical procedure consisting of the following steps. Firstly, a temperature annealing under an external electric field, $E_{[001]} = 10^8$ V/m, applied along the pseudocubic [001] direction. On reaching 15 K, the field is set to zero and the previously obtained low-temperature configuration is further relaxed. The resulting relaxed configuration features a spontaneous polarization along the axial direction of the nanowire, co-occurring with a flux-closure four-domain vortex structure of the cross-sectional polarization field, in-plane pattern, whereby the strength of the depolarization field is reduced. The corresponding cross-sectional polarization field is shown in Figure 4.10b, which displays the x and y components of the electric dipoles in an arbitrary (001)-plane of the shifted periodic supercell. This state will be referred to as $(V_{xy}|FE_z)$ in the following, to emphasize that it has a vortex state in the z-planes but also possesses an electrical polarization along the [001] direction. Moreover, in this $(V_{xy}|FE_z)$ state, in addition to the central vortex whose core is confined within the nanowire, the matrix exhibits a vortex at midway between second-nearest neighbor wires as well as two antivortices at midway between first-nearest neighbor wires of the periodic supercell. These punctual topological defects are associated with singularities in the 2D cross-sectional polarization field and are found to be anchored at the junction of DW_x and DW_y domain walls, which interpolate between domains of distinct in-plane (x and y) components of polarization (Figure 4.10b). The sum of the O(2) winding numbers, each defined as a line integral measuring the change in the angle of the 2D cross-sectional polarization field over a closed path and giving $n = +1$ for vortices and $n = -1$ for antivortices, yields a zero net topological charge.

In determining skyrmion configuration, the $(V_{xy}|FE_z)$ state is further subjected to a field pointing oppositely to $(V_{xy}|FE_z)$'s polarization. The evolution of the z-component of polarization, P_z, with $E_{[00\bar{1}]}$ is shown in Figure 4.11a. With decreasing $E[001]$, one can see that P_z first slightly decreases before exhibiting a steep drop. Microscopical insight into the behavior of the overall polarization is illustrated in Figure 4.11b–d, which shows the distribution of P_z within an arbitrary z-plane of the supercell for three different values of $E_{[00\bar{1}]}$. It is therein seen that as $E_{[00\bar{1}]}$ increases, the matrix transiently develops inhomogeneity and is thus primarily accountable for the observed reduction of P_z. After completion of this transient process, at a certain threshold value E^* of the field (see dashed vertical line in Figure 4.11a), the dipolar configuration is such that the z-component

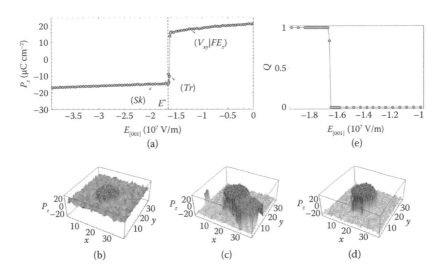

FIGURE 4.11 (a) Evolution of the P_z component of polarization with the gradually increasing external electric field at 15 K. Error bars indicate standard deviation and are less than or equal to the size of the points when not visible. Arrows indicate the state $(V_{xy}|FE_z)$ characterized by a vortex pattern in the z-planes co-occurring with an electrical polarization along the [001] direction as obtained for, comprising between -1.6×10^7 and 0 V/m, the transient state obtained over the narrow range extending from -1.65×10^7 to -1.6×10^7 V/m, and the skyrmionic state obtained under $E^* = -1.65 \times 10^7$ V/m, where E^* (dashed vertical line) is the threshold value of the field upon which the z-component of local dipoles is pointing along in the matrix and along [001] in the wire. (b–d) Distribution of P_z within an arbitrary z-plane of the supercell for the $(V_{xy}|FE_z)$ state, the transient state, and the skyrmionic state, respectively. The gray scale of (b–d) represents the magnitude of the out-of-plane polarization. (e) Dependence of the topological charge Q of the dipolar configurations on the external electric field. (From Nahas, Y., et al., *Nat. commun.*, 6, 8542, 2015. With permission.)

of the electric dipoles of the matrix is antiparallel to that of the dipoles comprised within the wire. By increasing the external electronic field, the topological charge Q jumps sharply from a topologically trivial state of $Q = 0$ to a skyrmion state of $Q = 1$ (Figure 4.11e).

4.3 Experimental Demonstration of Artificial 2D Skyrmion Lattices

Physics is a discipline based on experiments. Several different theoretical proposals for creating magnetic artificial skyrmion crystal are discussed above. The feasibilities of these proposals still require experimental examination. In the following, the experimental realization of the artificial skyrmion will be discussed.

Chapter 4

4.3.1 Sample Fabrication Process and Hysteresis Loop Characterization

[Pt(0.5 nm)/Co(0.5 nm)]$_5$/Pt(5 nm) has been deposited on a Si(001) substrate by dc magnetron sputtering, where the 5-nm Pt film acts as a buffer layer. The hysteresis loop confirms that the film has perpendicular anisotropy with the switching field of ~1.3 kOe. After that, the sample is transferred for patterning by means of an ultraviolet lithography system. A Co film of 32 nm is then deposited by electron beam evaporation onto the pre-patterned sample. After lift-off, edge-clipped circular disk arrays with the diameter of ~2 μm and a cutting angle of ~90° are obtained (Figure 4.12). The center-to-center separation between the neighboring Co disks is ~2.7 μm. As will be demonstrated below, the Co disk in this major-segment geometry has a vortex ground state, and its circulation can be controlled by in-plane magnetic field pulses.

A vortex has four degenerate states with CW/CCW circulation, and up or down polarity. The following manipulation sequences were enforced in order to format all Co disks into uniform polarity and circulation. Firstly, a weak in-plane magnetic field (0.2 kOe) along the clipped edge is applied to saturate the Co disks into a single domain. After releasing it, most of the disks would deform into vortices with the same circulation. Secondly, a perpendicular field of 10 kOe (which is slightly smaller than the saturation field of 13 kOe) is applied to format the vortex cores and surrounding material into a parallel configuration. Upon turning off this field, the disks remain in a vortex state with all the core magnetization pointing to the field direction. At last, a smaller

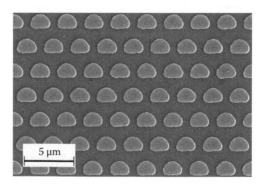

FIGURE 4.12 Typical scanning electron microscopy image of the fabricated structure: disk diameter, 2 μm; lattice constant, 2.7 μm. The major-segment disk is made of 32-nm-thick Co, while the perpendicular anisotropy material beneath is a [Pt(0.5 nm)/Co(0.5 nm)]$_5$ multilayer. (From Miao, B.F., et al., *Phys. Rev., B*, 90, 174411, 2014. With permission.)

perpendicular field of −1.5 kOe is applied to align the disk periphery antiparallel with the vortex cores. In such cases, the vortex core polarities remain unchanged and they are antiparallel with the surrounding Co/Pt multilayers. Thus, a skyrmion lattice is stabilized.

Figure 4.13a presents a hysteresis loop measured by a superconducting quantum interference device (SQUID) for Co/Pt multilayers (black curve, with a dimension of ~2.5 mm × 3.5 mm), which exhibits a sizable perpendicular anisotropy with a coercivity of ~1.3 kOe. For comparison,

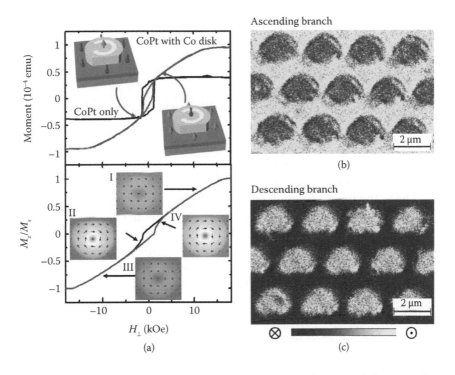

FIGURE 4.13 (a) Upper panel: hysteresis loops along the vertical direction for [Pt(0.5 nm)/Co(0.5 nm)]$_5$/Pt(5 nm)/Si substrate with (light gray) and without (black) Co disks. The insets show two skyrmions under different magnetic fields. The arrows represent the magnetization orientation of the local moments. Lower panel: the calculated hysteresis loop obtained using parameters defined in the text. The insets show the top view of magnetic configuration of the Co/Pt layer in a calculated artificial skyrmion crystal at different stages. In order to clearly show the contrast of the vortex core and surrounding material, the Co-disk diameter is set as 120 nm with a 150-nm separation. Polar Kerr microscopy images under perpendicular field: (b) and (c) present the magnetic contrast along the ascending branch and descending branch, respectively. Cross (dot) in a circle denotes the magnetization component pointing into (out of) the plane, respectively. The seemingly visual difference between the disks in (b) and (c) is only caused by the different (bright/dark) surrounding background. (From Miao, B.F., et al., *Phys. Rev.*, B, 90, 174411, 2014. With permission.)

Chapter 4

the hysteresis loop after the Co-disk deposition is plotted in the same figure. After coating the Co disks, the switching field of the film is sizably reduced to ~0.9 kOe. The reduced coercivity demonstrates that there is ferromagnetic coupling between the Co disks and the underneath Co/Pt multilayer as the in-plane magnetization of the Co disk acting as an effective transverse field which softens the Co/Pt multilayer. Meanwhile, the switching amplitude is also reduced to 60% of the value that, prior to the Co disk deposition, was in agreement with the estimated area ratio (55%). The reduced amplitude in the jump evidences that the Co/Pt multilayers underneath the Co disks is no longer perpendicular due to the strong coupling between them. This is consistent with the previous calculation that the vortex state penetrates into the Co/Pt multilayers underneath the disks.

4.3.2   The Kerr Microscopy Study of 2D Skyrmion Crystal

The polar Kerr microscopy measurements are performed to depict the out-of-plane magnetization component. The original image typically contains both the topographic and magnetic contrasts. To remove the topographic contrast, a background image is taken. In the case studied here, the magnetic contrast along the ascending branch is obtained by subtracting image at ~1.0 kOe (switched) by that at ~0.9 kOe (pre-switched) (Figure 4.13b). Similarly, one can obtain the image for the descending branch of the hysteresis loop (Figure 4.13c). The surrounding CoPt film changes from bright to dark, reflecting the switching of the magnetization direction. The sharp contrast between Figure 4.13b and c demonstrates that the surrounding Co/Pt multilayer can be controlled by a perpendicular magnetic field with the switching field amplitude of ~0.95 kOe.

To confirm that the disks are in the vortex state, the Kerr microscopy measurements were performed in two longitudinal geometries (Figure 4.14a and b). In these geometries, the contrast is sensitive to the in-plane magnetization along the optical plane. As shown in the sketches in Figure 4.14, the optical planes in these two geometries are aligned perpendicular to each other, i.e., along or perpendicular to the cutting-edge direction. In such a case, the Kerr signal is sensitive to the magnetization along and perpendicular to the clipped edge in Figure 4.14a and b, respectively. In order to remove the morphology contrast and the residual polar Kerr contributions, a background subtraction technique is again used. The disks were found to be saturated into a uniform single-domain state upon applying an in-plane field of 0.2 kOe. The average image obtained with

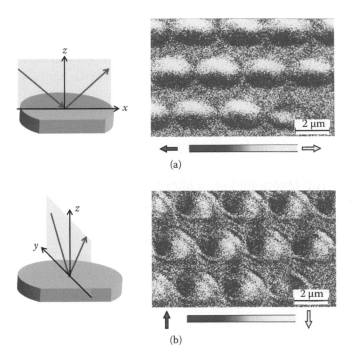

(a)

(b)

FIGURE 4.14 Kerr microscopy images with the sensitivities along the x-axis (a) and the y-axis (b). Blue (yellow) arrow denotes the magnetization component along the left (right) in (a) and up (down) in (b), respectively. Both images are obtained with longitudinal magneto-optical Kerr effect geometry with the corresponding schematic setups shown on the left. (Note that the weak stripe contrast found near the disk edges in (b) was caused by the imperfect background subtraction process.) (From Miao, B.F., et al., *Phys. Rev., B*, 90, 174411, 2014. With permission.)

± 0.2 kOe is chosen as the background. Figure 4.14a depicts the magnetization component along the clipped edge in zero-field after the background subtraction. The Co disks have a contrast between up and down portions, reflecting opposite magnetization components along the x-direction. Remarkably, most of the disks, except the one at the bottom right, show bright contrast on top and dark contrast at the bottom, suggesting they are mostly in the same magnetic configurations. The up/down contrast suggests that the disks can be either in two-domain or vortex states. In order to identify the magnetic configuration of the Co disks, similar measurements were further performed with the optical plane rotated by 90°. If the disks are in vortex states, the contrast should also rotate 90° accordingly (Ding et al. 2005). Because the domain wall typically has a width on the order of tens of nanometers—which makes it unresolved within the limitations of

Chapter 4

Kerr microscopy—there will be no contrast for the two-domain configuration. Figure 4.14b shows the result of the same spot after rotating the optical plane by 90°. The image has contrast between left and right portions; that is, the contrast rotates following the rotation of the optical plane. This unambiguously proves that the disks are in the vortex state. From their contrasts, 12 of 13 disks shown in the image were identified to have the same CW circulation. We note that the weak contrast near the disk edge in Figure 4.14b originates from the imperfect background subtraction.

4.3.3   The Magnetic Force Microscopy Study of 2D Skyrmion Crystal

To further confirm that the disks are in the vortex state, the magnetic force microscopy (MFM) measurements have been used. For a perfect circular vortex disk, the MFM image typically exhibits a dark or bright spot at the disk center only (Figure 4.15a) (Shinjo et al. 2000). For a nonperfect disk such as an elliptical one, the vortex is no longer perfectly circular and the stray field also exists in the surrounding part. In such a case, a pattern of four quadrants with alternating dark and bright contrasts is expected (Figure 4.15b) (Okuno et al. 2002). Figure 4.16a shows the typical MFM image of our patterned structure. Except for several disks which have fine structures (and may possess a multi-vortex/multi-domain state

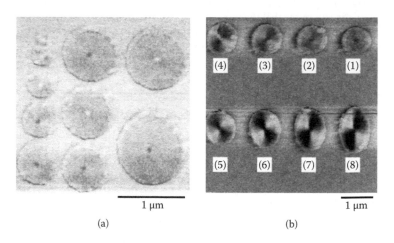

(a)   (b)

FIGURE 4.15   (a) MFM image of an ensemble of 50-nm-thick permalloy dots with diameters varying from 0.1 to 1 mm after applying an external field of 1.5 T along an in-plane direction. (b) An MFM image of elliptical dots, where the width is 1 μm along the short axis and increases gradually from 1 μm (1) to 2 μm (8) along the long axis. (Adapted from Shinjo, T., et al., *Science*, 289, 930–932, 2000; Okuno, T., et al., *J. Magn. Magn. Mater.*, 240, 1–6, 2002. With permission.)

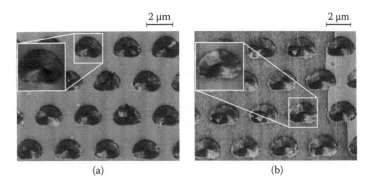

FIGURE 4.16   MFM image of two skyrmion states. The alternating dark and bright contrasts indicate the Co disks are in a vortex state. The insets in (a) and (b) present the zoomed-in image of a single vortex. A small portion of the underlying CoPt multilayer in the right edge did not switch, and thus, acted as a reference point. (From Miao, B.F., et al., *Phys. Rev., B*, 90, 174411, 2014. With permission.)

[Prejbeanu et al. 2002]), most of the disks (23 out of 28 disks examined) clearly show four quadrants with alternating dark and bright contrasts evidencing that they are in the vortex state since the edge-clipped circular disk naturally has a shape asymmetry. Two kinds of patterns of four quadrants in the disks correspond to CW or CCW circulation. From the patterns, we find most of the vortices have the same circulation, which are consistent with Kerr microscopy images in Figure 4.14. Figure 4.16b represents the MFM image with opposite vortex circulation with respect to Figure 4.16a after the field operation. An opposite perpendicular field close to the switch filed of surrounding CoPt multilayers, ~0.9 kOe, was chosen as the imaging condition. In this case, a noticeably small portion of CoPt film in the right side did not switch and shows opposite magnetization with respect to the left part.

The relative orientation of the vortex core and the surrounding perpendicular film can be further resolved with disks in almost perfect circular geometry. Fraerman et al. (2015) performed these measurements and confirmed that the vortex core and the surrounding perpendicular film can be configured to align oppositely. The formation of artificial skyrmion is therefore firmly confirmed.

4.3.4 Photoemission Electron Microscopy Study of the Epitaxial Artificial Skyrmion

The artificial skyrmion can also be realized in an epitaxial system. By growing an epitaxial Co vortex disk on top of an out-of-plane magnetized Ni film, Li et al. (2014) demonstrated a controllable skyrmion

Chapter 4

topological index and topological effect in a skyrmion core annihilation process. The authors fabricated 30-nm-thick Co circular disks with a radius of 1 μm on top of an epitaxially grown 30 monolayer thick Ni film on Cu(001) substrate. The epitaxial Ni/Cu(001) has an out-of-plane magnetization when the thickness of Ni is above ~7 monolayer. Co/Cu(001) film has an in-plane magnetization, and Co/Ni(30 monolayer)/Cu(001) magnetization undergoes a spin re-orientation transition from out-of-plane to in-plane direction as the Co thickness increases above ~1 nm. Therefore, the Co(disk)/Ni(30 monolayer)/Cu(001) sample should consist of two distinct regions in terms of magnetization direction: (1) the Co/Ni disk has an in-plane magnetization and (2) the Ni surrounding the disk has an out-of-plane magnetization. Element-specific X-ray magnetic circular dichroism (XMCD) measurements (Figure 4.17) confirm this spin configuration that the Co exhibits an in-plane magnetic hysteresis loop with a full remanence, and the surrounding Ni exhibits an out-of-plane magnetic hysteresis loop with a full remanence (the Ni XMCD measures only the Ni region surrounding the Co disks because of the surface sensitivity of the XMCD measurement).

The Co magnetic domain image from photoemission electron microscopy (PEEM) measurement clearly shows the formation of a magnetic vortex state (spins curling around the center of the vortex). The vortex contrast changes accordingly after changing the in-plane projection of the X-ray beam by 90° (Figure 4.17d), confirming that the vortex state is an in-plane curling of the spin texture (the observation of the out-of-plane vortex core is beyond the PEEM spatial resolution). After applying a $H = 2.0$ T magnetic field in the out-of-plane direction and then turning off the field, the central vortex core polarity will be parallel to the surrounding Ni spins with skyrmion number $Q = 0$ state (Figure 4.18a). If an 800-Oe magnetic field opposite to the core polarity direction is further applied and then turned off, the $Q = -1$ skyrmion state can be obtained (Figure 4.18b). As the magnetic coercivity of the surrounding Ni is 500 Oe and the core polarity reversal field for a permalloy disk is >3000 Oe, the 800-Oe field will only reverse the surrounding Ni spin direction without reversing the central vortex core polarity. Therefore, the ending state has an opposite orientation between the central vortex core polarity and the surrounding Ni spins, that is, the $Q = -1$ skyrmion state (Figure 4.18b).

The authors also use an in-plane field pulse to annihilate the skyrmion/vortex core by pushing it out of the disk region in $Q = -1$/ $Q = 0$ state. Figure 4.19 shows the PEEM images of the central vortex

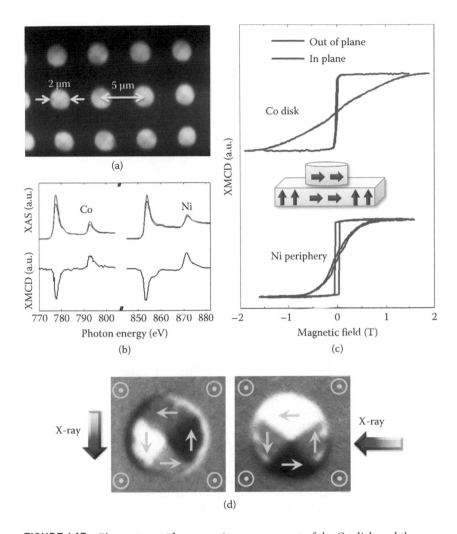

(a)

(b)

(c)

(d)

FIGURE 4.17 Element-specific magnetic measurement of the Co disk and the surrounding Ni. (a) PEEM image of the Co(disk)/Ni(30 monolayer)/Cu(001) sample fabricated using shadow mask. (b) X-ray absorption spectra (XAS) of Co and Ni at 2p levels for left circular polarized X-ray. The difference of the XAS for magnetization parallel (red color) and antiparallel (blue color) to the X-ray beam represents the element-specific XMCD signal. (c) Element-specific hysteresis loops shows that the Co disk (together with the Ni below the disk) has an in-plane magnetization, and the Ni surrounding the disk has an out-of-plane magnetization. (d) Co PEEM images with different X-ray incident directions show that the Co disk forms a magnetic vortex state. The central vortex plus the surrounding out-of-plane Ni spins correspond to a magnetic skyrmion. (From Li, J., et al., *Nat. Commun.*, 5, 4704, 2014. With permission.)

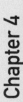

Chapter 4

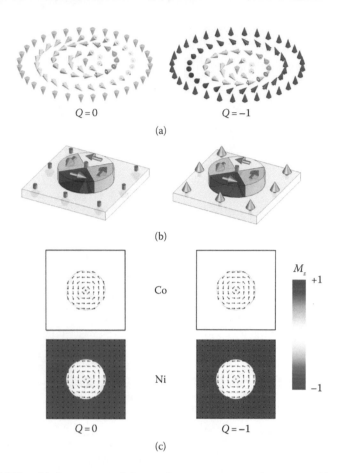

FIGURE 4.18 (a) $Q = 0$ state and $Q = -1$ skyrmion state are characterized by a parallel and antiparallel alignment of the central vortex core polarity relative to the surrounding out-of-plane spins, respectively. (b) $Q = 0$ state and $Q = -1$ skyrmion state in Co(disk)/Ni(30 monolayer)/Cu(100) can be prepared by switching the surrounding Ni spin direction without switching the central vortex core polarity using an 800-Oe out-of-plane magnetic field. (c) Micromagnetic simulation confirms the formation of $Q = 0$ state and $Q = -1$ skyrmion state by switching the surrounding Ni spin direction without changing the core polarity (blue dot at the center). (Adapted from Li, J., et al., *Nat. Commun.*, 5, 4704, 2014. With permission.)

state after application of an in-plane magnetic field pulse of different strengths. The Co spin texture is imaged, as discussed above, is also representative of the strong exchange coupled Ni spins. For the $Q = 0$ state, the central vortex state remains after the application of an in-plane magnetic field pulse up to ~100 Oe and switches to the single domain state for field greater than ~110 Oe (Figure 4.19a). In contrast, the $Q = -1$ skyrmion central vortex state remains up to a field strength of ~140 Oe and switches to the single domain state for field greater than

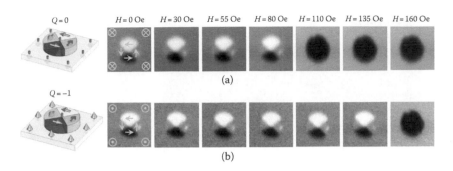

FIGURE 4.19 The central vortex is surrounded by out-of-plane Ni spins as indicated by the yellow symbols. (a) $Q = 0$ state for parallel alignment between the central vortex core and the surrounding Ni spins and (b) $Q = -1$ skyrmion state for antiparallel alignment between the central vortex core and the surrounding Ni spins. The critical field needed to switch the central vortex disk to the single-domain disk ($Q = 0$ state) is weaker for $Q = 0$ state than $Q = -1$ skyrmion, suggesting a topological effect in the skyrmion core annihilation process. (Adapted from Li, J., et al., *Nat. Commun.*, 5, 4704, 2014. With permission.)

~160 Oe (Figure 4.19b). As the PEEM images in Figure 4.19a and b are taken from the same disk, the different critical fields in annihilating the skyrmion core can be attributed to the different topologies of the skyrmion: a lower critical field for $Q = 0$ state as opposed to a greater field for $Q = -1$ skyrmion.

4.3.5 Scanning Electron Microscopy with a Polarization Analysis Study of Artificial Skyrmion Crystals

Gilbert et al. (2015) patterned the vortex-state Co nanodots in hexagonal arrays via electron-beam lithography on top of a Co/Pd thin film with perpendicular magnetic anisotropy, as illustrated in Figure 4.20a. A Co/Pd film was first sputter deposited and subsequently spin-coated with a polymethyl methacrylate film; arrays of asymmetric edge-cut anti-dots were patterned into this polymer layer by e-beam lithography. Next, the film was irradiated with energetic Ar$^+$ ions, thus modifying the multilayer structure in the exposed dot areas and tilting their easy axis toward the in-plane direction. The irradiation step of Ar$^+$ ions was attributed to be one of the crucial steps for forming skyrmion crystal in the system of Co disks on top of Co/Pd film. Finally, Co was sputtered over the irradiated anti-dot arrays, forming edge-cut Co dots after a lift-off process. The as-grown Co/Pd films exhibit strong PMA with an out-of-plane remanence of essentially unity and

Chapter 4

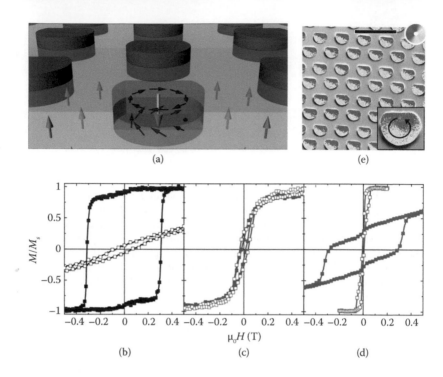

FIGURE 4.20 (a) The hybrid structure consists of Co dots (red) on top of Co/Pd PMA underlayer (grey) where the in-plane spin texture of the Co dots (purple arrows) is imprinted into an irradiated Co/Pd region (light blue) underneath the dots (tilted blue arrows). Green and yellow arrows indicate the moments in the Co/Pd underlayer and the core region of the (imprinted) vortex, respectively. Major in-plane (open symbols) and perpendicular (solid symbols) hysteresis loops are shown for (b) the Co/Pd underlayer as grown, (c) the irradiated Co/Pd witness sample, and (d) the hybrid Co+Co/Pd sample. Remanent-state (e) vectorial SEMPA image, after saturating the dots in an in-plane field parallel to the flat edge of the dots to the right, indicates circularity control. (Adapted from Gilbert, D.A., et al., *Nat. Commun.*, 6, 8462, 2015. With permission.)

a coercivity of $\mu_0 H_c = 320$ mT (Figure 4.20b). In contrast, for the Co/Pd film irradiated by Ar^+ over the entire area, the in-plane and perpendicular hysteresis loops are nearly identical (Figure 4.20c), confirming that the PMA has been successfully suppressed. The hybrid sample in the perpendicular geometry retains the characteristics of the unirradiated Co/Pd film (Figure 4.20d), illustrating that the processing did not damage the perpendicular layer other than as designed. These results demonstrate that indeed the hybrid sample has realized the designed magnetic configurations, that is, the Co/Pd underlayer is perpendicularly magnetized in a single domain state, while the Co dots are in an in-plane vortex state.

The skyrmion texture was then prepared by a designed field sequence similar as described above for the system of Co disks on top of Co/Pt film. After the field sequence, the scanning electron microscopy with polarization analysis (SEMPA) shows that all the disks have the same circulation providing direct evidence of circularity control over the dot arrays, as shown in Figure 4.20e.

The skyrmion lattice (SL) requires the core polarity opposite to the underlayer magnetization. By applying a perpendicular bias field parallel to the underlayer magnetization, the polarity can be aligned with the underlayer, which results in a vortex lattice (VL) on top of a PMA underlayer (skyrmion number $Q = 0$). The SL and VL will have different perpendicular remanent magnetizations as a result of the vortex core orientation. In addition, zero-bias field during the vortex nucleation in the Co dots would result in a random distribution of the core polarity. This mixed lattice (ML) is expected to have a remanence in between the cases with ordered polarity and a reversal behavior indicative of both features. The authors configure the remanent state of the hybrid structure into the SL, VL, and ML. As the perpendicular field was swept from 0 to negative saturation, the magnetization curves reveal clear differences, as shown in Figure 4.21a, with the difference between SL and VL highlighted in Figure 4.21b. In the early stage of the magnetization reversal (illustrated in Figure 4.21c), the case of parallel alignment between the core and underlayer (VL) exhibits the largest magnetization. Antiparallel alignment (SL) shows the smallest, while the ML is midway between the VL and SL curves. The magnetization difference between the VL and SL configurations remains almost constant along the field sweep until the underlayer's reversal field at $B \sim -0.3$ T, indicating the stability of the skyrmions. As the Co/Pd underlayer starts its reversal (shown in Figure 4.21d), the abrupt magnetization drop occurs first in the SL configuration and last in the VL. The difference may originate from the fact that in the SL the oppositely oriented cores will facilitate the nucleation of reversal domains, whereas in the VL the parallel cores will not. The difference in the nucleation field for SL versus VL is indicative of the topological effect on the nucleation process of Co/Pd reversal domains. This behavior is consistent with the presence of both types of pole configurations with opposite polarities. The magnetization difference near remanence and the field sweep thus demonstrates the polarity control.

The polarized neutron reflectometry and transport measurement are used to confirm that the chiral texture in the vortex-state Co dots is imprinted into the Co/Pd, thus an SL forms with the underlying layer.

Chapter 4

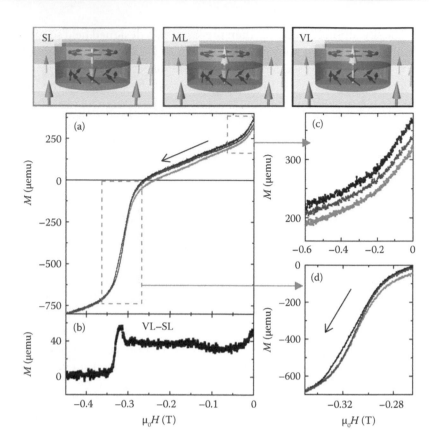

FIGURE 4.21 Top row shows schematic illustrations of the SL, VL, and ML states. The light gray arrows mark the core direction in the vortex and the imprinted region, while the other arrows represent the magnetic moments in other parts of the structure. (a) Magnetization curves, with the field sweeping from zero to negative saturation, for the hybrid structure prepared into the SL (light gray), VL (black), and ML (dark gray) states at remanence. (b) The image highlights the magnetization difference between the VL and SL, and zoomed-in views of the magnetization curves in dashed boxes are shown in (c) near zero field, and (d) ~320 mT where the Co/Pd underlayer starts its reversal. (Adapted from Gilbert, D.A., et al., *Nat. Commun.*, 6, 8462, 2015. With permission.)

4.4 Micromagnetic Study of Excitation Modes in an Artificial Skyrmion Crystal

4.4.1 Calculation Method

The excitation modes of the 2D artificial skyrmion crystal were studied using the NIST OOMMF code (Donahue and Porter 1999). Co is chosen as the vortex-disk material with zero crystalline anisotropy.

And for the perpendicular magnetic film, CoPt is used, which has a uniaxial perpendicular anisotropy $K_1^{CoPt} = 4.0 \times 10^5$ J/m^3 (Maret et al. 1997). The material parameters used in calculations are exchange constant: $A = 1.9 \times 10^{-11}$ J/m; saturation magnetization: $M_s^{Co} = 1.4 \times 10^6$ A/m for Co (Munzer et al. 2010); and $M_s^{CoPt} = 5.0 \times 10^5$ A/m for CoPt (Maret et al. 1997; Eyrich et al. 2012). The thickness of the Co disk and the CoPt perpendicular film is 24 and 8 nm, respectively. The Co disk diameter is $D = 90$ nm, with a center-to-center distance $S = 100$ nm. In the calculations, we apply 2D periodical boundary conditions within the plane and a grid size of $2 \times 2 \times 2$ nm^3. The Gilbert damping constant α is set to 0.02 in all the calculations.

In order to excite resonance modes in the skyrmion state ($Q = 1$) at a zero static field, a Gaussian-shaped pulse field is applied with an amplitude $h_0 = 10$ mT and width $w = 20$ ps along the x/z-direction, respectively. After this perturbation, the time evolution of the magnetization of each cell is calculated and stored. Fourier transform is then performed on the time evolution of the out-of-plane component of magnetization (M_z) in each cell, which yields the amplitude A_i and phase ϕ_i of each cell as a function of the frequency. The frequency resolution in our study is $1/(T_{end}-T_{start}) = 1/7$ ns $= 0.14$ GHz. The Fourier spectrum usually consists of a certain number of sharp peaks. Each of them indicates a resolved eigenmode of the system (Zhu et al. 2005; Mochizuki 2012). By plotting the real part of the Fourier transform $A_i \cos \phi_i$ at a certain resonance frequency determined from the spectrum, we can reconstruct the spatial profile of the corresponding eigenmode, where $A_i \cos \phi_i$ represents a snapshot of the dynamical out-of-plane component of the magnetization M_z at the specific frequency (Buess et al. 2004; Buess et al. 2005; Yan et al. 2006). Figure 4.22a shows the calculated spectra for the skyrmion state ($Q = 1$) excited by in-plane (black curve) and out-of-plane (light gray curve) Gaussian-shaped pulse field. The data have been normalized by the maximum value of each curve. Three resonance peaks can be observed under in-plane pulse field, located at 1.72, 10.44, and 12.88 GHz, respectively, while the out-of-plane pulse perturbation only yields two peaks at 1.72 and 18.45 GHz. As will also be discussed in the following part, the highest frequency peak at 18.45 GHz is an out-of-plane mode which can only be excited by an out-of-plane field. And due to the fact that OOMMF uses rectangular mesh to construct the Co disks, the circular symmetry of each Co disk near the edge is inevitably broken. Thus, the out-of-plane distortion of the magnetization also disturbs the in-plane component, causing the excitations of the in-plane modes at 1.72 GHz as well.

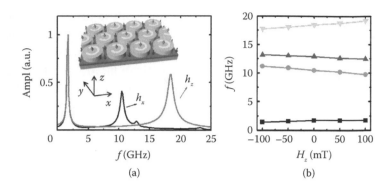

FIGURE 4.22 (a) Spin-wave excitation spectrum for a skyrmion crystal with skyrmion number $Q = 1$ under in-plane (black curve) and out-of-plane (light gray curve) excitation. The inset shows the schematic of a 2D artificial skyrmion crystal. Ordered arrays of vortex disks are prepared on top of a film with perpendicular anisotropy. The arrows represent the orientation of the local magnetization. The x-axis is defined along the cutting edge of vortex disk, with z-axis normal to the film surface. (b) Dependence of the four resonant frequencies on the perpendicular magnetic field H_z. (From Miao, B.F., et al., *Appl. Phys. Lett.*, 107, 222402, 2015. With permission.)

Figure 4.22b represents the evolution of four characteristic frequencies under different DC perpendicular fields, H_z. The lowest mode has a weak field dependence and its resonance frequency increases slowly with H_z. Both two medium modes are red-shifted as H_z increases, while the highest one is blue-shifted.

To identify each resonance mode, we present the snapshot of dynamical M_z (Figure 4.23a, left column) and phase ϕ (Figure 4.23a, right column) for each of the four eigenmodes of the artificial skyrmion. In the scale bar of M_z, red (blue) represents the positive (negative) absolute value of maximum. In the scale bar of ϕ, red, green, and yellow represent $-\pi/2$, $\pi/2$, $\pm\pi$ for the phase, respectively. Since for each mode, every individual skyrmion in the crystal behaves the same way, we focus on only one skyrmion unit hereafter. Figure 4.23a shows the first mode, in which M_z has both a blue and a red spot separated by two azimuthal nodes, which would rotate around the center as a function of time. The phase changes from $-\pi$ to π in the CCW sense. This mode is the strongest and with the lowest frequency, which corresponds to the gyration mode of a magnetic vortex core (Park and Crowell 2005; Zaspel et al. 2005). The dynamic M_z at 10.44 and 12.88 GHz shows similar features as that at 1.72 GHz, while the phase winds in the sense of CW and CCW. The two medium-frequency modes correspond to the CW and CCW rotational modes of skyrmion crystal. This point is further clarified by applying oscillating fields at given frequencies in the following

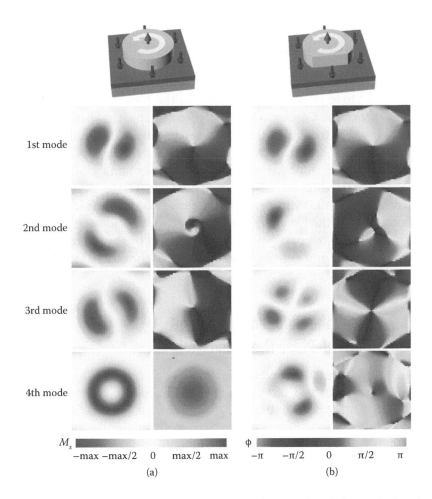

M_z | −max −max/2   0   max/2   max

ϕ | −π   −π/2   0   π/2   π

(a)   (b)

FIGURE 4.23 (a) Snapshots of the dynamic M_z (left column) and phase ϕ (right column) at different eigenfrequencies for a skyrmion state with $Q = 1$. The first, second, and third are azimuthal-like modes, and the fourth is a radial-like mode. (b) For comparison, we also present the results for a skyrmion crystal consisting of edge-cut Co disks. The color bars on the bottom are applicable for both (a) and (b). (From Miao, B.F., et al., *Appl. Phys. Lett.*, 107, 222402, 2015. With permission.)

discussion. For the highest frequency mode at h_z pulse perturbation, the dynamic M_z exhibits a ring pattern with concentric nodes only in the center of the disk and at its border, and the phase is almost uniform over the skyrmion (Figure 4.23a for the fourth mode). This mode corresponds to the breathing mode of skyrmion crystal.

We also calculated the dynamic behavior of skyrmion crystal with Co disk in the edge-cut geometry. Due to the fact that skyrmion is a topologically protected object, it is inertial under small perturbation.

Chapter 4

Thus, the skyrmion crystal with edge-cutting Co disk and circular Co disk exhibits similar dynamic features, three in-plane rotational modes and one out-of-plane breathing mode are observed (Figure 4.23b). And in the two cases, the skyrmion crystal gyrates in the same way for the low-lying three in-plane modes. Due to the breaking of the circular symmetry of the Co disk with an edge-cut shape, the spatial distributions of m_z and ϕ are distorted compared to those of a circular Co disk, especially at higher frequencies. For instance, the blue spot in dynamic m_z of the second mode is very light, and m_z exhibits a distorted ring pattern in the highest frequency mode; the phase ϕ also changes accordingly. In the third mode, we observe two red spots located at the upper-left and the lower-right part, and two blue spots at the upper-right and lower-left part. This is mainly caused by the stronger interaction of in-plane and out-of-plane resonance as the circular symmetry is broken in the edge-cut disk.

4.4.2 In-Plane Rotational Mode and Out-Of-Plane Breathing Mode

The spin dynamics is further studied by applying a continuous oscillating magnetic field at the resonant frequency f_r to the skyrmion crystal with skyrmion number $Q = 1$. First, an in-plane ac magnetic field is applied to the crystal to activate in-plane resonant modes, which is set as $h(t) = (h_x^f \sin(2\pi ft), 0, 0)$, with $h_x^f = 10$ mT. Figure 4.24a through c demonstrate the time evolutions of the magnetization for the skyrmion state ($Q = 1$) under a 1.72, 10.44, and 12.88 GHz ac in-plane field, respectively. In the magnetic configuration, the light/dark gray color denotes the magnetization component coming out-of-/into-surface, while the black arrow represents the in-plane direction. We find that the skyrmion core rotates in the sense of CCW for the low-lying mode, CW for the medium-mode, and CCW for the higher in-plane mode. The directions of their rotation are consistent with the phase profile calculated from Fourier transformation in Figure 4.23a. Equally interesting, the rotation directions are independent of the winding direction of spins in the skyrmion (Figure 4.24d). Instead, they are determined by the skyrmion number, manifesting its topologic nature of dynamic behavior. For instance, the skyrmion of a skyrmion crystal with $Q = -1$ (Figure 4.24e) rotates in the sense of CCW at 10.44 GHz, opposite to the skyrmion with $Q = 1$ (Figure 4.24b and d). Under the out-of-plane ac field at 18.45 GHz, the skyrmion oscillates in the breathing mode; its core extends and shrinks periodically

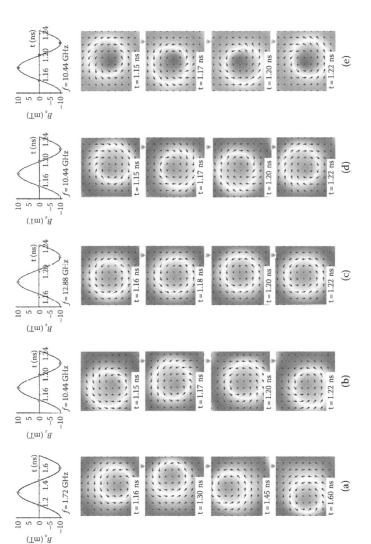

FIGURE 4.24 Time evolution of a skyrmion core ($Q = 1$) under continuous in-plane ac field at (a) 1.72 GHz, (b) 10.44 GHz, and (c) 12.88 GHz. The light/dark gray color denotes the magnetization coming out-of-the-surface/in-to-the-surface and the arrows represent the in-plane direction. (d) Time evolution of a skyrmion core ($Q = 1$) under continuous in-plane ac field at 1.72 GHz similar as (a) but with opposite (CW) circulation. (e) Time evolution of a skyrmion core with $S = -1$ under a 1.72 GHz in-plane ac field. (From Miao, B.F., et al., *Appl. Phys. Lett.*, 107, 222402, 2015. With permission.)

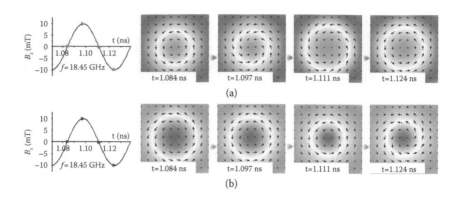

FIGURE 4.25 Breathing mode activated by a continuous out-of-plane ac field at 18.45 GHz. The skyrmion core extends and shrinks as a function of time. Skyrmion states (a) with $Q = 1$ and (b) $Q = -1$ breath out-of-phase with each other. (From Miao, B.F., et al., *Appl. Phys. Lett.*, 107, 222402, 2015. With permission.)

over time (Figure 4.25a). Similar to the in-plane rotational modes, the skyrmion state with opposite skyrmion numbers breathes out of phase with each other (Figure 4.25b).

4.4.3 Comparing the Spin–Wave Modes in Skyrmion with DMI

It is interesting to compare the spin-wave modes calculated herein for an artificial skyrmion with those of the skyrmion arising from the DMI, which were recently reported in Mochizuki (2012). In the previous study, it was found that two rotational modes exist under an in-plane ac field, with a CCW (CW) manner for the lower-lying (higher-lying) mode. And an out-of-plane ac field can excite the highest frequency breathing mode. It is important to note the skyrmion number $Q = -1$ in Mochizuki (2012), with the core pointing down and the periphery part pointing up. So the positive direction of the magnetic field H_z is opposite to our definition here. As already demonstrated in Figure 4.24, the directions of rotation modes of skyrmion crystal with $Q = 1$ are opposite with that of $Q = -1$. As a result, the CCW mode, CW mode, and breathing mode in Mochizuki (2012) correspond to the CW mode (10.44 GHz), CCW mode (12.88 GHz), and breathing mode (18.45 GHz) in this work, respectively. In addition, all these modes in artificial skyrmion have the same field dependence with those in a skyrmion with DMI. The additional resonant mode at around 1.72 GHz is the gyration mode of the vortex core.

4.5   Summary

In this chapter, we presented a joint theoretical and experimental exploration of a 2D artificial skyrmion crystal, which completely by-passes the need for DM interaction. The methodology is demonstrated with micromagnetic simulations and the computed skyrmion number per unit cell, and experimentally proved by Kerr microscopy, MFM, and hysteresis measurements. The created skyrmion crystal has a robust working regime including room temperature, much broader than that for DMI-driven skyrmion crystals. We have also studied the dynamics of artificial skyrmion crystal with micromagnetic simulations. In the absence of DMI, we do observe the in-plane CCW and CW rotational mode and out-of-plane breathing mode of skyrmion crystal, evidencing that the nontrivial magnetic structure is the intrinsic origin of skyrmion dynamics. We also find the skyrmion number determines the rotation direction of in-plane mode as well as the phase of out-of-plane breathing mode, manifesting the topological nature of skyrmion dynamics. Sidestepping the need for DMI, artificial skyrmion crystal greatly expands the pool of prospective materials in the future of skyrmion-based application devices.

Acknowledgments

This chapter is supported by NSFC (Grants No. 11304150, No. 11374145, and No. 51571109), and Natural Science Foundation of Jiangsu (Grant No. BK20150565).

References

Bogdanov, A. N., Rossler, U. K. and Shestakov, A. A. (2003). Skyrmions in nematic liquid crystals. *Physical Review E* **67**, 016602.

Brey, L., Fertig, H. A., Cote, R. and Macdonald, A. H. (1995). Skyrme crystal in a 2-dimensional electron-gas. *Physical Review Letters* **75**, 2562–2565.

Buess, M., Hollinger, R., Haug, T., Perzlmaier, K., Krey, U., Pescia, D., Scheinfein, M. R., Weiss, D. and Back, C. H. (2004). Fourier transform imaging of spin vortex eigenmodes. *Physical Review Letters* **93**, 077207.

Buess, M., Knowles, T. P. J., Hollinger, R., Haug, T., Krey, U., Weiss, D., Pescia, D., Scheinfein, M. R. and Back, C. H. (2005). Excitations with negative dispersion in a spin vortex. *Physical Review B* **71**, 104415.

Chen, G., Mascaraque, A., N'Diaye, A. T. and Schmid, A. K. (2015). Room temperature skyrmion ground state stabilized through interlayer exchange coupling. *Applied Physics Letters* **106**, 242404.

Chien, C. L., Zhu, F. Q. and Zhu, J. G. (2007). Patterned nanomagnets. *Physics Today* **60**, 40–45.

Chapter 4

Cowburn, R. P., Koltsov, D. K., Adeyeye, A. O., Welland, M. E. and Tricker, D. M. (1999). Single-domain circular nanomagnets. *Physical Review Letters* **83**, 1042–1045.

Dai, Y. Y., Wang, H., Tao, P., Yang, T., Ren, W. J. and Zhang, Z. D. (2013). Skyrmion ground state and gyration of skyrmions in magnetic nanodisks without the dzyaloshinsky-moriya interaction. *Physical Review B* **88**, 054403.

Ding, H. F., Schmid, A. K., Li, D. Q., Guslienko, K. Y. and Bader, S. D. (2005). Magnetic bistability of Co nanodots. *Physical Review Letters* **94**, 157202.

Donahue, M. J. and Porter, D. G. (1999). *Oommf user's guide, version 1.0.* National Institute of Standards and Technology, Gaithersburg, MD, NISTIR 6376.

Dumas, R. K., Gredig, T., Li, C. P., Schuller, I. K. and Liu, K. (2009). Angular dependence of vortex-annihilation fields in asymmetric cobalt dots. *Physical Review B* **80**, 014416.

Eyrich, C., Huttema, W., Arora, M., Montoya, E., Rashidi, F., Burrowes, C., Kardasz, B., Girt, E., Heinrich, B., Mryasov, O. N., From, M. and Karis, O. (2012). Exchange stiffness in thin film Co alloys. *Journal of Applied Physics* **111**, 07C919.

Fraerman, A. A., Ermolaeva, O. L., Skorohodov, E. V., Gusev, N. S., Mironov, V. L., Vdovichev, S. N. and Demidov, E. S. (2015). Skyrmion states in multilayer exchange coupled ferromagnetic nanostructures with distinct anisotropy directions. *Journal of Magnetism and Magnetic Materials* **393**, 452–456.

Gilbert, D. A., Maranville, B. B., Balk, A. L., Kirby, B. J., Fischer, P., Pierce, D. T., Unguris, J., Borchers, J. A. and Liu, K. (2015). Realization of ground-state artificial skyrmion lattices at room temperature. *Nature Communications* **6**, 8462.

Ha, J. K., Hertel, R. and Kirschner, J. (2003). Micromagnetic study of magnetic configurations in submicron permalloy disks. *Physical Review B* **67**, 224432.

Hertel, R., Gliga, S., Fahnle, M. and Schneider, C. M. (2007). Ultrafast nanomagnetic toggle switching of vortex cores. *Physical Review Letters* **98**, 117201.

Hertel, R. and Schneider, C. M. (2006). Exchange explosions: Magnetization dynamics during vortex-antivortex annihilation. *Physical Review Letters* **97**, 177202.

Huang, S. X. and Chien, C. L. (2012). Extended skyrmion phase in epitaxial FeGe(111) thin films. *Physical Review Letters* **108**, 267201.

Jiang, W. J., Upadhyaya, P., Zhang, W., Yu, G. Q., Jungfleisch, M. B., Fradin, F. Y., Pearson, J. E., Tserkovnyak, Y., Wang, K. L., Heinonen, O., te Velthuis, S. G. E. and Hoffmann, A. (2015). Blowing magnetic skyrmion bubbles. *Science* **349**, 283–286.

Jonietz, F., Muhlbauer, S., Pfleiderer, C., Neubauer, A., Munzer, W., Bauer, A., Adams, T., Georgii, R., Boni, P., Duine, R. A., Everschor, K., Garst, M. and Rosch, A. (2010). Spin transfer torques in mnsi at ultralow current densities. *Science* **330**, 1648–1651.

Klebanov, I. (1985). Nuclear-matter in the skyrme model. *Nuclear Physics B* **262**, 133–143.

Lee, M., Kang, W., Onose, Y., Tokura, Y. and Ong, N. P. (2009). Unusual hall effect anomaly in mnsi under pressure. *Physical Review Letters* **102**, 186601.

Li, J., Tan, A., Moon, K. W., Doran, A., Marcus, M. A., Young, A. T., Arenholz, E., Ma, S., Yang, R. F., Hwang, C. and Qiu, Z. Q. (2014). Tailoring the topology of an artificial magnetic skyrmion. *Nature Communications* **5**, 4704.

Maret, M., Cadeville, M. C., Poinsot, R., Herr, A., Beaurepaire, E. and Monier, C. (1997). Structural order related to the magnetic anisotropy in epitaxial (111) CoPt3 alloy films. *Journal of Magnetism and Magnetic Materials* **166**, 45–52.

Miao, B. F., Sun, L., Wu, Y. W., Tao, X. D., Xiong, X., Wen, Y., Cao, R. X., Wang, P., Wu, D., Zhan, Q. F., You, B., Du, J., Li, R. W. and Ding, H. F. (2014). Experimental realization of two-dimensional artificial skyrmion crystals at room temperature. *Physical Review B* **90**, 174411.

Miao, B. F., Wen, Y., Yan, M., Sun, L., Cao, R. X., Wu, D., You, B., Jiang, Z. S. and Ding, H. F. (2015). Micromagnetic study of excitation modes of an artificial skyrmion crystal. *Applied Physics Letters* **107**, 222402.

Mochizuki, M. (2012). Spin-wave modes and their intense excitation effects in skyrmion crystals. *Physical Review Letters* **108**, 017601.

Moritz, J., Rodmacq, B., Auffret, S. and Dieny, B. (2008). Extraordinary hall effect in thin magnetic films and its potential for sensors, memories and magnetic logic applications. *Journal of Physics D-Applied Physics* **41**, 135001.

Muhlbauer, S., Binz, B., Jonietz, F., Pfleiderer, C., Rosch, A., Neubauer, A., Georgii, R. and Boni, P. (2009). Skyrmion lattice in a chiral magnet. *Science* **323**, 915–919.

Munzer, W., Neubauer, A., Adams, T., Muhlbauer, S., Franz, C., Jonietz, F., Georgii, R., Boni, P., Pedersen, B., Schmidt, M., Rosch, A. and Pfleiderer, C. (2010). Skyrmion lattice in the doped semiconductor $Fe_{1-x}Co_xSi$. *Physical Review B* **81**, 041203.

Nagaosa, N. and Tokura, Y. (2013). Topological properties and dynamics of magnetic skyrmions. *Nature Nanotechnology* **8**, 899–911.

Nahas, Y., Prokhorenko, S., Louis, L., Gui, Z., Kornev, I. and Bellaiche, L. (2015). Discovery of stable skyrmionic state in ferroelectric nanocomposites. *Nature communications* **6**, 8542.

Neubauer, A., Pfleiderer, C., Binz, B., Rosch, A., Ritz, R., Niklowitz, P. G. and Boni, P. (2009). Topological hall effect in the a phase of MnSi. *Physical Review Letters* **102**, 186602.

Okuno, T., Shigeto, K., Ono, T., Mibu, K. and Shinjo, T. (2002). MFM study of magnetic vortex cores in circular permalloy dots: Behavior in external field. *Journal of Magnetism and Magnetic Materials* **240**, 1–6.

Onose, Y., Okamura, Y., Seki, S., Ishiwata, S. and Tokura, Y. (2012). Observation of magnetic excitations of skyrmion crystal in a helimagnetic insulator Cu_2OSeO_3. *Physical Review Letters* **109**, 037603.

Onose, Y., Takeshita, N., Terakura, C., Takagi, H. and Tokura, Y. (2005). Doping dependence of transport properties in $Fe_{1-x}Co_xSi$. *Physical Review B* **72**, 224431.

Pang, Z. Y., Yin, F., Fang, S. J., Zheng, W. F. and Han, S. H. (2012). Micromagnetic simulation of magnetic vortex cores in circular permalloy disks: Switching behavior in external magnetic field. *Journal of Magnetism and Magnetic Materials* **324**, 884–888.

Park, J. P. and Crowell, P. A. (2005). Interactions of spin waves with a magnetic vortex. *Physical Review Letters* **95**, 167201.

Petrova, O. and Tchernyshyov, O. (2011). Spin waves in a skyrmion crystal. *Physical Review B* **84**, 214433.

Prejbeanu, I. L., Natali, M., Buda, L. D., Ebels, U., Lebib, A., Chen, Y. and Ounadjela, K. (2002). In-plane reversal mechanisms in circular Co dots. *Journal of Applied Physics* **91**, 7343–7345.

Rossler, U. K., Bogdanov, A. N. and Pfleiderer, C. (2006). Spontaneous skyrmion ground states in magnetic metals. *Nature* **442**, 797–801.

Shinjo, T., Okuno, T., Hassdorf, R., Shigeto, K. and Ono, T. (2000). Magnetic vortex core observation in circular dots of permalloy. *Science* **289**, 930–932.

Skyrme, T. H. (1961). Particle states of a quantized meson field. *Proceedings of the Royal Society of London Series a-Mathematical and Physical Sciences* **262**, 237.

Skyrme, T. H. R. (1962). A unifield field theory of mesons and baryons. *Nuclear Physics* **31**, 556.

Sondhi, S. L., Karlhede, A., Kivelson, S. A. and Rezayi, E. H. (1993). Skyrmions and the crossover from the integer to fractional quantum hall-effect at small zeeman energies. *Physical Review B* **47**, 16419–16426.

Chapter 4

Sun, L., Cao, R. X., Miao, B. F., Feng, Z., You, B., Wu, D., Zhang, W., Hu, A. and Ding, H. F. (2013). Creating an artificial two-dimensional skyrmion crystal by nanopatterning. *Physical Review Letters* **110**, 167201.

Tokunaga, Y., Yu, X. Z., White, J. S., Ronnow, H. M., Morikawa, D., Taguchi, Y. and Tokura, Y. (2015). A new class of chiral materials hosting magnetic skyrmions beyond room temperature. *Nature Communications* **6**, 7638.

Tretiakov, O. A. and Tchernyshyov, O. (2007). Vortices in thin ferromagnetic films and the skyrmion number. *Physical Review B* **75**, 012408.

Van Waeyenberge, B., Puzic, A., Stoll, H., Chou, K. W., Tyliszczak, T., Hertel, R., Fahnle, M., Bruckl, H., Rott, K., Reiss, G., Neudecker, I., Weiss, D., Back, C. H. and Schutz, G. (2006). Magnetic vortex core reversal by excitation with short bursts of an alternating field. *Nature* **444**, 461–464.

Xie, K. X. and Sang, H. (2014). Three layers of skyrmions in the magnetic triple-layer structure without the dzyaloshinsky-moriya interaction. *Journal of Applied Physics* **116**, 223901.

Yan, M., Leaf, G., Kaper, H., Camley, R. and Grimsditch, M. (2006). Spin-wave modes in a cobalt square vortex: Micromagnetic simulations. *Physical Review B* **73**, 014425.

Yu, X. Z., Kanazawa, N., Onose, Y., Kimoto, K., Zhang, W. Z., Ishiwata, S., Matsui, Y. and Tokura, Y. (2011). Near room-temperature formation of a skyrmion crystal in thin-films of the helimagnet FeGe. *Nature Materials* **10**, 106–109.

Yu, X. Z., Kanazawa, N., Zhang, W. Z., Nagai, T., Hara, T., Kimoto, K., Matsui, Y., Onose, Y. and Tokura, Y. (2012). Skyrmion flow near room temperature in an ultralow current density. *Nature Communications* **3**, 988.

Yu, X. Z., Onose, Y., Kanazawa, N., Park, J. H., Han, J. H., Matsui, Y., Nagaosa, N. and Tokura, Y. (2010). Real-space observation of a two-dimensional skyrmion crystal. *Nature* **465**, 901–904.

Zaspel, C. E., Ivanov, B. A., Park, J. P. and Crowell, P. A. (2005). Excitations in vortex-state permalloy dots. *Physical Review B* **72**, 024427.

Zhu, X. B., Liu, Z. G., Metlushko, V., Grutter, P. and Freeman, M. R. (2005). Broadband spin dynamics of the magnetic vortex state: Effect of the pulsed field direction. *Physical Review B* **71**, 180408.

5. Imaging and Tailoring Chiral Spin Textures Using Spin-Polarized Electron Microscopy

Gong Chen and Andreas Schmid
Lawrence Berkeley National Laboratory, Berkeley, California

Yizheng Wu
Fudan University, Shanghai, People's Republic of China

Chapter 5

This chapter introduces recent work on imaging three-dimensional spin structures in ultrathin magnetic films using spin-polarized low-energy electron microscopy (SPLEEM). Magnetic chirality within domain walls or skyrmions is directly observed. Several examples of tailoring the chirality as well as its driving force, the Dzyaloshinskii–Moriya interaction (DMI), are reviewed. A multilayer system taking advantage of interlayer exchange coupling to stabilize a skyrmion ground state at room temperature and in absence of an applied field is described.

5.1   Introduction

Manipulation of magnetic domain walls by means of electric current alone, without applying external magnetic fields, is appealing for applications in memory and logic devices (Allwood et al. 2005; Parkin et al. 2008). In principle, passing electric current through a magnet usually exerts a so-called spin-transfer torque that results in domain-wall propagations in the direction of electron flow. This type of torque is relatively weak so that high current densities are required, which limits device applications (Slonczewski 1996; Katine et al. 2000; Thiaville et al. 2005). Recent discoveries of new mechanisms of current-driven domain-wall propagation dramatically enriched this field. It was found that spin accumulated through the spin Hall effect at the boundaries of current-carrying nonmagnetic conductors (Dyakonov and Perel 1971) can also exert a torque on the magnetization of adjacent magnetic films. This spin Hall effect-induced torque can be used to manipulate magnetizations (Jonietz et al. 2010; Liu et al. 2012a, 2012b; Pai et al. 2012; Emori et al. 2013; Haazen et al. 2013; Ryu et al. 2013). A key point is that the efficiency of spin Hall effect torques strongly depends on the micromagnetic structure of the domain walls. For instance, in a typical nanowire geometry, the torque might vanish in Bloch wall (see definition in Section 5.1.1). When magnetization rotates in a Néel-type texture (see definition in Section 5.1.1), then the spin Hall effect-induced torque reverses direction as a function of the handedness of the wall texture, pushing left- and right-handed domain walls in opposite directions. However, most ferromagnetic materials do not have a preferred magnetic chirality, thus application of the spin Hall effect-induced torques in memory and logic devices relies upon the development of experimental approaches to stabilize domain-wall chirality. This motivates the exploration of controlled domain-wall spin texture and homo-chiral handedness in engineered materials and structures. In this chapter, we review recent progress

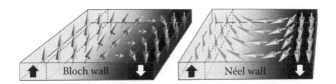

FIGURE 5.1   Sketches of Bloch-type and Néel-type domain wall in perpendicularly magnetized films; arrows indicate the orientation of local magnetization (From Chen, G. and Schmid, A.K.: Imaging and tailoring chirality of domain walls in magnetic films. *Advanced Materials*. 2015. 27. 5738–5743. Copyright Wiley-VCH Verlag GmbH & Co. KGaA. Reproduced with permission.)

employing spin-polarized electron microscopy to image and tailor domain-wall spin texture and chirality in epitaxial multilayers.

5.1.1   Ground States of Domain Walls

One common domain-wall type is called Bloch wall, where the magnetization within the domain wall rotates toward the domain-wall tangent, i.e., as a spin helix, as sketched in Figure 5.1. Another common domain-wall type is called Néel wall, where the magnetization within the domain wall rotates toward the domain-wall normal, i.e., as a spin cycloid, as sketched in Figure 5.1.

In the conventional picture, the spin structure of magnetic domain walls is governed by the interplay between exchange interaction, dipolar interaction, and magnetocrystalline anisotropy. Néel walls are often observed in films with in-plane magnetic anisotropy. However, in thin films with perpendicular magnetic anisotropy, Bloch walls are expected to be the ground state structure (Hubert and Schäfer 2008), because Néel walls are associated with an additional dipolar energy penalty, while the energy penalties associated with exchange and anisotropy interactions are equal for cycloidal and helical spin textures.

5.1.2   Origin of the Chirality

Exchange interaction, dipolar interaction, and magnetocrystalline anisotropy are all nonchiral interactions and therefore domain-wall spin textures with left-handed or right-handed rotation senses normally energetically degenerate. However, this degeneracy can be lifted in thin film systems. Because inversion symmetry breaks at interfaces, an additional asymmetric exchange interaction can occur at interfaces in the presence of spin–orbit coupling. Named the interfacial

Chapter 5

Dzyaloshinskii–Moriya interaction (DMI) (Dzyaloshinskii 1957; Moriya 1960; Fert 1990; Crépieux and Lacroix 1998), the energy contribution of the DMI is written as $E_{DM} = -D_{ij} (S_i \times S_j)$, where S_i and S_j are spins on neighboring atomic sites i and j. In most cases, the interfacial Dzyaloshinskii–Moriya vector D_{ij} is coplanar with the interface and is oriented perpendicular to the position vector $r_{ij} = r_i - r_j$ (Crépieux and Lacroix 1998), see Figure 5.2. As a result, Néel wall structures can lower (or raise) the DMI energy E_{DM} because $(S_i \times S_j)$ is parallel (or antiparallel) to D_{ij}, thus the DMI can stabilize homo-chiral spin textures with one handedness over the other, and the sign of D_{ij} determines whether right-handed or left-handed rotation sense between neighboring spins is the lower-energy configuration. In contrast, the DMI energy contribution vanishes in Bloch wall structures, where $(S_i \times S_j)$ always points orthogonally to D_{ij}. That said, Section 5.5 describes how the introduction of an additional uniaxial strain-induced anisotropy term can be used to stabilize chiral Bloch structures in the presence of the interfacial DMI. Note that the degeneracy of left-handedness and right-handedness can also be lifted in bulk systems where the lattice structure is chiral, i.e., B20 structures (Uchida et al. 2006). In B20 structures, D_{ij} is parallel to r_{ij} (Yu et al. 2010) and this kind of DMI will only stabilize

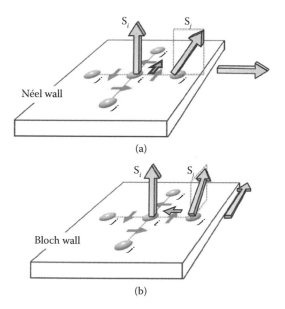

FIGURE 5.2 Sketch of the interfacial DMI vector (short/solid-gray arrows) for an atom i (circle) surrounded by four atoms j. Large arrows indicate the orientation of local spins in case of (a) Néel wall and (b) Bloch wall. Short/outlined arrows correspond to the direction of the cross-product $(S_i \times S_j)$.

Bloch-type chirality, while the DMI energy for Néel-type spin structures vanishes. More detailed discussion of bulk DMI systems exceeds the scope of this chapter, where we focus on the interfacial DMI.

5.2 Spin-Polarized Low-Energy Electron Microscopy

Experimental methods to determine Néel-type domain-wall chirality include real-space imaging using spin-polarized scanning tunneling microscopy (Bode et al. 2007) or scanning nitrogen-vacancy magnetometry (Tetienne et al. 2015), and testing of the dynamics of current-driven domain wall in the presence of an applied magnetic field (Emori et al. 2013; Ryu et al. 2013). In this chapter, we focus on spin-polarized low-energy electron microscopy (Rougemaille and Schmid 2010), which allows one to map the orientation and strength of the 3D magnetization vector in magnetic films in-situ with high special resolution of ~15 nm. This provides a powerful approach to explore magnetic chirality in multilayers as a function of varying sample design parameters.

In SPLEEM, a spin-polarized electron beam is elastically back-scattered from the sample surface. Magnetic contrast can be obtained by taking the difference of two low-energy electron microscopy (LEEM) images acquired with opposite spin polarization of the electron beam. The magnitude of the magnetic contrast scales with the scalar product $\mathbf{P} \cdot \mathbf{M}$ between the magnetization vector $\mathbf{M}$ and spin polarization vector $\mathbf{P}$, which means that magnetic contrast is a function of the magnetization component parallel/antiparallel to the spin polarization direction. The instrument is equipped with an electron-optical spin manipulator so that the orientation of the spin polarization $\mathbf{P}$ of the incident electron beam can be adjusted to point into any direction. By imaging samples with three orthogonal beam polarization directions, the 3D magnetization vector $\mathbf{M}$ can be mapped with high spatial resolution.

For instance, Figure 5.3a shows three SPLEEM images obtained at the surface of a Fe/Ni/Cu(001) bilayer film. Gray scales in these three images map the Cartesian components of the magnetization vector, and the symbols in the upper right of the three SPLEEM images indicate the directions of the spin polarization. In order to better visualize the magnetization configuration in a single graph, 3D magnetization vectors can be computed pixel-by-pixel from such sets of three images. Figure 5.3b represents a perspective view of the magnetization structure of the Fe/Ni/Cu sample, showing five in-plane magnetized domains that are separated by 180° domain walls.

Chapter 5

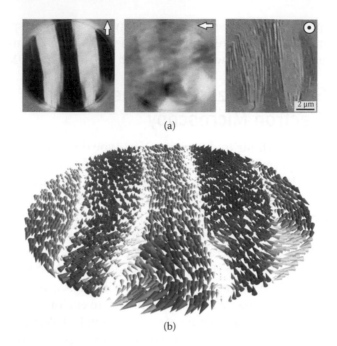

(a)

(b)

FIGURE 5.3 Three-dimensional magnetization vector measurement. (a) SPLEEM images of ~3 ML Fe/2 ML Ni/Cu(001) acquired with spin polarization of electron beams aligned along three Cartesian directions (symbols in the upper right of each image indicate the orientation of spin polarization.) (b) Spin texture computed from (a), rendering magnitude and orientation of local magnetization as cone symbols (From Chen, G. and Schmid, A.K.: Imaging and tailoring chirality of domain walls in magnetic films. *Advanced Materials.* 2015. 27. 5738–5743. Copyright Wiley-VCH Verlag GmbH & Co. KGaA. Reproduced with permission.)

5.3   Chiral Domain Walls

5.3.1   Imaging Magnetic Chirality

SPLEEM vector magnetometry can be applied to examine rotating spin textures within magnetic domain walls. Three images measuring the Cartesian components of the magnetization vector in a Fe/Ni/Cu(001) bilayer with slightly less thickness than in the sample of Figure 5.3 are shown in Figure 5.4a through c, and strong contrast in Figure 5.4a indicates that, at this thickness, the system is perpendicularly magnetized.

To represent the spin structure in the Fe/Ni bilayer in a single image that highlights the positions and magnetization directions of the domain walls, the information from the three Cartesian SPLEEM images is represented in shading in a single image, see Figure 5.4d (cf. Figure 1d in Chen et al. 2013a for color representation). This image plots perpendicular components of the magnetization within domains as light/dark gray (where magnetization points up/down, respectively), and the direction of in-plane components of the magnetization direction within the domain walls is represented in shading according to the gray wheel (inset).

Comparing colors at domain boundary in Figure 5.4d with the gray wheel indicates that the orientation of the magnetization within the domain walls always rotates toward the direction normal to the domain-wall tangent. This type of domain-wall texture, as sketched in Figure 5.1, is referred to as a Néel wall. Moreover, as highlighted by tracing the magnetization along the horizontal line in Figure 5.4d, the in-plane component of the magnetization of these domain walls always points from up domains (light gray) to down domains (dark gray). This indicates that these domain walls form a homo-chiral cycloidal structure: for instance, going left to right across three domains along the horizontal line marked in Figure 5.4d, the magnetization vector always rotates in a clockwise rotation sense, as sketched using shade-matched arrows in Figure 5.4e.

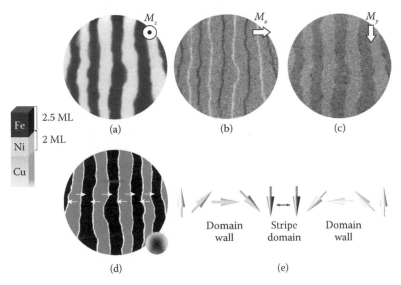

FIGURE 5.4 Imaging magnetic chirality in Fe/Ni/Cu multilayers. (a–c) SPLEEM images of 2.5 ML Fe/2 ML Ni/Cu(001), mapping orthogonal magnetization components: (a) M_z, (b) M_x, and (c) M_y. Top right symbols define x, y, z axes and spin directions. (d) Compound image constructed from the SPLEEM images (a–c), highlighting the domain walls. The gray wheel represents the direction of in-plane magnetization components. (e) Cross-sectional sketch of spin structure along gray line in (d), indicating right-handed chiral Néel walls. (From Chen, G., et al.: Novel chiral magnetic domain-wall structure in Fe/Ni/Cu(001) films. *Physical Review Letters*. 110. 177204. 2013a. Copyright APS. Reproduced with permission; Chen, G. and Schmid, A.K.: Imaging and tailoring chirality of domain walls in magnetic films. *Advanced Materials*. 2015. 27. 5738–5743. Copyright Wiley-VCH Verlag GmbH & Co. KGaA. Reproduced with permission.)

Chapter 5

5.3.2   Quantifying the Strength of the Dzyaloshinskii–Moriya Interaction

Both fundamental understanding and device applications depend on quantitative knowledge of the sign and magnitude of the DMI vector. One method to quantify the DMI is based on observations of film thickness-dependent transitions of the domain-wall type and chirality (Chen et al. 2013b).

The total magnetic free energy of a system can be expressed as the sum of exchange interaction, dipolar interaction, magnetocrystalline anisotropy, and the DMI. Comparing the free energy of Bloch- and Néel-wall textures, the energy cost of exchange interaction and magnetocrystalline anisotropy are equal. In addition, while the interfacial DMI does not affect the total energy of helical textures such as Bloch walls, it does contribute to the total energy of Néel walls. This is because the interfacial DMI vector D_{ij} is oriented perpendicular to the position vector r_{ij}, and the DMI energy term $= -D_{ij} (S_i \times S_j)$ vanishes in helical textures (here $(S_i \times S_j)$ points orthogonal to D_{ij}). Depending on the sign of the vector D_{ij}, the total energy of Néel walls is raised/lowered as a function of the walls' handedness. Since the dipolar interaction favors Bloch walls, as discussed above in Section 5.1.1, dipolar forces and the interfacial DMI are in competition in these structures and which of these two interactions dominates determines whether Bloch walls or Néel walls are the ground state in a given thin film system.

This provides an opportunity to measure the strength of the interfacial DMI in SPLEEM experiments in the following manner (Chen et al. 2013a, 2013b). The interfacial origin of the DMI implies that its energy contribution is independent of magnetic film thickness, whereas the contribution from the dipolar interaction increases with increasing magnetic film thickness. Therefore, one might expect a transition from Néel walls in thinner films to Bloch walls above a certain thickness threshold. This can be observed in the Fe/Ni/Cu(001) system. In bilayers with 7 ML Ni film thickness (Figure 5.5a), Néel walls with right-handed chirality are observed. In contrast, at larger Ni film thickness of 10 ML, magnetization within domain walls lies in the direction parallel to the domain-wall tangent, which means that these are Bloch-type domain walls (Figure 5.5b). This means that the Fe/Ni/Cu(100) system goes through a thickness-dependent transition of domain-wall type at a Ni film thickness somewhere between 7 and 10 ML.

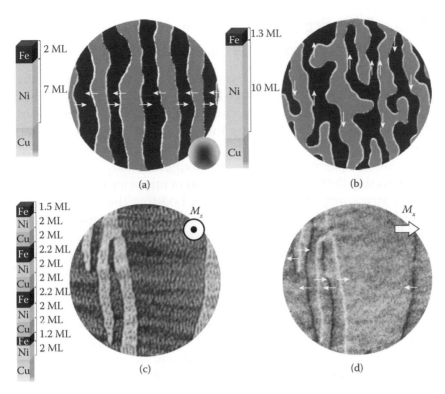

FIGURE 5.5 (a, b) Compound magnetic images of Fe/Ni/Cu(001) samples with different film thickness. (c, d) SPLEEM images in ternary superlattice with structure as sketched on left. White arrows in (d) highlight in-plane spin orientations in the domain walls. Field of view is 8 μm. (From Chen, G., et al.: Novel chiral magnetic domain-wall structure in Fe/Ni/Cu(001) films. *Physical Review Letters*. 110. 177204. 2013a. Copyright APS. Reproduced with permission; Chen, G. and Schmid, A.K.: Imaging and tailoring chirality of domain walls in magnetic films. *Advanced Materials*. 2015. 27. 5738–5743. Copyright Wiley-VCH Verlag GmbH & Co. KGaA. Reproduced with permission.)

Thus, in films where Néel walls are observed, the DMI energy contribution must be larger than the difference between the dipolar energy contributions of Bloch and Néel walls. At the transition film thickness, the energy difference between Bloch and Néel walls must equal the DMI contribution. This dipolar energy difference between Bloch and Néel wall can be calculated numerically (see details in Chen et al. 2013a, 2013b). Comparing these calculations and the two measurements shown in Figure 5.5a and b shows that the interfacial DMI energy in the Fe/Ni/Cu(100) system amounts to approximately $D_{DM} \sim 0.17$ meV/atom (Chen et al. 2013a).

Chapter 5

5.3.3   Enhancing the Dzyaloshinskii–Moriya Interaction in Ternary Superlattice

The picture described in the previous section is based on the fact that the contribution of the interfacial DMI per volume decreases with increasing magnetic film thickness, suggesting that domain-wall homo-chirality is limited to very thin layers. Extending domain-wall texture and chirality control to thicker magnetic structures might be an interesting requirement for some device applications.

Increasing the number of interfaces, as in preparing magnetic super-lattice structures, offers a route toward boosting the strength of the interfacial DMI. One important point, however, is that when the growth order of two elements is reversed, then the sign of the DMI is flipped at that interface. As a result, the total DMI strength in a binary superlattice is limited because contributions from neighboring interfaces tend to cancel out. This symmetry can be lifted by preparing ternary super-lattices. By inserting Cu spacer layers between a number of Fe/Ni bilayers, resulting in the [Fe/Ni/Cu]$_n$ multilayer structure sketched in the left of Figure 5.5c, the enhancement of the DMI strength was indeed observed. The out-of-plane (Figure 5.5c) and in-plane (Figure 5.5d) SPLEEM images show that magnetization within domain walls always points from up- to down-magnetized domains (light gray and dark gray, in Figure 5.5c). The observed domain-wall spin structure is the same as in the ultrathin Fe/Ni/Cu(100) bilayer shown in Figure 5.4a through e; that is, in both systems domain walls exhibit homo-chiral, right-handed Néel wall texture. Recalling that in Fe/Ni/Cu(100) bilayers the transition from DMI-stabilized homo-chiral Néel wall to nonchiral Bloch wall occurred at approximately 10 ML total film thickness, the observation that in the [Fe/Ni/Cu]$_n$ multilayer structure homo-chiral Néel walls persist up to 21 ML total thickness clearly indicates that the total DMI strength in the system is enhanced. This approach can be applied more generally as a way to amplify the DMI, another ternary multilayer structure is described in Section 5.4.2.

5.4   Tailoring the Interfacial Dzyaloshinskii–Moriya Interaction and the Magnetic Chirality

The dynamic properties of domain walls under conditions of applied fields, currents, etc., strongly depend on domain-wall type and chirality (Emori et al. 2013; Khvalkovskiy et al. 2013; Ryu et al. 2013); therefore, rich possibilities to influence domain-wall dynamics may arise if the sign and strength of the interface DMI can be adjusted. This section reviews one possible approach, based on adjusting a nonmagnetic spacer layer at the interface between a magnetic multilayer and a nonmagnetic substrate. This modification of this interface permits stabilization of either right- or left-handed chiral Néel walls, or nonchiral Bloch walls, within otherwise identical magnetic film structures (Chen et al. 2013b).

5.4.1   Substrate-Dependent Chirality

$[Co/Ni]_n$ superlattices are a prototypical perpendicularly magnetized system. Growing $[Co/Ni]_n$ multilayers either on Pt(111) or on Ir(111) substrates (the structures are sketched in Figure 5.6), it turns out that domain-wall structures are homo-chiral in both cases, but with an interesting twist. The compound SPLEEM image of $[Co/Ni]_n$/Pt(111) reproduced in Figure 5.6a shows that the magnetization within domain walls always points from bright gray (up) domains to dark gray (down) domains, indicating right-handed chiral Néel domain-wall texture. In contrast, when otherwise equivalent $[Co/Ni]_n$ multilayers are grown on Ir(111) substrates, then the domain-wall magnetization always points the opposite way, from dark gray domains to light gray domains (see Figure 5.6b), indicating left-handed chiral Néel wall texture. The observation that the domain-wall chirality in $[Co/Ni]_n$ multilayers switches—from right-handedness when grown on Pt(111) to left-handedness when grown on Ir(111)—shows that the magnetic layer/substrate interface plays a dominant role in the sign of the DMI vector and the handedness of the domain-wall chirality.

5.4.2   Tailoring the Dzyaloshinskii–Moriya Interaction by Interface Engineering

The experimental observation of opposite magnetic chirality at Ni/Pt(111) and Ni/Ir(111) interfaces raises interesting possibilities to tailor the DMI and the magnetic chirality: if an ultrathin Ir spacer layer is inserted between a $[Co/Ni]_n$ multilayer and a Pt(111) substrate, can domain-wall chirality in the magnetic film be tailored to be right-handed (as on a pure Pt substrate) or left-handed (as on a pure Ir substrate)?

This idea was tested by systematically tuning the thickness of an Ir spacer layer inserted in the $[Co/Ni]_n$/Ir/Pt(111) system. With this spacer layer, the domain-wall spin structure can be engineered to form ground

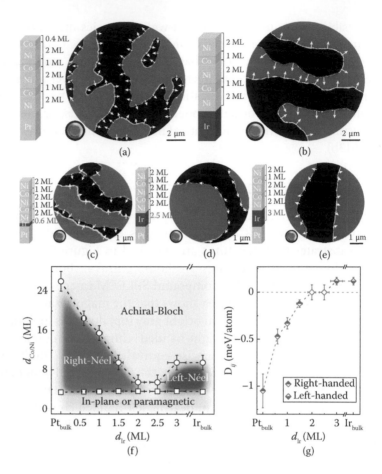

FIGURE 5.6 Real-space observation of chiral Néel walls in $[Co/Ni]_n$ multilayers. (a) (left) Sketch of $[Co/Ni]_3$ grown on Pt(111) substrate. (right) Compound SPLEEM image highlighting domain-wall magnetization, color represents the in-plane magnetization direction within domain walls, according to color wheel in inset. Additionally, white arrows highlight in-plane spin orientations in the domain walls. (b) Compound SPLEEM image of Ni/$[Co/Ni]_2$/Ir(111) and sketch of the structure. Compound SPLEEM images of Ni/$[Co/Ni]_2$/Ir/Pt(111) with different thickness of the Ir layer: (c) 0.6 ML, (d) 2.5 ML, and (e) 3 ML. (f) Experimental domain-wall type and chirality-phase diagram of $[Co/Ni]_n$ multilayer grown on Pt(111) substrate with Ir spacer layer. Square symbols indicate the transition from in-plane anisotropy to out-of-plane anisotropy, and circular symbols indicate the transition from chiral Néel wall to nonchiral Bloch wall. Label 'In-plane or paramagnetic' refers to films with either in-plane anisotropy or paramagnetic susceptibility, where no out-of-plane magnetic contrast was observed. (g) Estimated magnitude of the DMI vector D_{ij} deduced from the data (circles) plotted in (f). (From Chen, G., et al.: Tailoring the chirality of magnetic domain walls by interface engineering. *Nature Communications*. 4. 2671. 2013b. Copyright NPG. Reproduced with permission; Chen, G. and Schmid, A.K.: Imaging and tailoring chirality of domain walls in magnetic films. *Advanced Materials*. 2015. 27. 5738–5743. Copyright Wiley-VCH Verlag GmbH & Co. KGaA. Reproduced with permission.)

states ranging from right-handed chiral Néel wall when the Ir layer is very thin (see Figure 5.6c) to nonchiral Bloch wall when the Ir layer thickness is 2.5 ML (see Figure 5.6d) and finally to left-handed chiral Néel wall when the Ir layer thickness is 3 ML or more (see Figure 5.6e). To better understand this interface-engineered manipulation of the domain-wall chirality, the dependence of domain-wall spin textures on thickness of the $[Co/Ni]_n$ multilayers was measured for a selection of Ir-spacer layer thicknesses. A domain-wall spin texture phase diagram is summarized in Figure 5.6f, where the thickness of the Ir spacer layer is represented on the horizontal axis and the thickness of the $[Co/Ni]_n$ multilayer is represented on the vertical axis. Three different regimes of domain-wall texture can be clearly distinguished in the phase diagram; right-handed Néel walls, left-handed Néel walls, and nonchiral Bloch walls can be stabilized as a function of spacer layer and multilayer thickness.

The transition of the domain-wall type can be further analyzed as described in Section 5.3.2, in order to estimate the value of the DMI as a function of the Ir spacer layer thickness. Results of the analysis (Figure 5.6g) indicate that 3 ML of Ir are sufficiently thick to completely quench the contribution of the Pt substrate on the effective DMI at the $[Co/Ni]_n$ multilayer. Monte Carlo simulations successfully reproduce the experimental phase diagram, suggesting that rich possibilities to engineer domain-wall types with controlled chirality can be achieved by extending the choices of materials beyond the handful of elements used in this study (Chen et al. 2013b).

Note that in the $[Co/Ni]_n/Pt(111)$ system, due to cancellation of contributions from neighboring interfaces within the multilayer (see Section 5.3.3) the total strength of the DMI is dominated by the contribution from the Ni/Pt(111) interface. As a result, the magnetic chirality of the system is limited to relatively thin deposits. The experimental observation that the strength of the DMI at Co/Ir interfaces is three times stronger than the DMI at Ni/Ir interfaces (Chen et al. 2015a) suggests that the approach of growing ternary superlattice introduced in Section 5.3.3 may boost the total DMI strength also in this system; an extended thickness range of magnetic chirality was indeed observed in $[Co/Ir/Ni]_n$ multilayers (Chen et al. 2015b).

5.5 Unlocking Bloch Chirality through Uniaxial Strain

A numerical study predicts that Bloch walls in magnetic films can have chirality-dependent domain-wall propagation behaviors (Khvalkovskiy et al. 2013). If Bloch chirality can be stabilized, these phenomena may

Chapter 5

open up new possible device applications. However, on first sight it may seem that left-handed and right-handed Bloch walls should be energetically degenerating. This is because, as noted above, the interfacial DMI vector $\mathbf{D}_{ij}$ is perpendicular to the position vector $\mathbf{r}_{ij} = \mathbf{r}_i - \mathbf{r}_i$ (Dzyaloshinskii 1957; Moriya 1960; Fert 1990; Crépieux and Lacroix 1998; Vedmedenko et al. 2007), and in Bloch-type spin textures, the cross product $(\mathbf{S}_i \times \mathbf{S}_j)$ is parallel to the position vector $\mathbf{r}_{ij}$. For this reason, it is usually thought that the interfacial DMI usually cannot stabilize chiral Bloch wall structures.

It turns out, however, that the introduction of uniaxial strain into a system featuring interfacial DMI offers an approach to lift the left/right-handed degeneracy of Bloch-type spin structures. Prior work on the Ni/W(110) system (Sander et al. 1998) established that magnetoelastic contributions to the magnetic anisotropy from in-plane anisotropic strain due to lattice mismatch give rise to an in-plane uniaxial magnetic anisotropy K_u with easy axis along W[001]. By growing Fe/Ni bilayers on W(110) substrates (see Figure 5.7a), this uniaxial magnetic anisotropy is combined with significant DMI. In such a system, the spin structure inside the domain wall is not just a result of interplay between the interfacial DMI and the dipolar interaction; in addition, the contribution from K_u must be considered. The nature of twofold symmetry of the uniaxial anisotropy means that K_u does not influence the chirality of the domain walls, but it provides an additional force favoring alignment of the magnetization within domain walls toward the easy magnetization axis <001>.

Figure 5.7b shows a compound SPLEEM image of a Fe/Ni bilayer grown on W(110) substrate, where lime/violet indicates magnetization components along W[1–10]/W[–110] and cyan/red indicates magnetization components along W[001]/W[00–1]. In Figure 5.7b, all domain walls are either cyan or red, which means that the in-plane components of all domain walls are either parallel or antiparallel to W[001], even though the directions of domain walls are oriented along all possible directions within the film plane. This means that the domain-wall type depends on the angle φ (see sketch in Figure 5.7c) between the domain-wall normal and the W[001] easy-axis direction. For example, domain walls could be either pure Néel walls when φ = 0°, pure Bloch walls when φ = 90°, or mixed walls containing both Néel and Bloch components when −90° < φ < 90°. Figure 5.7d reproduces SPLEEM images cropped from Figure 5.7b to show domain-wall sections, highlighted by dashed lines, where certain domain-wall orientations are prevalent (red and blue arrows in Figure 5.7d indicate directions of domain-wall magnetization unit vector **m** and domain boundary normal vector **n**).

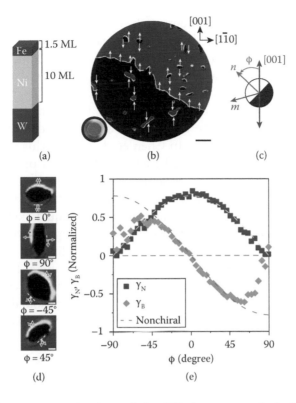

(a)   (b)   (c)

(d)   (e)

FIGURE 5.7   Anisotropic chirality and chiral Bloch component in the Fe/Ni/W(110) system. (a) Sketch of Fe/Ni bilayer grown on W(110) substrate. (b) Compound SPLEEM image; color within domain walls highlights the in-plane orientation of the magnetization inside the domain wall, according to color wheel in inset. Scale bar 1 μm. White arrows highlight in-plane spin orientations in the domain walls. (c) Definition of angles ϕ: blue arrow indicates domain boundary normal vector n, grey arrow indicates W[001] direction, and m indicates the in-plane component of the domain wall. (d) Examples of different domain-wall orientations; data are cropped from (b). (e) ϕ dependent averages of Néel-type (γ_N) and Bloch-type (γ_B) components, −1 corresponds to right-handedness, 0 corresponds to nonchirality and +1 corresponds to left-handedness. (From Chen, G., et al.: Unlocking Bloch-type chirality in ultrathin magnets through uniaxial strain. *Nature Communications*. 6. 6598. 2015b. Copyright NPG. Reproduced with permission.)

The dependence of magnetic chirality on the angle ϕ can be further quantified by decomposing the magnetization into Bloch and Néel components by projecting the magnetization unit vector onto the domain-wall tangent and normal directions (for details, see Chen et al. 2015b). Plotting averages of these components as a function of ϕ shows the dependence of average Bloch-type chirality γ_B and Néel-type chirality γ_N on ϕ, as shown in Figure 5.7e. The average Néel chirality γ_N is always positive and follows a cosine curve (light blue dashed line),

Chapter 5

indicating that Néel components in the Fe/Ni/W system are always left-handed, regardless of the local domain-wall orientation. The average Bloch chirality follows a sine curve (dark blue dashed line) in the middle region of the plot, where the magnitude of ϕ is less than approximately 60°; this Bloch chirality only vanishes when ϕ approaches ±90°. (The loss of Bloch chirality near ϕ = ±90° is due to the competition between the DMI and the dipolar interaction; for details, see analysis and supportive materials in Chen et al. 2015b.) This indicates that Bloch wall spin structure can be either left-handed or right-handed, depending on the direction of domain walls with respect to the easy-axis direction. The sinusoidal ϕ-dependence suggests that left-handed and right-handed Bloch components occur with equal likeliness in this system. This is similar to the case of nonchiral magnets where the DMI can be neglected. However, this system is clearly different from nonchiral magnets, in which case one would expect the quantities plotted in Figure 5.7e to scatter about a flat line where the average chirality vanishes, $\gamma = 0$ (purple dashed line). In contrast, the sinusoidal dependence of average Bloch-type chirality γ_B on ϕ is a result of the anisotropic DMI in the Fe/Ni/W(110) system.

5.6   Room-Temperature Skyrmion Ground State Stabilized by Interlayer Exchange Coupling

In general, the simultaneous presence of the Dzyaloshinskii–Moriya interaction and the exchange interaction tends to stabilize spin spiral structures (Uchida et al. 2006; Bode et al. 2007). A key question is how a magnetic system might be modified such that skyrmions can be stabilized (Mühlbauer et al. 2009; Neubauer et al. 2009; Jonietz et al. 2010; Yu et al. 2010, 2011; Seki et al. 2012; Li et al. 2013) and manipulated by means of spin-polarized current (Iwasaki et al. 2013; Romming et al. 2013; Sampaio et al. 2013). One approach is to introduce Zeeman energy into the system by applying an external magnetic field, which can break the energy degeneracy of "up" and "down" domains in spin spirals, resulting in skyrmion formation (Jonietz et al. 2010; Yu et al. 2010, 2011; Seki et al. 2012; Iwasaki et al. 2013; Li et al. 2013; Romming et al. 2013; Sampaio et al. 2013). In some specific systems, such as Fe/Ir(111), four-spin exchange interactions can stabilize skyrmions in the absence of an applied magnetic field (Heinze et al. 2011). Another experimental approach is to use planar confinement structures to stabilize skyrmionic vortices (Sun et al. 2013; Li et al. 2014; Miao et al. 2014).

Another method to stabilize magnetic skyrmions in Fe/Ni bilayer without applying an external magnetic field can be achieved by coupling the Fe/Ni layer through a nonmagnetic Cu spacer layer to a perpendicularly magnetized Ni film (Chen et al. 2015c). The structure of the thin film is illustrated in Figure 5.8a. A compound SPLEEM image reproduced in Figure 5.8b shows the spin structure of bubble domains in the Fe/Ni/Cu/Ni/Cu(001) multilayers.

As in the previous sections, image gray scale in Figure 5.8b represents the perpendicular magnetization component, and the magnetization component along in-plane directions is represented in the color wheel shown in the inset. The small circular-shaped domains all share the same spin structures: in the perimeter regions, the magnetization vector points toward the domain centers (dark gray domain). The perimeters are thus right-handed chiral Néel-type structures (Chen et al. 2013a, 2013b), and thus these bubble domains can be described as right-handed hedgehog skyrmions with a winding number 1.

A magnified color-coded SPLEEM image of one single skyrmion with a radius of approximately 200 nm is shown in more detail in Figure 5.8c. Figure 5.8d plots a pixel-by-pixel measurement of the angle θ between the magnetization **m** as a function of distance r from the center of the skyrmion shown in Figure 5.8c. The plot (Figure 5.8d) highlights how the magnetization canting angle tilts smoothly in almost linear proportion to r (blue circles are pixel-by-pixel measurements; a cyan solid line is a guide to the eye). The linear dependence of θ on r suggests that the spin rotates as a homogeneous spin spiral along any trajectory crossing the center of the skyrmion. The slope of this linear dependence shows that the spin rotates by ~0.8° per nm, which is about 0.14° between the two neighboring atoms. To display the spin texture more clearly, the array of vectors shown in Figure 5.8e illustrates the same data as the compound SPLEEM image from Figure 5.8c. Here, arrows show the magnetization direction of spin blocks averaged from 3×3 pixels in the original data, highlighting the measured spin texture of this hedgehog skyrmion.

5.7   Summary

This chapter summarizes recent progress in characterizing and tailoring chiral domain walls by combining molecular beam epitaxy sample growth with in-situ SPLEEM imaging. Experimental results on magnetic bilayers and multilayers demonstrate that the DMI can stabilize chiral domain walls; measuring the film thickness dependence of domain-wall spin textures provides an approach to estimate the

Chapter 5

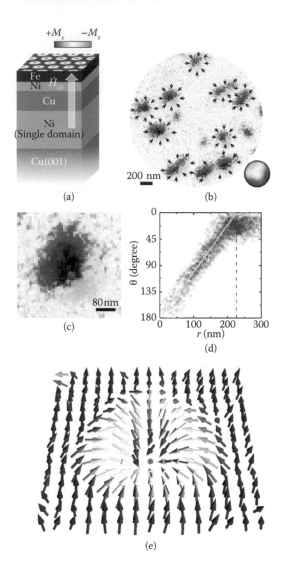

FIGURE 5.8   Stabilizing skyrmion through interlayer exchange coupling. (a) Sketch of the multilayer structure indicates how a virtual magnetic field $\vec{H}_{eff}$ is introduced by coupling Fe/Ni bilayers to buried single-domain Ni layers. (b) Compound SPLEEM image plots 3D magnetization vector components of skyrmions according to color wheel shown in inset. Additionally, black arrows highlight skyrmion perimeter magnetization directions. (c) Magnified image of a single skyrmion. (d) Pixel-by-pixel measurement of the angle θ between magnetization and the (001) surface normal, as a function of distance r from the center of skyrmion. (e) Arrows-array representation of the experimental data shown in (c). The orientation of arrows shows magnetization vector direction of spin blocks averaged from 3 × 3 pixels in the original data. (Reprinted with permission from Chen, G., et al. Room temperature skyrmion ground state stabilized through interlayer exchange coupling. *Applied Physics Letters*. 106. 242404. 2015c. Copyright (2015), American Institute of Physics.)

strength of the DMI. Tailored domain-wall chirality can be achieved by adjusting the thickness of an Ir spacer layer between $(Co/Ni)_n$ multilayers and a Pt substrate. The Ir spacer provides a way to manipulate the DMI, to stabilize either left-handed or right-handed Néel walls, or nonchiral Bloch walls. Moreover, chiral-Bloch wall components can be stabilized by adding uniaxial strain to chiral magnetic films. Finally, a method to stabilize skyrmions at ambient temperature through interlayer exchange coupling is reviewed. These results exemplify the rich physics of DMI effects in magnetic films and demonstrate how interface engineering opens up new degrees of freedom in controlling magnetic chirality, which may be useful for designing domain wall–based spintronic devices.

References

Allwood DA, et al. (2005). Magnetic domain-wall logic. *Science*. 309:1688–1692.

Bode M, et al. (2007). Chiral magnetic order at surfaces driven by inversion asymmetry. *Nature*. 447:190–193.

Chen G, et al. (2013b). Tailoring the chirality of magnetic domain walls by interface engineering. *Nat. Commun.* 4:2671.

Chen G, Mascaraque A, N'Diaye AT and Schmid AK. (2015c). Room temperature skyrmion ground state stabilized through interlayer exchange coupling. *Appl. Phys. Lett.* 106:242404.

Chen G, et al. (2015b). Unlocking Bloch-type chirality in ultrathin magnets through uniaxial strain. *Nat. Commun.* 6:6598.

Chen G, N'Diaye AT, Wu Y and Schmid AK. (2015a). Ternary superlattice boosting interface-stabilized magnetic chirality. *Appl. Phys. Lett.* 106:062402.

Chen G and Schmid AK. (2015). Imaging and tailoring chirality of domain walls in magnetic films. *Adv. Mater.* 27:5738–5743.

Chen G, et al. (2013a). Novel chiral magnetic domain wall structure in Fe/Ni/Cu(001) films. *Phys. Rev. Lett.* 110:177204.

Crépieux A and Lacroix C. (1998). Dzyaloshinskii-Moriya interactions induced by symmetry breaking at a surface. *J. Magn. Magn. Mater.* 182:341–349.

Dyakonov MI and Perel VI. (1971). Current-induced spin orientation of electrons in semiconductors *Phys. Lett. A.* 35:459–460.

Dzyaloshinskii IE. (1957). Thermodynamic theory of "weak" ferromagnetism in antiferromagnetic substances. *Sov. Phys. JETP.* 5:1259–1272.

Emori S, Bauer U, Ahn SM, Martinez E and Beach GSD. (2013). Current-driven dynamics of chiral ferromagnetic domain walls. *Nat. Mater.* 12:611–616.

Fert A. (1990).Magnetic and transport properties of metallic multilayers. *Metallic Multilayers.* 59–60:439.

Haazen PPJ, et al. (2013). Domain wall depinning governed by the Spin Hall Effect. *Nat. Mater.* 12:299–303.

Heinze S, et al. (2011). Spontaneous atomic-scale magnetic skyrmion lattice in two dimensions. *Nat. Phys.* 7:713–718.

Hubert A and Schäfer R. (2008). *Magnetic Domains: The Analysis of Magnetic Microstructures.* FL: Springer Berlin Heidelberg: New York.

Chapter 5

Iwasaki J, Mochizuki M and Nagaosa N. (2013). Current-induced skyrmion dynamics in constricted geometries. *Nat. Nanotechnol.* 8:742–747.

Jonietz F, et al. (2010). Spin transfer torques in MnSi at ultralow current densities. *Science.* 330:1648–1651.

Katine JA, Albert FJ, Buhrman RA, Myers EB and Ralph DC. (2000). Current-driven magnetization reversal and Spin-wave excitations in Co/Cu/Co pillars. *Phys. Rev. Lett.* 84:3149.

Khvalkovskiy AV, et al. (2013). Matching domain-wall configuration and spin-orbit torques for efficient domain-wall motion. *Phys. Rev. B.* 87:020402(R).

Li J, et al. (2014). Tailoring the topology of an artificial magnetic skyrmion. *Nat. Commun.* 5:4704.

Li Y, et al. (2013). Robust formation of skyrmions and Topological Hall Effect anomaly in epitaxial thin films of MnSi. *Phys. Rev. Lett.* 110:117202.

Liu L, Lee OJ, Gudmundsen TJ, Ralph DC and Buhrman RA. (2012a). Current-induced switching of perpendicularly magnetized magnetic layers using spin torque from the Spin Hall Effect. *Phys. Rev. Lett.* 109:096602.

Liu L, et al. (2012b). Spin-torque switching with the giant Spin Hall Effect of tantalum. *Science.* 336:555–558.

Miao BF, et al. (2014). Experimental realization of two-dimensional artificial skyrmion crystals at room temperature. *Phys. Rev. B.* 90:174411.

Moriya T. (1960). Anisotropic superexchange interaction and weak ferromagnetism. *Phys. Rev.* 120:91–98.

Mühlbauer S, et al. (2009). Skyrmion lattice in a chiral magnet. *Science.* 323:915–919.

Neubauer A, et al. (2009). Topological Hall Effect in the A phase of MnSi. *Phys. Rev. Lett.* 102:186602.

Pai CF, et al. (2012). Spin transfer torque devices utilizing the giant Spin Hall Effect of tungsten. *Appl. Phys. Lett.* 101:122404.

Parkin SSP, Hayashi M and Thomas L. (2008). Magnetic domain-wall racetrack memory. *Science.* 320:190–194.

Romming N, et al. (2013). Writing and deleting single magnetic skyrmions. *Science.* 341:636–639.

Rougemaille N and Schmid AK. (2010). Magnetic imaging with spin-polarized low-energy electron microscopy. *Eur. Phys. J. Appl. Phys.* 50:20101.

Ryu KS, Thomas L, Yang SH and Parkin SSP. (2013). Chiral spin torque at magnetic domain walls. *Nat. Nanotechnol.* 8:527–533.

Sampaio J, Cros V, Rohart S, Thiaville A and Fert A. (2013). Nucleation, stability and current-induced motion of isolated magnetic skyrmions in nanostructures. *Nat. Nanotechnol.* 8:839–844.

Sander D, Schmidthals C, Enders A and Kirschner J. (1998). Stress and structure of Ni monolayers on W(110): The importance of lattice mismatch. *Phys. Rev. B.* 57:1406.

Seki S, Yu XZ, Ishiwata S and Tokura Y. (2012). Observation of skyrmions in a multiferroic material. *Science.* 336:198.

Slonczewski JC. (1996). Current-driven excitation of magnetic multilayers. *J. Magn. Magn. Mater.* 159:L1–L7.

Sun L, et al. (2013). Creating an artificial two-dimensional skyrmion crystal by nanopatterning. *Phys. Rev. Lett.* 110:167201.

Tetienne JP, et al. et al. (2015). The nature of domain walls in ultrathin ferromagnets revealed by scanning nanomagnetometry. *Nat. Commun.* 6:6733.

Thiaville A, Nakatani Y, Miltat J and Suzuki Y. (2005). Micromagnetic understanding of current-driven domain wall motion in patterned nanowires. *Europhys. Lett.* 69:990–996.

Uchida M, Onose Y, Matsui Y and Tokura Y. (2006). Real-space observation of helical spin order. *Science*. 311:359–361.

Vedmedenko EY, Udvardi L, Weinberger P and Wiesendanger R. (2007). Chiral magnetic ordering in two-dimensional ferromagnets with competing Dzyaloshinskii-Moriya interactions. *Phys. Rev. B*. 75:104431.

Yu XZ, et al. (2011). Near room-temperature formation of a skyrmion crystal in thinfilms of the helimagnet FeGe. *Nat. Mater*. 10:106–109.

Yu XZ, et al. (2010). Real-space observation of a two-dimensional skyrmion crystal. *Nature (London)*. 465:901–904.

Chapter 5

6. Epitaxial Thin Films of the Cubic B20 Chiral Magnets

Sunxiang Huang
University of Miami, Coral Gables, Florida

Chia-Ling Chien
Johns Hopkins University, Baltimore, Maryland

Bulk B20 magnets exhibit an exotic skyrmion phase in a narrow phase region at a temperature close to Curie temperature within a magnetic field. The investigation and exploitation of these exotic phases, and especially potential devices, require thin films. In this chapter, we discuss the epitaxial growth, the unusual magnetic and transport properties, with the strong thickness dependence of the B20 films. We have also realized an extended skyrmion phase in thin films.

6.1   Introduction

Common magnets, such as Fe, Ni, and Co, have ferromagnetic ground state, where all the moments are aligned with a saturation magnetization M_S. This well-known fact is due to symmetric Heisenberg exchange interaction as a result of the centrosymmetric crystal structures of the common magnets.

Chapter 6

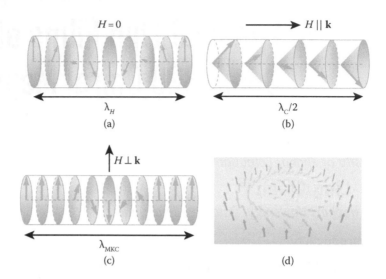

FIGURE 6.1   (a) Helical, (b) conical, (c) MKC, and (d) skyrmion spin textures.

In cubic B20 chiral magnets (e.g., MnSi, FeGe, and MnGe), the atoms within the cubic unit cell have noncentrosymmetric arrangements with the loss of inversion symmetry. As a result, in addition to the symmetric Heisenberg interaction, there is also a nontrivial antisymmetric Dzyaloshinskii–Moriya interaction (DMI) (Dzyaloshinskii 1958; Moriya 1960), which favors perpendicular alignment between neighboring spins. As a result of the competing Heisenberg interaction and DMI, the B20 magnets have a spin helical ground state (Bak and Jensen 1980) (Figure 6.1a). In a spin helix, spins form ferromagnetic sheets. The spins in adjacent sheets maintain a fixed advancing angle thus forming a spin helix with a helical wavelength λ_H along the propagation vector $\mathbf{k}$. The helical wavelength $\lambda_H = 4\pi A/D$ is determined by the ratio of Heisenberg interaction with the exchange stiffness A and the DMI with the Dzyaloshinskii constant D. In the absence of DMI, λ_H would be an infinite thus ferromagnetic alignment. Among various B20 magnets, λ_H varies from a few nanometers to several hundred nanometers.

Under a magnetic field parallel to the $\mathbf{k}$ direction, the spins are tilted towards the field direction, and the helical spin texture is transformed to a conical spin texture (Figure 6.1b). All the spins are then aligned in the field direction at the saturation field of $H_D \approx D^2 M_S/2A$, where M_S is the saturation magnetization. When the field is perpendicular to the $\mathbf{k}$ direction, a magnetic kink crystal (MKC, or helicoid spin texture, Figure 6.1c) may be realized (Yurii 1984; Ulrich et al. 2011). However, due to the small magnetic anisotropy in the bulk B20 magnets, a modest magnetic field may align the **k** vector along the field direction no matter what the initial **k** vector may be (Grigoriev et al. 2007; Huang et al. 2014). The spin helix, the conical spin texture, and the aligned state, under an increasing magnetic field, occur at low temperatures, sufficiently below the Curie temperature T_C.

Most remarkably, at temperature close to Curie temperature T_C and with the presence of a magnetic field, the bulk B20 magnets exhibit a new and exotic phase, known since the 1980s as "A-phase," A for anomalous. During the last few years, neutron diffraction and Lorentz TEM have revealed that the A-phase is the skyrmion phase, where magnetic skyrmions (Figure 6.1d) can be stabilized by thermal fluctuations in bulk B20 magnets (Mühlbauer et al. 2009; Yu et al. 2010). In a magnetic skyrmion, whose diameter is about λ_H, the spins form an intricate double-twist texture along both the angular and the radial directions. Magnetic skyrmions have a topological charge of 1 in real space. They exhibit unusual topological and dynamic properties (Nagaosa and Tokura 2013) and may be used in ultra-dense memory or logic devices (Fert et al. 2013).

However, the skyrmion phase in bulk B20 magnets only exists in a small temperature-field region, a few degrees in temperature and a few tenths of tesla in field. As such, the magnetic skyrmions in bulk B20 materials are unlikely to be useful. However, theories indicate that the skyrmion phase may be extended by reduced dimensionality and/or uniaxial distortion as in thin films (Bogdanov and Yablonsky 1989; Bogdanov and Hubert 1994, 1999; Butenko et al. 2010; Yu et al. 2010; Rößler et al. 2011: Karhu et al. 2012), as observed in thin crystal platelets (Yu et al. 2010; Huang and Chien 2012; Li et al. 2013). The first epitaxial B20 FeGe films have been reported by Huang and Chien (2012), revealing the much extended skyrmion phase in thin films. Theories also suggest that new soliton states may be realized when the film thickness is comparable with the helical wavelength (Rybakov et al. 2015). Of course, to explore the spintronics applications of skyrmions, thin films are essential for patterned structures.

In what follows, we will discuss the realization of epitaxial growth of B20, mostly FeGe films. We will elaborate on the unusual magnetic and transport properties of the B20 films, the extended skyrmion phase, and their strong thickness dependence.

6.2 Epitaxial Growth of B20 Films on Si(111) Substrate

As shown in Figure 6.2a, B20 magnets, such as FeGe, have a cubic unit cell with one lattice constant $a = b = c$, but within the unit cell, the atomic

Chapter 6

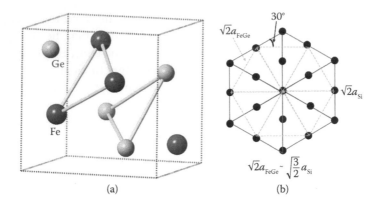

FIGURE 6.2 (a) Cubic B20 FeGe crystal structure. (b) Epitaxial relationship between the FeGe(111) plane (dotted lines) and the Si(111) plane (solid lines).

Table 6.1 B20 Magnets Studied in This Work

B20 Magnets	MnSi	FeGe	MnGe	$Fe_{0.5}Co_{0.5}Si$
T_C (K)	29	280	170	35
a (Å)	4.56	4.7	4.795	4.47
λ_H (nm)	18	70	3	90
Lattice mismatch (%)	−3	−0.06	+2	−5

arrangement with the unit cell lacks the inversion and fourfold rotation symmetries. Within each unit cell, the eight atoms do not occupy lattice points. Upon closer examination, all the Fe and the Ge atoms are situated in the planes parallel to the (111) plane, suggesting that epitaxial B20 films tend to grow in the (111) orientations. The lattice constants of B20 magnets (Table 6.1) are quite different from those of common substrates including Si. For example, FeGe has a lattice constant of 4.7 Å, which is 15% smaller than the lattice constant of 5.43 Å of Si. This large lattice mismatch will not support epitaxial growth. However, if we rotate the FeGe(111) plane (dotted lines in Figure 6.2b) by 30°, FeGe(111) plane matches perfectly onto the Si(111) plane (solid lines in Figure 6.2b) with a lattice mismatch of 0.06%. In this manner, the FeGe[1$\bar{1}$0] direction is parallel to the Si[11$\bar{2}$] direction. The actual lattice mismatch is determined by the difference between $\sqrt{2}\,a_{B20}$ and $\sqrt{3/2}\,a_{Si}$ (Figure 6.2b). Table 6.1 shows the actual lattice mismatch between various B20 magnets and Si(111). Among the B20 materials, FeGe has the smallest lattice mismatch, whereas $Fe_{0.5}Co_{0.5}Si$ has a large lattice mismatch of 5%.

We have grown epitaxial FeGe(111) with various thicknesses by magnetron sputtering in a high vacuum system with a base pressure of

3.5×10^{-8} torr onto HF etched undoped Si(111) substrates at a substrate temperature of 500°C. Figure 6.3a shows the $\theta/2\theta$ X-ray diffraction (XRD) pattern of the FeGe films. It shows a very strong FeGe(111) peak with a much smaller FeGe(210) peak and other tiny impurity peaks (< 0.5% impurity phases). It should be noted that the growth conditions (substrate temperature, Ar sputtering gas pressure, and deposition rate) can significantly affect the phase of the FeGe films. The out-of-plane lattice constant d_{111} is about 0.5% less than that in bulk B20 FeGe, indicating a tensile strain of the films. The in-plane φ scan (Figure 6.3b) of the (111) peaks of the FeGe film and the Si substrate reveals the 30° rotation of the aforementioned epitaxial relationship with a twinning.

In contrast to FeGe, both MnSi and MnGe have a much larger lattice mismatch and are difficult to grow directly on Si(111) by sputtering. However, we find that a thin (1–2 nm) FeGe(111) buffer layer greatly promotes the epitaxial growth of MnGe(111) and MnSi(111) films. As shown in the XRD phase scan (Figure 6.4a), the MnGe film

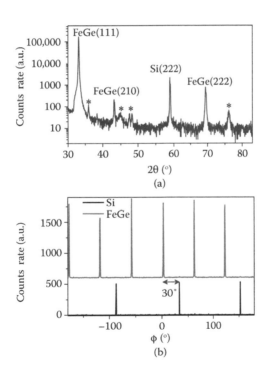

FIGURE 6.3 (a) X-ray diffraction phase scan of the FeGe(111) film. Asterisk (*) indicates impurity phases. (b) In-plane φ scan about the (111) peaks of FeGe film and Si substrate.

Chapter 6

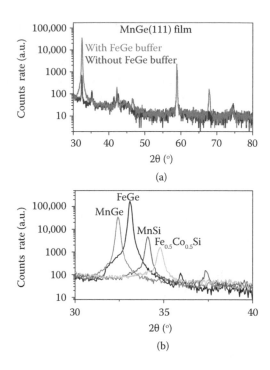

FIGURE 6.4 X-ray diffraction phase scan of (a) MnGe(111) films with FeGe buffer layer (light gray lines) and without FeGe buffer layer (dark gray lines), and (b) MnGe, FeGe, MnSi, and $Fe_{0.5}Co_{0.5}Si$ films.

with FeGe buffer layer has a much larger diffraction intensity and less impurity peaks, compared with the MnGe film without the buffer layer. With the FeGe buffer layer, MnSi(111) films (Figure 6.4b) can also be grown by sputtering. $Fe_{0.5}Co_{0.5}Si$ has a lattice mismatch of 5% with Si but has a lattice mismatch of 2% with MnSi. Therefore, double buffer layers, which consist of a thin FeGe(111) film and a thin MnSi(111) film, are used to grow $Fe_{0.5}Co_{0.5}Si$(111) films (Figure 6.4b).

In addition to magnetron sputtering, molecular beam epitaxy (MBE) has also been used to grow MnSi(111) (Karhu et al. 2010, Li et al. 2013), FeGe(111) (Kanazawa et al. 2015, Porter et al. 2015), and $Fe_{1-x}Co_xSi$(111) (Porter et al. 2012) epitaxial films on Si(111) substrates.

6.3 Magnetic and Transport Properties

Magnetic hysteresis loops (Figure 6.5) of a 300-nm FeGe(111) film show that the magnetization is easier to saturate under an in-plane field ($H_\parallel$) than under an out-of-plane field ($H_\perp$), indicating that the in-plane magnetization is favored by the shape anisotropy. The saturated

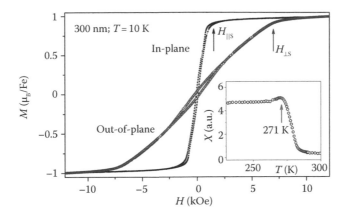

FIGURE 6.5 Hysteresis loop of 300-nm FeGe film at 10 K for in-plane field (black) and out-of-plane field (gray). Inset: In-plane ac susceptibility of FeGe(111) film as a function of temperature. (Adapted from Huang, S.X., and Chien, C.L., *Phys. Rev. Lett.*, 108, 267201, 2012. With permission.)

magnetic moment is about 1 μ_B per Fe, which is consistent with the value of bulk B20 FeGe. The in-plane M-H loop shows a negligible magnetic remanence, a small coercivity, and a slanted curve with a saturation field ($H_{\|S}$), defined at the knee point in the M–H curve, of about 1.3 kOe. The $H_{\|S}$ is much smaller than the saturation field H_D of about 3 kOe for bulk FeGe (Lebech and et al. 1989), suggesting more complex spin textures under the in-plane field in the thin films. The $H_{\perp S}$ is about 7 kOe and is close to $H_D + 4\pi M_S$ (4.5 kOe), indicating a negligible uniaxial anisotropy for 300-nm films. Note that both $H_{\|S}$ and $H_{\perp S}$ depend systematically on film thickness. The Curie temperature, as indicated in the ac susceptibility measurement with a 973 Hz in-plane ac field of 10 Oe, is about 271 K (inset of Figure 6.5), slightly lower than 280 K of bulk FeGe.

Resistivities of films with thicknesses from 18 to 300 nm increase with temperature (Figure 6.6), demonstrating metallic behavior with a systematic dependence on film thicknesses. The resistivity at room temperature is around 300 $\mu\Omega\cdot$cm for films with a thickness above 30 nm and increases rapidly to 1200 $\mu\Omega\cdot$cm for the 18-nm film when the interfacial scattering becomes significant, as shown in the inset in Figure 6.6. The resistivity ratio γ, defined as the $R(300K)/R(10K)$, decreases systematically as the thickness increases. The value γ is as high as 7 for the 300-nm film, suggesting high quality of the epitaxial films.

The Hall resistivity at a field larger than H_S has a nonmonotonic dependence on temperature as shown in Figure 6.7a. The Hall resistivity

Chapter 6

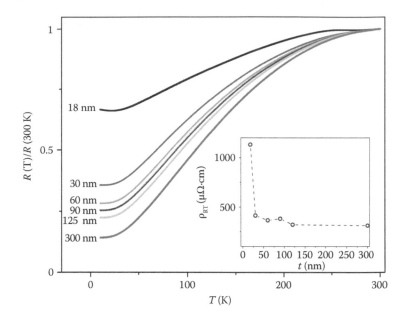

FIGURE 6.6 Resistivity ratio R(T)/R (300 K) as a function of temperature for films with various thicknesses indicated. Inset: resistivity at 300 K as a function of thickness. (Adapted from Huang, S.X., and Chien, C.L., *Phys. Rev. Lett.*, 108, 267201, 2012. With permission.)

$\rho_H = R_0 H + R_S M$ at $H > H_S$ include two contributions of ordinary Hall resistivity (OHR) $R_0 H$ and a much larger anomalous Hall resistivity (AHR) $R_S M$, where R_0 is the ordinary Hall coefficient and R_S is the anomalous Hall coefficient. The ordinary Hall coefficient R_0 depends on the carrier density and is roughly the same for most metals. In general, R_S increases with resistivity but may be a complex function of resistivity ρ (e.g., $a\rho + b\rho^2$), depending on the detailed scattering mechanisms (Nagaosa et al. 2010). The measured ρ_H (dominated by $R_S M$) increases monotonically from low temperature to around 210 K due to the increasing resistivity (Figure 6.6). It begins to decrease from 210 K since the saturated magnetization decreases rapidly when the temperature approaches T_C. As shown in Figure 6.7b, ρ_H has a good linear relationship with $\rho^2 M_S$ below 210 K (light gray curve), suggesting that the dominating scattering mechanism of AHR in epitaxial FeGe films is the intrinsic contribution ($R_S = S_A \rho^2$). As the temperature approaches T_C when the resistivity increases, it deviates from the linear relation (dark gray curve, Figure 6.7b), suggesting that the extrinsic contributions, such as impurity scattering, set in.

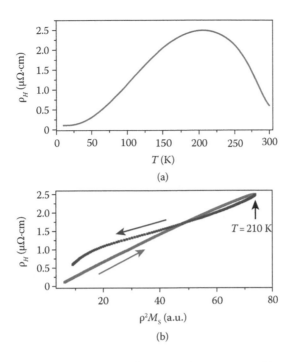

FIGURE 6.7 (a) Hall resistivity ρ_H as a function of temperature. (b) ρ_H as a function of $\rho^2 M_S$. Hall measurements are performed at a field larger than H_S.

6.4 Extended Skyrmion Phase

Magnetic skyrmions carry a spin Berry phase in real space and have a topological charge of 1. When an electron traverses through a skyrmion, its spin follows local moments, thus picking up the Berry phase. This in turn generates a fictitious gauge field B_{eff}, which deflects the electron motion, much like the deflection by a magnetic field in ordinary Hall effect. This effect is called the topological Hall effect (Ye et al. 1999; Bruno et al. 2004; Lee et al. 2009; Neubauer et al. 2009), a hallmark for the existence of skyrmions in conducting materials. As a result, the total Hall resistivity is:

$$\rho_H = R_0 H + R_S M + \rho_{TH}, \tag{6.1}$$

where ρ_{TH} is the topological Hall resistivity due to the skyrmion spin textures. As discussed above, at temperatures below 200 K, the dominant scattering mechanism of the AHE is intrinsic contribution which gives $R_S = S_A \rho^2$. At higher temperatures, R_S deviates from ρ^2. Since the

Chapter 6

magnetoresistance (MR) of the FeGe films is less than 0.5% at any temperatures, R_S is weakly dependent on magnetic field. For simplicity, we adapt $R_S = S_A \rho^2$ as in previous works, where S_A is independent of magnetic field (Lee et al. 2007; Kanazawa et al. 2011). At a large field when all the spins are aligned, ρ_{TH} is 0 and $\rho_H = R_0 H + R_S M$. Following the analysis in Kanazawa et al. (2011), by a linear fitting of ρ_H/H versus R_S/H at large fields, the coefficients R_0 and S_A can be obtained. The value of S_A obtained from the fitting is positive for the FeGe epitaxial thin films, whereas it is negative for MnSi (Lee et al. 2007) and positive for $Fe_{1-x}Co_xSi$ (Onose et al. 2005).

As shown in Figure 6.8a, at high fields $H > H_{Skx}$ (~14 kOe for the 60-nm film) when all the spins are aligned, the experimental data (black lines) agree well with the calculated values. Below H_{Skx}, the measured ρ_H clearly deviates from $R_0 H + R_S M$ (red lines) and includes contribution of THR. The ρ_{TH} (blue lines) can then be obtained by subtraction of $R_0 H + R_S M$ from the total Hall resistivity. The field dependence of ρ_{TH} shows a dome shape with a peak at a magnetic field H_m. The dome-shaped THR has been observed in previous experiments (Lee et al. 2009; Neubauer et al. 2009, Kanazawa et al. 2011) and described in theories (Yi et al. 2009). The THR values are proportional to the density of the skyrmions and are consistent with the Lorentz transmission electron microscopy (TEM) measurements on a thin plate of B20 magnets (Yu et al. 2010). At zero field, it has a helical ground state with the wave vector **k** lying in the plane. By applying a small out-of-plane magnetic field, the skyrmions start to emerge and co-exist with the spin helixes (or the helicoids). Increasing the magnetic field suppresses the spin helixes and promotes the formation of the skyrmion lattices. Further increasing the magnetic field suppresses the skyrmions and eventually aligns all the spins to the saturated ferromagnetic phase due to the large reduction of Zeeman energy at high fields. The field dependence of small-angle neutron scattering (SANS) on FeGe thin films, which corroborates with the field dependence of THR, will be discussed later. It is interesting to note that the H_{Skx} is much larger than H_S. This is likely due to the formation of free isolated (repulsive) skyrmions. The isolated skyrmions are excitations of the saturated state and exist as topologically stable 2D solitons at field far above H_S as predicted in theory (Rößler et al. 2011).

For all the FeGe films, ρ_{TH} shows a similar dome behavior. The maximum ρ_{TH} is over 0.1 $\mu\Omega\cdot cm$, much larger than that of several $n\Omega \cdot cm$ found in MnSi (Neubauer et al. 2009). The maximum THR ratio (THR_m), defined as $\rho_{TH}(H = H_m)/\rho_H(H = 25 \text{ kOe})$, is more than 20% at 10 K and decreases to 6% at 150 K (Figure 6.8b). At $T = 280$ K, which

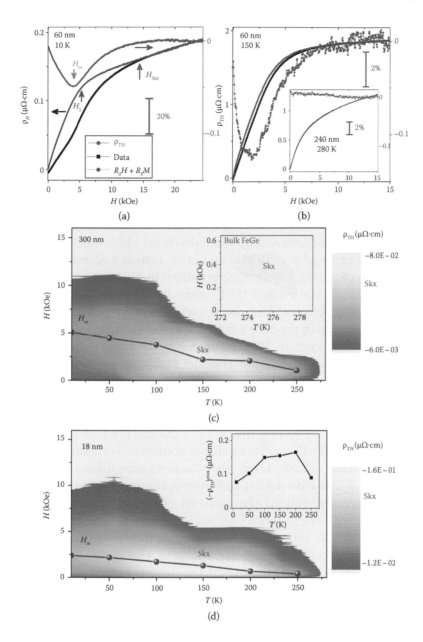

FIGURE 6.8   Hall resistivity as a function of field for (a) 60-nm film at 10 K and (b) 60-nm film at 150 K, inset: 240-nm at 280 K, showing experimental ρ_H (black), calculated (red) values of $R_0H + R_SM$ according to Equation 6.1, and the topological Hall resistivity ρ_{TH} (blue). Color map of topological Hall resistivity in H–T plane for (c) 300-nm and (d) 18-nm film and the values of H_m. Inset in (c): skyrmion region for bulk FeGe. (From Wilhelm, H., et al., *Phys. Rev. Lett.*, 107, 127203, 2011. Inset in (d): maximum values of $(-\rho_{TH})$ as a function of temperature. (Adapted from Huang, S.X., and Chien, C.L., *Phys. Rev. Lett.*, 108, 267201, 2012. With permission.)

is slightly above T_C, THR is zero and the total Hall resistivity equals to $R_0H + R_SM$ within 1% error for all fields (inset in Figure 6.8b).

The THR, originating from a nonzero topological charge of skyrmions, provides strong evidence of the skyrmion phase, as observed in numerous experiments (Lee et al. 2009; Neubauer et al. 2009; Kanazawa et al. 2011). The measured ρ_{TH} at different temperatures and fields are then used to determine the skyrmion phase diagram in epitaxial FeGe films (Figure 6.8c). In bulk FeGe, the skyrmion phase only exists in a very narrow region with a temperature range of 5 K at magnetic fields from 0.1 to 0.5 kOe (inset in Figure 6.8c) (Wilhelm et al. 2011). In contrast, in FeGe films, the skyrmion phase is greatly extended to cover the *entire* temperature range below T_C of 271 K (Figure 6.8c and d). In addition, the skyrmion phase exists in a much wider field range of over 10 kOe and exists even at zero magnetic field. The values of ρ_{TH} show systematic variations in T–H phase diagram and reach a maximum at around 150 K, where they are 0.16 and 0.08 $\mu\Omega\cdot$cm for the 18- and 300-nm films, respectively. The ρ_{TH}, due to the skyrmion spin textures which produce a fictitious magnetic field B_{eff}, reads $\rho_{TH} = n_{\text{Skx}}PR_0B_{\text{eff}}$ (Bruno et al. 2004; Neubauer et al. 2009), where n_{Skx} is the relative skyrmion density (for compact arrays, $n_{\text{Skx}} = 1$) and P is the local spin polarization. The fictitious magnetic field $B_{\text{eff}} = 2\phi_0/A$, where ϕ_0 is the flux quantum and A is the area of a skyrmion, is about 1 T if the skyrmion size of 70 nm for FeGe is adapted. The R_0, measured at $T = 380$ K for the 300-nm film, is 0.072 $\mu\Omega\cdot$cm/T, which gives a value of $R_0B_{\text{eff}} = 0.072$ $\mu\Omega\cdot$cm. This value is consistent with the maximum measured value of 0.08 $\mu\Omega\cdot$cm when the films have the maximum n_{Skx}, assuming R_0 and P do not depend on temperature. The variation of ρ_{TH} with temperature reflects the variation of n_{Skx} with temperature, in agreement with the Lorentz TEM measurement for the 15-nm region of a wedged FeGe thin specimen (15- to 75-nm wedge) (Yu et al. 2011). However, the skyrmion phase only exists in a small T–H region when the thickness is around the helical wavelength λ_H (the 75-nm region of the wedge) (Yu et al. 2011). For all the epitaxial FeGe films, the skyrmion phase is still greatly extended to cover the entire temperature range below T_C, even for the thick films, with thickness several times that of λ_H.

The maximum ρ_{TH} for the 18-nm film is larger than that in the 300-nm film, possibly due to a smaller skyrmion size in thinner films which produces larger B_{eff} (Kiselev et al. 2011). In addition, H_m of 18 nm is significantly smaller than that of 300 nm. The thicker film has a larger skyrmion region, in contrast with that from Yu et al. (2011). This feature is systematically demonstrated in Figure 6.9. The field dependence

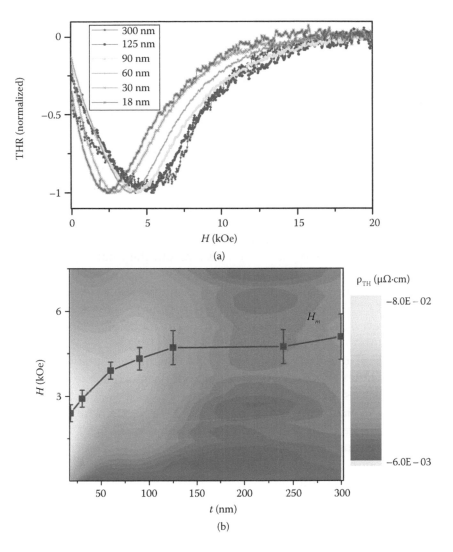

FIGURE 6.9 (a) Normalized THR = ρ_{TH}/ρ_H (25 kOe) as a function of magnetic field for various thicknesses at 10 K. (b) Color map of topological Hall resistivity in H–t plane at $T = 10$ K; square symbols are the values of H_m. (Adapted from Huang, S.X., and Chien, C.L., *Phys. Rev. Lett.*, 108, 267201, 2012. With permission.)

of THR shows a systematic variation with thickness, and H_m progressively shifts to lower values when the thickness decreases. The thickness dependence of the skyrmion phase is likely due to the induced uniaxial anisotropy which varies systematically with thickness as discussed in detail (Huang and Chien 2012). Nevertheless, the thickness provides a special knob to tune the skyrmion phase with an out-of-plane magnetic field.

Chapter 6

As shown in Figures 6.8 and 6.9, the ρ_{TH} has nonzero value at zero magnetic field when the magnetic field sweeps from positive to zero, suggesting the skyrmions can exist at zero field in the epitaxial FeGe thin films. The stabilization of the skyrmions at zero field may come from the pinning effect by the grain boundary, such as the pinned 'half-skyrmion'-like structure at the grain boundary at zero field, observed in the Lorentz TEM experiments (Yu et al. 2010). The nonzero ρ_{TH} is further revealed in the hysteresis loop of the Hall resistivity shown in Figure 6.10. The magnetization of FeGe films shows a usual counterclockwise hysteresis loop with positive remanence $M_r > 0$ when field sweeps from a large positive H to zero (denoted as $H = 0^+$). At $H = 0^+$, AHR gives $S_A \rho^2 M_r > 0$ since $S_A > 0$. However, the hysteresis loop of the Hall resistivity shows a clockwise hysteresis loop with negative ρ_H at $H = 0^+$. As a result, the total negative Hall resistivity ($\rho_H(H = 0^+) < 0$) must include a negative THR to compensate for the positive AHR at $H = 0^+$. By the same token, there is a positive THR at $H = 0^-$, indicating that the residue skyrmions at $H = 0^-$ reverse their cores and have opposite topological charge compared with the skyrmions at $H = 0^+$.

The Hall resistivity hysteresis loops remain robust down to the 50-nm wide devices in Hall-bar microdevices, while the hysteresis loop collapses for the 30-nm wide device (Kanazawa et al. 2015). For Hall-bar devices with width from 50 to 250 nm, stepwise profiles of the topological Hall resistivity are observed at small fields before the field-increasing and field-decreasing branches are overlapped, suggesting discontinuous motion and the creation/annihilation of skyrmions (Kanazawa et al. 2015).

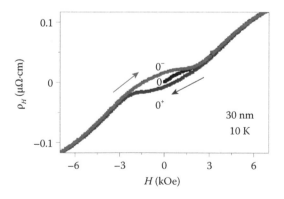

FIGURE 6.10   Hall resistivity hysteresis loop for 30-nm films at 10 K. (Adapted from Huang, S.X., and Chien, C.L., *Phys. Rev. Lett.*, 108, 267201, 2012. With permission.)

6.5 SANS on Epitaxial FeGe Films

In bulk B20 magnets, skyrmions form regular hexagonal lattices which are perpendicular to the applied field. As a result, they show a sixfold diffraction pattern (Mühlbauer et al. 2009) in SANS measurements when the magnetic field is parallel with the direction of the incident neutron beam (Figure 6.11). SANS measures the wave vectors which are perpendicular to the beam direction. As discussed earlier, due to the negligible magnetic anisotropy in bulk B20 magnets, the wave vectors of helical (conical) spin textures are aligned to the field direction. In the geometry shown in Figure 6.11 for bulk B20 magnets, the diffraction pattern on the detector screen shows no features for helical/conical spin textures whose **k** vectors are aligned to the field direction, but shows six spots for the skyrmion lattices whose **k** vectors are perpendicular to the field direction. Therefore, SANS measures skyrmion lattices in the momentum space and provides an indirect evidence of the existence of the skyrmion phase in B20 magnets, as first demonstrated in 2009 (Mühlbauer et al. 2009).

In collaboration with Dr. Colin Broholm at Johns Hopkins University, Dr. Lisa Debeer-Schmitt, and Dr. Ken Littrell at Oak Ridge National Laboratory, we performed the SANS measurements on epitaxial FeGe thin films. To increase the diffraction signals by the weak neutron beam, 10 pieces of 2 cm^2 300-nm epitaxial FeGe films were used. The neutron beam and the magnetic field are perpendicular to the film planes (Figure 6.11). Figure 6.12 shows the integrated intensities as a function of |**k**| at various magnetic fields at 150 K. The diffraction intensity at zero field indicates spin ordered states with in-plane **k** at zero field. Lorentz TEM measurements show that the ground states are helical spin textures with in-plane **k** at zero field in the thin plates of B20 magnets (Yu et al. 2010). In the FeGe thin films, the spin ordered states

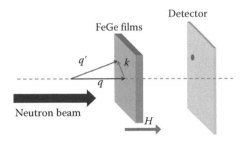

Detector

FeGe films

q' k

q

Neutron beam

H

FIGURE 6.11   Experimental set-up for SANS on FeGe films.

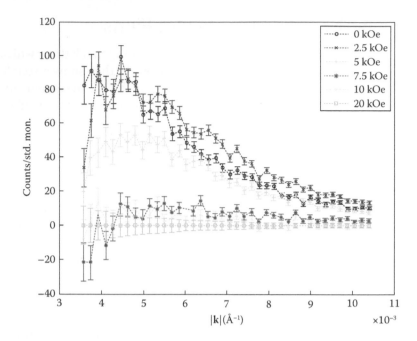

FIGURE 6.12 Integrated intensity of neutron scattering as a function of **k** at different magnetic fields at 150 K.

with in-plane **k** are likely to be helical/helicoid spin textures, but also include small portion of the residue skyrmion spin textures embedded in the helical spin textures since the THR is nonzero at zero field. When the magnetic field is increased to 2.5 kOe where the THR is maximum in the transport measurement (Figure 6.8c), the diffraction intensity increases. If the spins in the helical state are simply aligned by the magnetic field (i.e., more ordered along the field direction), the diffraction intensity should decrease. The increase in intensity suggests that the density of the skyrmions increases by suppressing the spin helixes. Further increasing the field aligns more spins along the field direction, thus reducing the skyrmion density and the diffraction intensity. As all the spins are aligned at large fields, the diffraction intensity drops to zero. The SANS results shown in Figure 6.12 are well corroborated with the dome-shaped THR results shown in Figure 6.8.

Now we use diffraction intensity at zero field as a background and plot intensity map $I = I(H = 2.5 \text{ kOe}) - I(H = 0 \text{ Oe})$ in the **k** plane. In this method, the differential diffraction intensity reflecting the skyrmion phase is shown in Figure 6.13. In contrast to the sixspots, the intensity map shows a ring-like feature, indicating disordered skyrmions in the film plane of the FeGe thin films, instead of regular skyrmion lattices

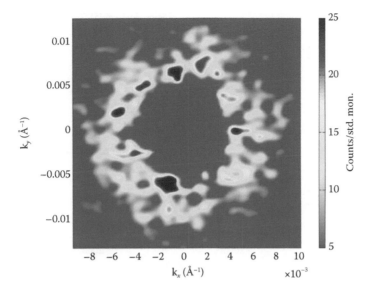

FIGURE 6.13   Neutron scattering intensity map for $I = I(H = 2.5 \text{ kOe}) - I(H = 0 \text{ Oe})$.

observed in bulk single crystals. The disordered skyrmions are likely due to the defects in the thin films and are also observed in the Lorentz TEM measurements of epitaxial MnSi thin films which are grown by MBE (Li et al. 2013).

6.6   Spin Textures under In-Plane Magnetic Fields

As discussed in the previous sections, the skyrmion phase in epitaxial FeGe films is greatly extended under an out-of-plane magnetic field. The film thickness has a strong effect on the stabilization of the skyrmion phase. In the FeGe thin films, the **k** wave vectors tend to lie in the film plane as shown in the SANS measurements. It is interesting to study if/how the skyrmion phase can be stabilized in thin films by the in-plane field. Moreover, how will the film thickness affect the spin textures under in-plane fields when it is smaller or larger than the helical wavelength of 70 nm for FeGe?

Ordinary Ferromagnetic (FM) films such as $Co_{40}Fe_{40}B_{20}$ films always exhibit anisotropy for the in-plane fields. As shown in Figure 6.14a, the MH loop of the $0°$ field is distinct from that of the $90°$ field in the plane. The remanence ratio M_r/M_S for the $0°$ field is nearly 1 while it is close to zero for the $90°$ field. In contrast, the FeGe films show completely isotropic MH loops when the in-plane fields are applied. The MH loops

Chapter 6

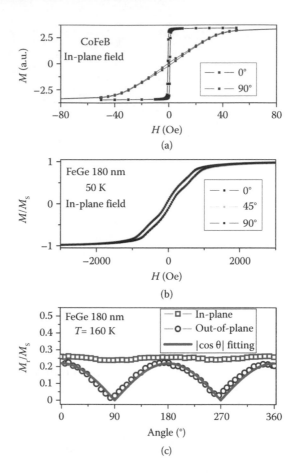

FIGURE 6.14 (a) Magnetic hysteresis loops of $Co_{40}Fe_{40}B_{20}$ films for in-plane field parallel with (black) and perpendicular to (dark gray) the easy axis. (b) Magnetic hysteresis loops of FeGe films for in-plane field with different orientations. (c) Normalized angular remanence as a function of angle for in-plane and out-of-plane orientations.

for 0°, 45°, and 90° fields are identical, and the M_r/M_S is only about 0.25 (Figure 6.14b). The isotropy of the *MH* loops under the in-plane fields is further illustrated in the angular-dependent remanence (Figure 6.14c). The remanence ratio is nearly a constant of 0.25 for the in-plane fields along any directions. The isotropy of the *MH* loops is also observed in the B20 $Fe_{1-x}Co_xSi$ single crystals (Huang et al. 2014). However, for the out-of-plane fields, the remanence shows anisotropic behavior due to the shape anisotropy. The remanence ratio is 0 for $\theta = 90°$ and its angular dependence can be fitted by $|\cos\theta|$, as shown in Figure 6.14c, where θ is the out-of-plane angle ($\theta = 90°$ when the field is perpendicular to the film plane).

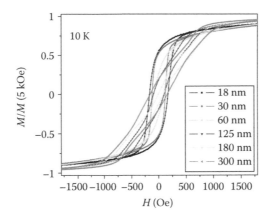

FIGURE 6.15 Magnetic hysteresis loop of FeGe films with different thicknesses at 10 K.

The *MH* loops are independent of the directions of the in-plane magnetic fields. But their shapes vary systematically on the film thickness (Figure 6.15). When the thickness t is larger than λ_H (~70 nm), the *MH* loops show slanted features with small remanences. For the films with thicknesses less than λ_H, the *MH* loops become squarer at low fields with much larger remanences, but the magnetization is more difficult to fully saturate. These results suggest that the film's thickness has a strong effect on the spin textures of the FeGe films under the in-plane fields.

For thicker films ($t > \lambda_H$, $t = 125$, 180 and 300 nm), the *MH* loop shows multiple features (Figure 6.15). The derivative of the *MH* loop (Figure 6.16a) clearly shows two distinct features. H_1 is defined as the transition field between the two features and H_S is the saturation field defined above. When temperature increases, these two features persist and the values of H_1 and H_S gradually decrease (Figure 6.16b). Above T_C, there is no feature as expected. We can then plot the phase diagram according to H_1 and H_S for FeGe films under the in-plane fields (Figure 6.16c). Below H_1, the helical spin textures are aligned along the field direction. Above H_S, it is the saturated state. Between H_1 and H_S, there may be elliptic skyrmion states since the film thickness is larger than the skyrmion size, as proposed in the studies of *MH* loops of MnSi films under the in-plane fields (Wilson et al. 2012). The in-plane skyrmions have also been suggested by the magnetic and planar Hall measurements on epitaxial MnSi films with thickness of several times the helical wavelength (Yokouchi 2015). For thinner films ($t < \lambda_H$, $t = 18$, 30, and 60 nm), the *MH* loop is much simpler and behaves like ordinary FM films (Figure 6.15), suggesting simpler spin textures and the in-plane

Chapter 6

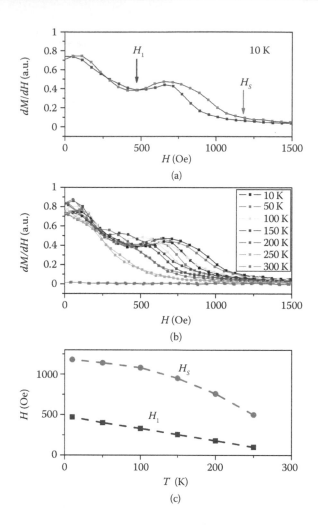

FIGURE 6.16 (a) Derivative of magnetic hysteresis loop for 180-nm FeGe films. (b) Derivatives of magnetic hysteresis loops as a function of temperature. (c) Phase diagram of 180 nm FeGe films under in-plane fields.

skyrmions cannot survive as a result of the thickness limitation. These results indicate that the film thickness can be used to tune the spin textures under the in-plane fields.

6.7 Magnetoresistance and Planar Hall Resistance

In ordinary conducting ferromagnets, such as Fe and Ni, a characteristic MR is the well-known anisotropic magnetoresistance (AMR) due to spin–orbit interaction and *s–d* scattering (Smit 1951; McGuire and Potter 1975). For polycrystalline FMs, the resistivity ρ at $H > H_S$ depends on the angle (ϕ) between the directions of the electric current (**I**) and the saturated magnetization (**M**$_S$) with an axial symmetry of (McGuire and Potter 1975):

$$\rho = \rho_\perp + (\rho_\| - \rho_\perp)\cos^2\phi, \tag{6.2}$$

where $\rho_\perp$ and $\rho_\|$ are the resistivities with **M**$_S$ parallel and perpendicular to **I**, respectively. For single crystals of centrosymmetric FMs such as Fe and Ni, the AMR also exhibits the fourfold rotation symmetry. As a result, the AMR in Fe or Ni single crystals has a combined symmetry of two- and fourfold symmetries (Berger and Friedberg 1968). However, in the cubic B20 magnets, the fourfold rotation symmetry is broken. This nontrivial broken rotation symmetry in cubic magnets can be directly revealed by the AMR measurements in $Fe_{1-x}Co_xSi$ single crystals which show only twofold symmetry (Huang et al. 2016). It should be noted that the maximum resistivity value is not necessary at $\phi = 0°$, but depends on the angle between the current and the crystalline axis.

In the epitaxial FeGe(111) thin films, the resistivities at $H > H_S$ for in-plane fields show twofold symmetry with maximum values at $\phi = 0°$ and $180°$. The AMR data can be well fitted by Equation 6.2 and the AMR ratio (($\rho_\| - \rho_\perp$)/$\rho_\perp$) is about 0.36%. When the transverse voltage is measured, the planar Hall resistance (PHR), which has the same origin of AMR, reads (McGuire and Potter 1975):

$$\rho_{PHR} = (\rho_\| - \rho_\perp)\cos\phi \cdot \sin\phi. \tag{6.3}$$

The PHR shows $\sin 2\phi$ dependence while the AMR shows $\cos 2\phi$ dependence. As shown in Figure 6.17b, the measured PHR is described well by Equation 6.3 and its maximum value is about 0.1 $\mu\Omega\cdot$cm at $\phi = 45°$ and $225°$. The PHR has both positive and negative values and reflects the magnetization. It can be used to measure the skyrmion phase in B20 magnets (Yokouchi 2015).

When $H < H_S$, in ordinary magnets, domain structures are formed. In each domain, all the spins are aligned. Domains are separated by narrow domain walls. Consequently, for $\phi = 0°$, the resistance at $H < H_S$ is smaller than $R_\|$ at $H > H_S$ when some domains have magnetization deviating from the current direction, assuming the contribution of domain wall resistance is negligible. On the other hand, for $\phi = 90°$, the resistance at $H < H_S$ is larger than $R_\perp$ at $H > H_S$. This is the typical field dependence of resistance in ordinary FMs as shown in Figure 6.18a.

Chapter 6

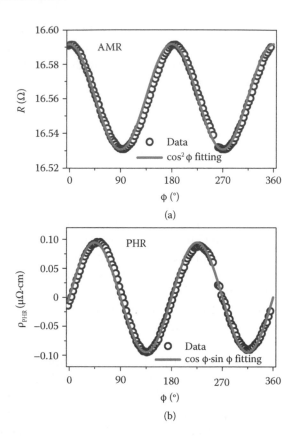

FIGURE 6.17 Angular dependence of (a) anisotropic magnetoresistance (AMR), and (b) planar Hall resistivity (PHR) of a 120-nm FeGe film at 10 K. Solid lines are $\cos^2 \phi$ (a) and $\cos \phi \cdot \sin \phi$ fitting (b) to the data ($T = 10$ K and $H = 15$ kOe).

In contrast, the field dependence of resistance in 120-nm epitaxial FeGe films shows very different behaviors. The MR has a positive background above H_S, which is nearly independent of field directions. The background may have several contributions such as ordinary MR due to Lorentz force (Porter et al. 2014) and requires further studies. For $\phi = 0°$, the MR shows a hysteresis loop which correlates with the magnetic hysteresis loop in Figure 6.15, suggesting that it originates from the spin textures. Interestingly, for $\phi = 0°$, the resistance at $H < H_S$ is larger than $R_{\parallel}$ (background subtracted) at $H > H_S$. The high resistance at $H < H_S$ cannot originate from AMR which should give rise to a low resistance. When the spins are not aligned, such as that in a domain wall and a spin valve with antiparallel spin alignment, the spin-dependent scattering due to the spin textures results in a higher resistance compared with the aligned spin configuration. This is the collective spin-dependent

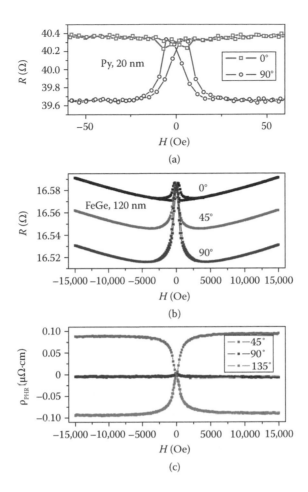

FIGURE 6.18 (a) Resistance (measured at room temperature) of 20-nm Py films as a function of in-plane magnetic fields for field angle of 0° and 90°. (b) Resistance ($T = 10$ K) of 120-nm FeGe films as a function of in-plane magnetic fields for field angle of 0°, 45°, and 90°. (c) Planar Hall resistivity ($T = 10$ K) of 120-nm FeGe films as a function of in-plane magnetic fields for field angle of 45°, 90°, and 135°.

scattering, inherent to the spin textures. For example, the domain wall resistance originates from the scattering of the spin textures of the domain wall. Spin helix, shown in Figure 6.1a, is a macroscopic Bloch domain wall and has an intrinsic helical resistance (Huang et al. 2014). In the epitaxial FeGe films, at $H < H_S$, the spin textures are more complicated and may include spin helix, elliptic skyrmions, and helical domain structures. Nevertheless, all the spin textures, which occupy the entire films, will give rise to collective spin-dependent scattering which increases the resistance. As a result, for fields along any directions, the resistance at $H < H_S$ is larger than R_S (background subtracted)

Chapter 6

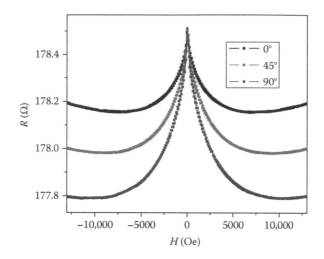

FIGURE 6.19 Resistance (T = 10 K) of 30-nm FeGe films as a function of in-plane magnetic fields for field angle of 0°, 45°, and 90°.

at $H > H_S$. Interestingly, the resistances at zero field (switching from a large field) are the same no matter what the field directions are. The PHR, on the other hand, only picks the contribution of AMR. Figure 6.18c shows the field dependence of PHR at ϕ = 45°, 90°, and 135°. The 45° PHR and the 90° PHR have opposite signs while the 90° PHR is near zero for all the fields. The PHR has a small asymmetry on field, possibly due to a small contribution from AHR from misalignment between the sample plane and the field direction.

For thinner films (Figure 6.19), the resistances at high fields have the same AMR behavior which reads $R(\phi = 45°) \approx (R(\phi = 0°) + R(\phi = 90°))/2$. The resistance at low fields is larger than that at high fields regardless of field directions. The resistances at zero field (switching from a large field) are the same for all the original field directions. However, when the film thickness is reduced, the interfacial scattering sets in and the field dependence of resistance (Figure 6.19) does not correlate with the magnetic hysteresis loop. It includes contributions from AMR, resistance due to spin textures, and other mechanisms, such as weak localization, which require further investigations.

6.8 Summary

Bulk B20 magnets host a helical ground state and the exotic skyrmion state in a small region in the phase space. Epitaxial B20 thin films can be deposited on Si(111) substrates by exploiting a special epitaxial growth.

In the epitaxial FeGe(111) films, the skyrmion phase, determined by the topological Hall resistance and SANS has been greatly expanded to essentially the entire temperature range below T_C under a perpendicular field. The extended skyrmion phase can be manipulated by the film' thickness. Moreover, under an in-plane field, the films thicknesses also control the spin structures, and the elliptic skyrmion phase may exist in films thicker than the helical wavelength. The FeGe(111) films show unconventional MR due to the spin textures. The unusual MR depends on the film thickness. These controllable extended skyrmion phases and the thickness-dependent magnetic and transport properties can be used to explore the spintronics applications of skyrmions.

Acknowledgment

This work has been supported by NSF Grant DMR1262253. We are grateful for the helpful discussions with JD Zang, now at University of New Hampshire. A portion of this research used resources at the High Flux Isotope Reactor, a DOE Office of Science User Facility operated by the Oak Ridge National Laboratory.

References

Bak P and Jensen MH (1980). Theory of helical magnetic structures and phase transitions in MnSi and FeGe. *Journal of Physics C: Solid State Physics* 13:L881.

Berger L and Friedberg SA (1968). Magnetoresistance of a permalloy single crystal and effect of 3d orbital degeneracies. *Physical Review* 165:670–679.

Bogdanov A and Hubert A (1994). Thermodynamically stable magnetic vortex states in magnetic crystals. *Journal of Magnetism and Magnetic Materials* 138:255–269.

Bogdanov A and Hubert A (1999). The stability of vortex-like structures in uniaxial ferromagnets. *Journal of Magnetism and Magnetic Materials* 195:182–192.

Bogdanov A and Yablonsky D (1989). Thermodynamically stable "vortices" in magnetically ordered crystals. The mixed state of magnets. *Soviet Physics JETP* 68:101.

Bruno P, Dugaev VK and Taillefumier M (2004). Topological Hall effect and Berry phase in magnetic nanostructures. *Physical Review Letters* 93:096806.

Butenko AB, Leonov AA, Rößler UK, and Bogdanov AN (2010). Stabilization of skyrmion textures by uniaxial distortions in noncentrosymmetric cubic helimagnets. *Physical Review B* 82:052403.

Dzyaloshinskii I (1958). A thermodynamic theory of "weak" ferromagnetism of antiferromagnetics. *Journal of Physics and Chemistry of Solids* 4:241–255.

Fert A, Cros V and Sampaio J (2013). Skyrmions on the track. *Nature Nano technology* 8:152–156.

Grigoriev SV, et al. (2007). Magnetic structure of $Fe_{1-x}Co_x$ Si in a magnetic field studied via small-angle polarized neutron diffraction. *Physical Review B* 76:224424.

Huang SX, et al. (2016). Unusual magnetoresistance in cubic B20 Fe0.85Co0.15Si chiral magnets. *New Journal of Physics* 18:065010.

Chapter 6

Huang SX and Chien CL (2012). Extended skyrmion phase in epitaxial FeGe(111) thin films. *Physical Review Letters* 108:267201.

Huang SX, et al. (2014). Universal ratio of intrinsic resistivities of spin helix in B20 (Fe-Co)Si magnets. *ArXiv* 1409.7869.

Kanazawa N, et al. (2015). Discretized topological Hall effect emerging from skyrmions in constricted geometry. *Physical Review B* 91:041122.

Kanazawa N, et al. (2011). Large topological Hall effect in a short-period helimagnet MnGe. *Physical Review Letters* 106:156603.

Karhu E, et al. (2010). Structure and magnetic properties of MnSi epitaxial thin films. *Physical Review B* 82:184417.

Karhu EA, et al. (2012). Chiral modulations and reorientation effects in MnSi thin films. *Physical Review B* 85:094429.

Kiselev NS, Bogdanov AN, Schäfer R and Rößler UK (2011). Chiral skyrmions in thin magnetic films: New objects for magnetic storage technologies? *Journal of Physics D: Applied Physics* 44:392001.

Lebech B, Bernhard J and Freltoft T (1989). Magnetic structures of cubic FeGe studied by small-angle neutron scattering. *Journal of Physics: Condensed Matter* 1:6105.

Lee M, Kang W, Onose Y, Tokura Y and Ong NP (2009). Unusual Hall effect anomaly in MnSi under pressure. *Physical Review Letters* 102:186601.

Lee M, Onose Y, Tokura Y and Ong NP (2007). Hidden constant in the anomalous Hall effect of high-purity magnet MnSi. *Physical Review B* 75:172403.

Li Y, et al. (2013). Robust formation of skyrmions and topological Hall effect anomaly in epitaxial thin films of MnSi. *Physical Review Letters* 110:117202.

McGuire T and Potter R (1975). Anisotropic magnetoresistance in ferromagnetic 3D alloys. *IEEE Transactions on Magnetics* on 11:1018–1038.

Moriya T (1960). Anisotropic superexchange interaction and weak ferromagnetism. *Physical Review* 120:91–98.

Mühlbauer S, et al. (2009). Skyrmion lattice in a chiral magnet. *Science* 323:915–919.

Nagaosa N, Sinova J, Onoda S, MacDonald AH and Ong NP (2010). Anomalous Hall effect. *Reviews of Modern Physics* 82:1539.

Nagaosa N and Tokura Y (2013). Topological properties and dynamics of magnetic skyrmions. *Nature Nano technology* 8:899–911.

Neubauer A, et al. (2009). Topological Hall effect in the A phase of MnSi. *Physical Review Letters* 102:186602.

Onose Y, Takeshita N, Terakura C, Takagi H and Tokura Y (2005). Doping dependence of transport properties in $Fe_{1-x}Co_xSi$. *Physical Review B* 72:224431.

Porter NA, Creeth GL and Marrows CH (2012). Magnetoresistance in polycrystalline and epitaxial $Fe_{1-x}Co_xSi$ thin films. *Physical Review B* 86:064423.

Porter NA, Gartside JC and Marrows CH (2014). Scattering mechanisms in textured FeGe thin films: Magnetoresistance and the anomalous Hall effect. *Physical Review B* 90:024403.

Porter NA, et al. (2015). Manipulation of the spin helix in FeGe thin films and FeGe/Fe multilayers. *Physical Review B* 92:144402.

Rößler UK, Leonov AA and Bogdanov AN (2011). Chiral skyrmionic matter in non-centrosymmetric magnets. *Journal of Physics: Conference Series* 303:012105.

Rybakov FN, Borisov AB, Blügel S and Kiselev NS (2015). New type of stable particlelike states in chiral magnets. *Physical Review Letters* 115:117201.

Smit J (1951). Magnetoresistance of ferromagnetic metals and alloys at low temperatures. *Physica* 17:612–627.

T. Yokouchi NK, et al. (2015). Formation of in-plane skyrmions in epitaxial MnSi thin films as revealed by planar Hall effect. *ArXiv* 1506.04821.

Ulrich KR, et al. (2011). Chiral skyrmionic matter in non-centrosymmetric magnets. *Journal of Physics: Conference Series* 303:012105.

Wilhelm H, et al. (2011). Precursor phenomena at the magnetic ordering of the cubic helimagnet FeGe. *Physical Review Letters* 107:127203.

Wilson MN, et al. (2012). Extended elliptic skyrmion gratings in epitaxial MnSi thin films. *Physical Review B* 86:144420.

Ye J, et al. (1999). Berry phase theory of the anomalous Hall effect: Application to colossal magnetoresistance manganites. *Physical Review Letters* 83:3737–3740.

Yi SD, Onoda S, Nagaosa N and Han JH (2009). Skyrmions and anomalous Hall effect in a Dzyaloshinskii-Moriya spiral magnet. *Physical Review B* 80:054416.

Yu XZ, et al. (2011). Near room-temperature formation of a skyrmion crystal in thin-films of the helimagnet FeGe. *Nature Materials* 10:106–109.

Yu XZ, et al. (2010). Real-space observation of a two-dimensional skyrmion crystal. *Nature* 465:901–904.

Yurii AI (1984). Modulated, or long-periodic, magnetic structures of crystals. *Soviet Physics Uspekhi* 27:845.

7. Formation and Stability of Individual Skyrmions in Confined Geometries

Haifeng Du, Chiming Jin, and Mingliang Tian

Chinese Academy of Sciences, Hefei, People's Republic of China
Nanjing University, Nanjing, People's Republic of China

Magnetic skyrmion is a topologically stable whirlpool-like spin texture that offers great promise as an information carrier for future memory and logical devices owing to its prominent features, including small size, topological stability, and the small critical current required to move it. To enable such applications, it is essential to understand skyrmion properties in highly confined geometry. In this section, we seek to provide a fundamental insight into this issue. Theoretical progresses ranging from the novel magnetic vortex to field-driven skyrmion dynamics in helimagnetic nanodisks are covered. Furthermore, experimental achievements regarding the formation and stability of individual skyrmions in nanowires (NWs) and nanostripes are outlined.

Chapter 7

7.1   Introduction

Traditional magnetic storage in hard disk drive technology is based on the controllable manipulation of magnetic domains by a magnetic field. Magnetic recording devices require high coercivity materials to insure thermal stability at high storage densities. However, the requirements of small volume (for high densities), low coercivity (for good write-ability), and high thermal stability cannot be optimized at the same time. The capacity of the conventional hard disk storage is thus almost approaching its limit [1]. The emergence of topological phenomena and topological materials offers immense opportunity to extend the storage device roadmap [2–5]. In magnetic materials, a notable example of a topologically stable object is the skyrmion, a swirl-like spin texture that is characterized by all the magnetic moments pointing in all directions wrapping a unit sphere. The peculiar twists of the magnetization within the skyrmion enable efficient couplings within the spin current, leading to topological Hall effects, and strong spin-transfer torque effects [6,7]. Accordingly, the critical current density to move skyrmions is several orders of magnitude lower than that needed to drive ferromagnetic domain walls [8–10]. Meanwhile, the size of a single skyrmion is typically on the order of 5–100 nm and can be continuously tuned by doping [11]. This high mobility, small size, and topological stability are all advantages to building skyrmion-based memory or logical devices, wherein the designs are all based on the formation and controlled manipulations of the spin textures in magnetic nanostructured elements [12]. It is therefore an important issue to understand the formation and stability of skyrmions in highly confined geometries.

A key ingredient of stabilizing the skyrmion phase is the Dzyaloshinskii–Moriya (DM) interaction [13]. In addition to this coupling, it has also been addressed that magnetic dipolar interaction [14], frustrated exchange interaction [15], and four-spin exchange interactions [16] are all able to create skyrmions. Among them, the DM-induced skyrmions have evident advantages including a small and tunable size [11], fixed unique chirality [13], and extra stability even in highly confined geometries [17–19]. DM couplings come from symmetry breaking in the crystal lattice or interface and surface. In this chapter, we will focus on skyrmions induced by the DM couplings originated from the symmetry breaking in the crystal lattice. Other types of skyrmions and their features can be found in other chapters. The typical materials are the noncentrosymmetric B20 compounds including MnSi, FeGe, and $Fe_xCo_{1-x}Si$ [13], which are named as helical magnets.

The competition between ferromagnetic exchange and DM couplings leads to the spin helix with a fixed wave vector, $\mathbf{Q}$, along the high symmetry crystal axis. If a finite magnetic field, $\mathbf{B}$, is applied at the temperatures (T) lower than the Curie temperature T_C, it becomes energetically favorable to form a conical phase with $\mathbf{Q} \| \mathbf{B}$. At an even higher field, the conical phase transfers into ferromagnetic spin alignment. Both helical and conical states belong to single-twist magnetic structure since the rotation of their magnetization is only in one direction. In contrast, the magnetization within the skyrmion rotates in two directions, forming double-twist modulations. However, skyrmions occupy only a tiny pocket in the $T\sim B$ phase diagram as the temperature is slightly below T_C in bulk compounds. The evolution of spin configurations with varied B and T in FeGe is shown in Figure 7.1 and reflects the common behaviors in helical magnets [13].

The emergent phenomena in helical magnets have been well explained regarding the above-mentioned physical model concerning the DM, ferromagnetic exchange, and external magnetic interaction. The formation of the skyrmion state out of the conical states in bulk materials is explained by the thermal fluctuation effect [20]. For the low-dimensional helical magnets, the skyrmion state has been

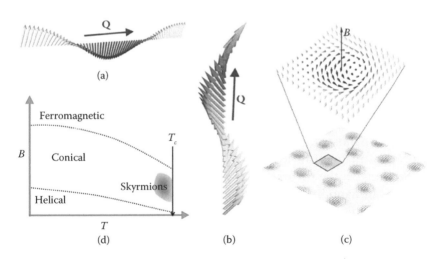

FIGURE 7.1 Common magnetic phase diagram of bulk helimagnets. (a), (b), and (c) schematically represent the spin configurations of helical, conical, and skyrmion phases, respectively, which develop below the Curie temperature, T_C. (d) Phase diagram of magnetic structure in bulk helical magnets. (From Du, H. F., et al., *Nature Commun.*, 6, 8504, 2015. With permission.)

Chapter 7

observed to show high stability [21]. Up to now, three mechanisms were proposed to explain the extended skyrmion state. The first is based on the spatial confinement effects [22]. When the thickness of the film is below a threshold, a type of a three-dimensional (3D) skyrmion, characterized by a superposition of conical modulations along the skyrmion axis and double-twist rotation in the perpendicular plane, is thermodynamically stable in a broad T–B range, with B normal to the film plane. This mechanism may explain the experimental results in two-dimensional (2D) materials including mechanically thinned flakes and homogeneous films, even quasi-2D MnSi nanostripes [19,23–25]. But this mechanism makes sense only if the film thickness is comparable to or less than the skyrmion lattice constant L_D. The second mechanism invokes the uniaxial anisotropy originating from the surface, interface, or pressure [26]. When applying the external magnetic field along the easy axis, the combination of the uniaxial anisotropy and the external magnetic field leads to a significant increase in conical phase energy and thus stabilizes the skyrmion state. The last mechanism also involves uniaxial anisotropy [27], but with its hard axis perpendicular to the film plane. Extended elliptic skyrmion gratings are suggested as a thermodynamically stable state when applying the external magnetic field in the film plane. Actually, the last two mechanisms can be unified in one argument that an effective uniaxial anisotropy may stabilize the skyrmion state if the magnetic field is applied parallel to the easy axis of the effective uniaxial anisotropy. In the highly confined helical magnets, the experimental observations, as discussed below, cannot be explained by independently using the three mechanisms.

This section provides a general introduction to the highly confined helical magnets concerning the formation and stability of skyrmions in these nanostructured elements, trying to relate fundamental aspects of magnetism to the very important potential applications. In Section 7.2, the theoretical progress, mainly focusing on the spin textures in nanodisks, is discussed, while Section 7.3 is reversed for a more detailed demonstration on the experimental observation of magnetic states in NWs and nanostripes by using electrical probing and direct Lorentz transmission electron microscopy (LTEM) observations. In addition to the present materials, most of the theoretical efforts have been devoted to the device design by using a single skyrmion confined in the nanostripe as an information bit, where the spin-polarized current is used to drive the motion of the single skyrmion [9,10,17]. These relevant contents have been discussed in other chapters and are not considered in this chapter.

7.2 Theoretical Progress of Skyrmions in a Helimagnetic Nanodisk

For the theoretical investigation on helical magnets, a generally accepted simplest Hamiltonian is written as [13]:

$$w = A(\nabla\mathbf{m})^2 + D\mathbf{m}\cdot(\nabla\times\mathbf{m}) - \mathbf{m}\cdot B, \qquad (7.1)$$

where $\mathbf{m}$ is the unit vector of magnetization, A is a ferromagnetic exchange constant, and B is the external magnetic field. The three terms in Equation 7.1 are, in turn, exchange, DM, and Zeeman energy, respectively. The equation-minimizing energy (7.1) include axisymmetric localized solutions, which are obtained by solving the corresponding Euler equation, as discussed in Section 7.2.1. More generally, minimizing energy (7.1) can be achieved by the Monte Carlo (MC) simulation, where the sample is divided into discrete blocks. This process was performed in Section 7.2.2.

7.2.1 Magnetic Vortex with Skyrmionic Core in Nanodisk

The equation-minimizing functional w in Equation 7.1 includes solutions for one-dimensional (1D) helical or conical states and 2D isolated skyrmions. Because the disk shares the same symmetry with the axisymmetric localized solution, the magnetic structures in the thin nanodisk of helical magnets are investigated by only considering axisymmetric localized solution $\psi = \varphi + \pi/2$, $\theta = \theta(\rho)$ [28–31]. In this case, the spatial variable $r = (\rho \cos \varphi, \rho \sin \varphi, z)$ is written in cylindrical coordinates and the magnetization in spherical coordinates $m = (\sin \theta \cos \psi, \sin \theta \sin \psi, \cos \theta)$. With reduced length unit A/D, and energy unit D^2/A, the vortex energy density W of the disk with radius R can be expressed as $E = (2/R^2)\int_0^R w(\rho)\rho\, d\rho$, with

$$w(\rho) = \left[\left(\frac{d\theta}{d\rho}\right)^2 + \frac{1}{\rho^2}\sin^2\theta\right] - \left(\frac{d\theta}{d\rho} + \frac{1}{\rho}\sin\theta\cos\theta\right). \qquad (7.2)$$

In the thin disk limit, the edge-pinning effect imposed by the surface energy at the lateral disk edge is neglected. The Euler equation for $\theta(\rho)$ with certain conditions is deduced as

$$\left[\frac{d^2\theta}{d\rho^2} + \frac{1}{\rho}\frac{d\theta}{d\rho} - \frac{1}{\rho^2}\sin\theta\cos\theta\right] - \frac{1}{\rho^2}\sin^2\theta = 0. \qquad (7.3)$$

Chapter 7

Equation 7.3 is the simple one without considering the effect of an external magnetic field. For an isolated skyrmion, the equation has the boundary conditions: $\theta(0) = \pi$ and $\theta(\infty) = 0$.

For a disk without the edge limitation, it belongs to the natural boundary conditions with its expression:

$$\theta(0) = 0, \ d\theta/d\rho|_{\rho=R} = \theta_\rho(R) = 0.5. \tag{7.4}$$

Equation 7.4 belongs to a two-point boundary-value problem in ordinary differential equations, which at the starting point do not determine a unique solution to start with. A random choice among the solutions that satisfy these incomplete starting boundary conditions $\theta(0) = 0$ is almost certain not to satisfy the boundary conditions at the end point $\theta_\rho(R) = 0.5$. The two-point boundary value problem can be solved by using the standard initial value problem [28]:

$$\theta(0) = 0, \ \theta_\rho(0) = \alpha. \tag{7.5}$$

In this way, Equation 7.3 may be solved just by numerical iteration from initial point to its end. For different values of α, the trajectories $\theta_\rho(\rho,\alpha)$ with $0 < \alpha < \infty$ define a family of solutions, which include the required solutions with boundary conditions in Equation 7.4. In Ref. [28], a MC method is performed to find the desired solutions, where the desired initial value depends on the disk size. Figure 7.2 shows the required $\alpha(R)$ curve, from which Equation 7.3 with boundary conditions in Equation 7.4 is solved.

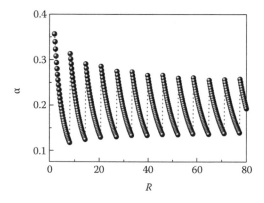

FIGURE 7.2 The solution of Equation 7.3 with boundary conditions in Equation 7.4. (From Du, H.F., et al., *EPL*, 101, 37001, 2013. With permission.)

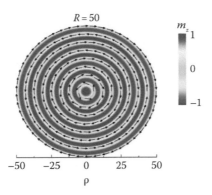

FIGURE 7.3 The featured parameters and spin arrangements of the magnetic vortex for a fixed radius $R=50$; the magnetization profile of the disk and the spin texture are characterized by a skyrmionic core and a serial of circle spin stripes, respectively. (From Du, H.F., et al., *EPL*, 101, 37001, 2013. With permission.)

The typical magnetization profiles for randomly chosen radius $R = 50$ are illustrated in Figure 7.3. The spin texture is characterized by a vortex core with a series of circle spin stripes. The period of stripes is $T_m \approx 4\pi$, which approximately equals to that of a helix state. The final spin stripe with an incomplete helical period is only partially owing to the limitation of the boundary conditions in Equation 7.4.

The dependence of the size of the skyrmion core, R_S, on the disk radius, R, is displayed in Figure 7.4a. When $R < R_c$ with $R_c = 8.0$ as the critical disk size, a vortex, but not a skyrmion, is formed. Just above the threshold value R_c, a complete skyrmion core with an incomplete edge twist. When $R < 14.4$, R_S increases linearly with increasing R with the slope 0.5. When $R > 14.4$, R_S jumps again to a small value. $R_S(R)$ shows oscillation behavior. These data suggest that by controlling the size of the disk, one can realize the effective manipulation of the size of the skyrmionic core.

Figure 7.4b and c represents the dependence of the energy density E of the disk on its size R, the dotted line $E_h = -0.25$ is the energy density of the helix state [20]. The horizontal axis is divided into two intervals: $R \in (0,30)$ and $R \in (30,100)$. In the former interval, $E(R)$ oscillates around E_h with a period of $T_E \approx 2\pi$, suggesting a possible vortex ground state since the energy of a vortex is lower than that of a helix structure in some interval of disk radius R, marked by double arrows. The helix state would be distorted at the edge of the disk due to the spatial confinement effect, lowering the energy of the system, so that it could not allege that the spontaneous vortex ground state may form within this region,

Chapter 7

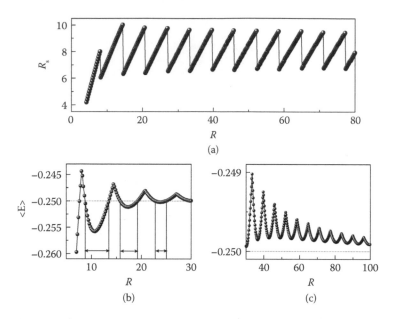

FIGURE 7.4 (a) The radius R_s of a skyrmionic core as a function of the disk radius R. Energy density as the function of size of disk: (b) $R \in (0,30)$ and (c) $R \in (30,100)$. The dotted lines represent the energy density of the helix state; in the interval marked by double arrows, the energy density of the vortex state is lower than that of the helix state, indicating a big possibility for the formation of a spontaneous vortex ground state even without the help of external field and thermal fluctuation. (From Du, H.F., et al., *EPL*, 101, 37001, 2013. With permission.)

but it gives a route to find this spontaneous vortex ground state even without the help of external fields and thermal fluctuation with high probability. In the latter interval, the energy difference between the helix state and vortex state first increases, then decreases, with the increase of R. Especially when $R > 60.0$, the relative difference of energy between two states is only 0.1%, indicating an increased opportunity to form this metastable vortex structure.

This theoretically predicated circle spin stripe has been observed by Lorentz microscopy in 2D chiral magnet FeGe [32] (Figure 7.5a), where it is only partially displayed when one edge of the thin sample is nearly round. Interestingly, another spin structure, shown in Figure 7.6, might be also the possible axisymmetric solution for the helimagnetic nanodisk. The contour lines of any fixed m_z are all well fitted by Archimedian spirals [33]. This special spin structure has also been observed in FeGe (Figure 7.5a), and the authors there named it a Swiss scroll state.

In addition to two main energy scales, i.e., ferromagnetic exchange and DM couplings, some weak energy contributions, such as dipolar energy,

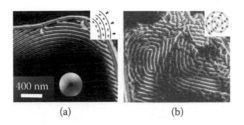

(a)   (b)

FIGURE 7.5   Spontaneous ground state in FeGe thin films. (a) The circle spin stripe with a skyrmionic core. (b) Swiss scroll state. The images are obtained by using high-resolution Lorentz TEM. (From Uchida, M., et al., *Phys. Rev. B*, 77, 184402, 2008. With permission.)

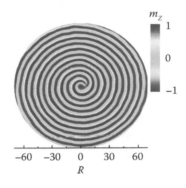

FIGURE 7.6   The polar magnetization profile of Swiss scroll state.

crystal field energy, and uniaxial anisotropy energy, also play important roles in determining the properties of helimagnets. In analog to soft magnetic vortex in nanodots and bubble domain in film, dipolar interaction may stabilize the vortex state since it is a magnetic charge free spin microstructure [34]. Crystal field energy favors the spin along its easy axis direction, promoting a helix state. It is the reason why the vortex state is found only in FeGe with weak crystal field and dislocation-like defect often observed in $Fe_{0.5}Co_{0.5}Si$ with stronger crystal field energy.

In this section, numerical calculations, based on a fundamental model, suggested a new type of spin vortex texture with a skyrmionic core existence in a thin nanodisk of helimagnets. The size of the skyrmionic core is able to be tuned by controlling the disk size. Moreover, this special spin texture may form spontaneous ground state even without external field and thermal fluctuation in some certain intervals of disk size.

Chapter 7

7.2.2 Field–Driven Evolution of Spin Textures in the Nanodisk

In Section 7.2.1, a novel magnetic vortex is suggested, where the external magnetic field H is not considered. In this section, the MC method is used to calculate the field-driven evolution of spin textures in heli-magnetic nanodisk with the energy including dipolar–dipolar interaction, uniaxial anisotropy, and Zeeman energy. The discrete expression for the energy is written as [31,35]:

$$E = -J \sum_{i<j} \hat{S}_i \cdot \hat{S}_j - \sum_{i,j} D_{R_{ij}} \cdot \left(\hat{S}_i \times \hat{S}_j \right) - K \sum_i \left(\hat{S}_i \cdot \hat{e}_i \right)^2$$

$$+ d \sum_{i<j} \left(\frac{\hat{S}_i \cdot \hat{S}_j}{\left| R_{ij} \right|^3} - \frac{3(\hat{S}_i \cdot \hat{R}_{ij}) \cdot (\hat{S}_j \cdot \hat{R}_{ij})}{\left| R_{ij} \right|^5} \right) - B \cdot \sum_i \hat{S}_i. \tag{7.6}$$

The five terms in Equation 7.6 denote the ferromagnetic exchange coupling with exchange constant J, DM interaction with constant $\left| D_{R_{ij}} \right| = D$ and $D_{R_{ij}}$ pointing along the vector at sites i and j, uniaxial anisotropy energy with constants K, the dipolar–dipolar interaction between the blocks with dipolar strength constant d, and Zeeman energy with an external field perpendicular to the disk plane, respectively. The ratio D/J was chosen to yield the spiral propagation wavelengths of certain lattice constants. In a discrete model, both direct and DM–exchange interaction in an inhomogeneous state will have a slightly different energy. Depending on their orientation with respect to the underlying discrete bond orientation of the lattice, the propagation direction of the helix state was locked into the [11] directions even without the weak crystal field energy [3,31], which determined the propagation direction of the helix state in the continuous model. A detailed investigation into the effect of crystalline field interactions on the spin arrangement in 2D film of the chiral magnets has been performed by Yi et al. [31].

Generally, a high-temperature annealing metropolis algorithm is used to obtain the equilibrium spin configurations [31]. After obtaining the final equilibrium magnetic configurations, the relative local chirality τ_r at lattice r can be calculated by:

$$\tau_r = \hat{S}_r \cdot \left(\hat{S}_{r+\hat{x}} \times \hat{S}_{r+\hat{y}} \right) + \hat{S}_r \cdot \left(\hat{S}_{r-\hat{x}} \times \hat{S}_{r-\hat{y}} \right). \tag{7.7}$$

The total chirality τ of the disk is calculated by $\tau = a \sum \tau_r$ with the constant a, which is chosen to guarantee the unit τ of one skyrmion.

The magnetic phase transition in the simulation is determined by monitoring the total chirality, which is equivalent to topological charge if an integral τ is obtained. A skyrmion is distinguished from a ferromagnet or other trivial state by the topological charge.

Figure 7.7 shows a typical field-driven evolution of the chiral modulations for the disk with radius $R_d = 10$. Discontinuous jumps in $\tau(B)$ and $m_z(B)$ curves mark the system with different spin textures. In bulk materials, the theoretical calculation has proved only two first-order transition lines exist, named as B_h and Bs, corresponding to the evolutions from helix state to skyrmion lattice and subsequently to ferromagnetic state, respectively [13]. In the nanodisk system, the situation is more complex. For $B < B_h$ with the critical field $B_h = 0.2$, the helix state is distorted at the edge, favoring their in-plane spin

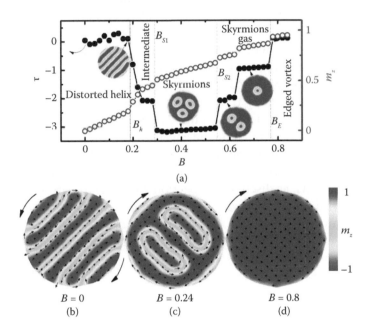

FIGURE 7.7   (a) The normalized magnetization m_z and chirality τ of disk for $R = 10$ as the functions of magnetic field B and discontinuous curves $m_z(B)$ and $\tau(B)$, indicating different spin textures; the corresponding snapshots of out-of-plane magnetization m_z is shown in the inset. The full spin orientation for three typical magnetic fields: (b) for the zero field $B = 0$, the complex spin textures may be regarded as the superposition of the edged state with the in-plane spin orientation perpendicular or parallel to the edge and bulk state with the features of helix. (c) For the moderate field $B = 0.24$, the edged state and the helix state are transferred into edged vortex and bimeron, respectively. (d) For high field $B = 0.8$, the skyrmions at the center of the disk are transferred into a ferromagnetic state, while the edged vortex still exists. (From Du, H.F. et al., *Phys. Rev. B*, 87, 014401, 2013. With permission.)

orientation perpendicular or parallel to the edge, as illustrated in Figure 7.7b. A similar phenomenon has been observed in the stripe domain of microsized Ni wires with high perpendicular anisotropy [36,37]. The distortion of spin textures at the edge results in nonzero chirality. In rigorous continuous model (Equation 7.1 just including direct and DM–exchange interactions), the distorted helix state is expected to be isotropic and continuously degenerate with respect to propagation directions in nanodisk. However, the discretization of the isotropic model in the *MC* simulation fixed the propagation direction of the helix state into the [11] directions, leading to fourfold degenerate ground states with every in-plane 90° rotation. In addition, two types of spin configurations with different symmetry are observed. For $B < 0.1$, in-plane spins m_{xy} show mirror symmetry and m_z displays two fold rotational symmetry with respect to the <11> axis, as displayed in Figure 7.7b. If $B > 0.1$, the symmetry of m_z and m_{xy} exchanged each other, as illustrated in the inset in Figure 7.7a ($B = 0.18$).

For $H_h \leq H < H_{SI}$ with the critical field $B_{SI} = 0.29$, the system undergoes an intermediate state before entering Skyrmions. With increasing B, the proportion of the parallel spins of the edge-state increases, and the spin textures finally transfer into an edged magnetic vortex. The typical intermediate state in Figure 7.7c is characterized by two compact bimerons surrounded by the edged vortex. Indeed, the edged vortex, favoring their in-plane spin orientation completely parallel to the edge, related to the rotational symmetric solutions of Euler equations of the helimagnets, as discussed in Section 7.2.1. The bimeron, composed of two half-disk domains with nonvanishing 1/2 topological charge and a rectangular stripe domain with zero topological charge, has been predicted to exist in 2D easy-plane ferromagnets with DM interactions at a finite temperature [38].

For $B_{SI} \leq B < B_E$ with the critical field $B_E = 0.77$ corresponding to the transition line from skyrmions to edged vortex, skyrmions have a countable number. Both skyrmions in the interior of the disk and the edged vortex have the same sense of rotation, which is determined by the sign of D[13], but the spins in the center of two spin textures point to opposite directions, resulting in opposite rotation directions of in-plane spins. According to Equation 7.7, the total chirality of the system is lower than that of pure skyrmions. To describe the transformation in detail, this region is further divided into two parts separated by the transition line $B_{S2} = 0.55$, i.e., the region with maximal skyrmion numbers are called skyrmions, and the rest is named as skyrmion gas. Finally, for $B > B_E$ only the edged vortex exists, as displayed in Figure 7.7d. Numerical calculation has proved that Equation 7.1 includes axisymmetric solutions with certain boundary conditions [29,30]. In the case of disk, neglecting the edge-pinning effect, the global minimal energy corresponds to the natural boundary conditions [28], which results in the nonzero in-plane projects of spins at the edge even at a high field. This was clearly observed by nonsaturated magnetization m_z, i.e., the edged vortex is protected by the natural boundary conditions.

The normalized $\tau(H,R)$ as the functions of external field and disk size is shown in Figure 7.8a. Some representative snapshots of m_z distribution are shown in Figure 7.8b. For $R \leq R_c$ with the critical size $R_c \approx 5 = 0.8Th$, rather than skyrmions an incomplete helix state with a "U" shape (Figure 7.8b, $R=5$, $B=0.2$) formed. For $6 \leq R \leq 8$, the transformations from a helix state to skyrmions with the maximal skyrmions number $N_S^m = 1$ (Figure 7.8b, $R=8$, $B=0.4$) and subsequently to edged vortex are all first order. For $9 \leq R \leq 12$, the evolution of spin textures in the external field is similar to that of $R=10$, where the intermediate state and skyrmions gas arise. Especially, for $R=9$, different initial

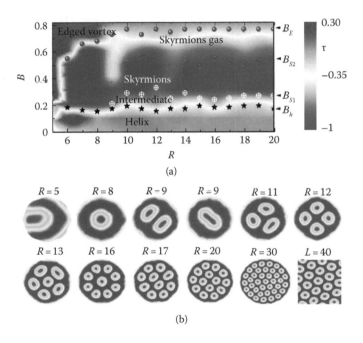

(a)

(b)

FIGURE 7.8 (a) The phase diagram as the functions of external field B and disk size R based on the normalized curves $\tau(B,R)$. (b) The dependence of the arrangements of skyrmions on the disk size, while $R \leq 5$ indicates no skyrmions are formed. With increasing R, the skyrmions' number increases discontinuously and sensitively depending on the disk size. For comparison, the spin arrangement of skyrmions in 2D films is also displayed ($L=40$, $B=0.4$). (From Du, H.F. et al., *Phys. Rev. B*, 87, 014401, 2013. With permission.)

Chapter 7

simulated parameters bring out different spin textures with nearly identical energy under the same field (Figure 7.8b, $R=9$, $B=0.38$). It indicates that the system may degenerate with different skyrmion numbers at some certain fields. For $13 \leq R \leq 17$, skyrmion arrangement is characterized by one in the center surrounded by the rest. For $R > 17$, $<\tau(B)>$ curves gradually transferred into continuous ones. Instead of forming ideal hexagonal skyrmion crystals observed in bulk or 2D film, the skyrmions shift like those of a liquid, but a freezing frame would reveal that the system has long-range order like a solid. Each skyrmion in the interior of a disk has on average six neighbors, forming a hexatic phase [39].

The effect of the uniaxial anisotropy and dipolar–dipolar interaction on the magnetic structure is investigated by varying d and K for fixed D and J. The resulting magnetic phase diagram without the external field in Figure 7.9 for $R = 7$ is composed of four regions: (I) distorted helix state, (II) chiral soft magnetic vortex for stronger dipolar couplings, (III) chiral stripe domain for stronger uniaxial anisotropy, and (IV) special magnetic vortex with a skyrmion core. The formation of spin textures in regions II and IV is easily understood as the counterparts of the vortex in microsized soft magnetic nanodisk [22] and stripe domain in 2D films with high uniaxial anisotropy [27], respectively. Antisymmetry DM couplings lift the degeneracy and put the chiral feature into the systems. Recent experiments provide clear evidence for the

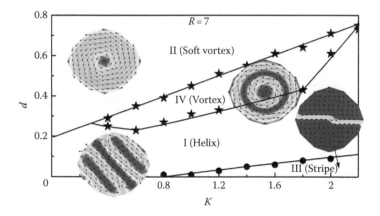

FIGURE 7.9 The magnetic phase diagram as the functions of d and K for fixed D and J and $R=7$; it is composed of four regions: (I) distorted helix state, (II) chiral soft magnetic vortex for stronger dipolar couplings, (III) chiral stripe domain for stronger uniaxial anisotropy, and (IV) special magnetic vortex with skyrmions' core. The corresponding snapshots of out of plane are illustrated in the inset. (From Du, H.F. et al., *Phys. Rev. B*, 87, 014401, 2013. With permission.)

chiral magnetic vortex in FeNi alloy [34] and chiral stripe domain in Fe(2 ML)/Ni/Cu(001) films [35]. The special vortex with a skyrmion core in region IV has also been analyzed in detail by numerical methods in Ref. [28], where the special vortex corresponds to the axisymmetric solutions of Euler equation of Equation 7.1 without magnetostatic energy. In particular, for certain disk sizes, the energy density of the special vortex is lower than that of the helix state in 2D films of chiral magnets, indicating a big chance to form spontaneous vortex ground state even without the help of uniaxial distortions and dipolar couplings. Here, the formation of this vortex ground state required the combination of the uniaxial anisotropy and dipolar couplings. The discrepancy between a numerical method and MC simulation probably results from the discretization of Equation 7.1, leading to some inaccuracy of MC simulations [36]. Indeed, this special vortex has partially been observed in FeGe films by Lorentz microscopy, where an incomplete circle helix state formed in the irregular inhomogeneous sample [37]. Recently, the homogeneous thickness-controllable FeGe films with skyrmion states have been fabricated by magnetron sputtering [38]. The advance provides a good basis to fabricate nanodisk of helimagnets, and then to realize the special vortex ground state.

Using MC simulations, Leonov et al. [40] reported in magnetic NWs unusual skyrmions with a doubly twisted core and a number of concentric helicoidal undulations are thermodynamically stable even in the absence of single-ion anisotropies. They named it as target skyrmion (Figure 7.10a), which is actually the novel magnetic vortex in Ref. [28].

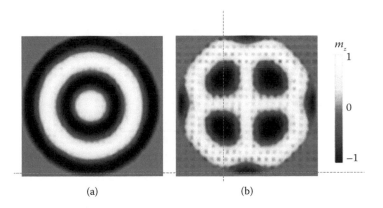

(a)   (b)

FIGURE 7.10 The spin textures in the nanowire with circular cross-section. (a) Target skyrmion, which are the same as the magnetic vortex in Fig. 7. (b) Skyrmion cluster states when a moderate magnetic field is applied along the long axis of the nanowire. (From Leonov, A.O., et al., *EPJ Web Conf.*, 75, 05002, 2014. With permission.)

Such skyrmions are free of magnetic charges and carry a noninteger skyrmion charge s. This state competes with clusters of $s = 1$ skyrmions (Figure 7.10b). For very small radii, the target skyrmion transforms into a skyrmion with $s < 1$, which resembles the vortex-like state stabilized by surface-induced anisotropies.

It must be noted that all the present discussion is only about the equilibrium state. In real materials, magnetization process depends on the magnetization history due to the existence of a metastable state, leading to hysteresis. By contrast, micromagnetism simulation is also used to calculate the spin configurations in nanostructured elements [41]. The ground state in these two nanostructures is the target skyrmions

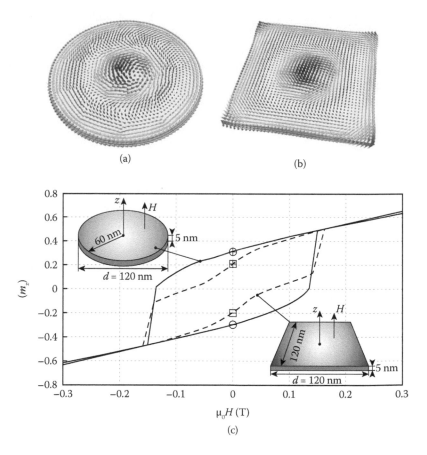

(a)   (b)

(c)

FIGURE 7.11 Nanostructure geometries and examples of obtained skyrmion ground states. The simulated nanostructures are FeGe thin film samples with a thickness $t = 5$ nm. (a) Disk with diameter 120 nm and (b) square with edge length 120 nm. The external magnetic field is applied out-of-plane in the positive z-direction. (c) Hysteresis loops. (From Beg, M., et al., *Sci. Rep.*, 5, 17137, 2015. With permission.)

(Figure 7.11a and b). As the OOMMF software calculates the dynamical magnetization process, a hysteresis is obtained (Figure 7.11c). Hysteresis loops of a skyrmion in the disk with 5 nm thickness and square thin film elements with 120 nm size are shown in Figure 7.11c. The loops are simulated by evolving the system to a steady state after changing the applied field, and then using the resulting state as the starting point for a new evolution after changing the field. In this way, a magnetization loop takes into account the history of magnetic configurations.

In this section, the equilibrium magnetization process of the helimagnets with uniaxial anisotropy and dipolar coupling in thin nanodisk is calculated by means of MC simulation. The complex spin textures may simply be regarded as the superposition of an edged state with their orientation of in-plane spin perpendicular or parallel to the edge and bulk state with the features analogous to 2D chiral magnetic films. With the increase in the external field, the parallel spins of the edge state increase and, finally, transfer into an edged magnetic vortex, which is protected by the natural boundary condition. The arrangement of skyrmions strongly depends on the disk size, in some aspect, similar to the vortex lattice in a confined Type-II superconductor [42]. In addition, other methods including micromagnetism are also introduced, where a hysteresis forms if the dynamical process is considered.

7.3 Experimental Demonstration of Individual Skyrmions

So far, we have discussed the spin textures in quasi-two-dimensional or zero-dimensional nanodisk by using numerical methods. But, the nanodisk is experimentally hard to achieve. By contrast, 1D helimagnetic NWs, such as MnSi, were synthesized almost at the same time as the discovery of the magnetic skyrmions in 2009 [43–45]. In this section, we will first demonstrate the skyrmion state in MnSi NWs with the external magnetic field applied along the long axis of the NWs [46,47]. Then, the formation and stability of skyrmions in 1D nanostripes are discussed with the external magnetic field normal to the stripe plane [48].

7.3.1 Highly Stable Skyrmion State in MnSi NWs

MnSi NWs have been synthesized by several different groups [43,44,49]. A representative scanning electron microscopy (SEM) image of a single wire with a smooth (111) surface is shown in Figure 7.12a.

Chapter 7

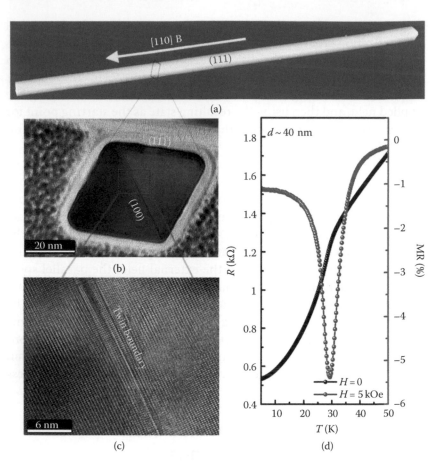

FIGURE 7.12 Crystal morphology and transport properties of MnSi nanowires. (a) A typical SEM image of a MnSi NW with a smooth (111) surface and [110] growth direction. (b) A cross-sectional TEM image of a *NW* with a merohedral twin boundary, where the (001) twin plane is parallel with the [110] growth direction. (c) High-resolution TEM image of the cross-section at the twin boundary. (d) Temperature dependence of the resistance (black) and magnetoresistance (MR) (colored dots) at 0 and 5 kOe, respectively, showing a T_C of 29 K. (From Du, H.F., et al. *Nat. Commun.*, 6, 7637, 2015. With permission.)

The cross-section of the *NW* shows a merohedral twinning structure with the (001) twin plane parallel with the [110] growth direction (Figure 7.12b). This is a common feature of *NW*s of noncentrosymmetric B20 compounds, in which the unique (001) twin plane partitions the *NW* into two parts with opposite handedness [50]. A high-resolution TEM image (Figure. 7.12c), together with previous TEM diffraction results [43], confirms the perfect B20 crystal structure. MR measurements were carried out by a standard four-probe technique on individual NWs. Both resistance (R) versus temperature (T) and MR–T curves are shown in Figure 7.12d, where MR at field, B, is defined as MR = $[R(B) - R(0\ Oe)]/R(0\ Oe)$. The residual resistivity ratio of the wire is $R(300\ K)/R(5\ K) \sim 7.5$. The Curie temperature T_C of the NW is determined to be ~ 29 K from Figure 7.12, which is almost the same as that of bulk MnSi (29.5 K). The room temperature resistivity, $\rho \approx 198\ \mu\Omega$ cm, is quite close to the value of 180 $\mu\Omega$ cm reported for bulk single-crystal MnSi [51,52].

For this NW, initial MR measurements are performed as the magnetic field is applied perpendicular to the long axis of the NW. Figure 7.13 shows MR as a function of applied field at two temperatures below T_C. For scans at 5 and 10 K, a sharp change in the MR slope is observed near 0.7 and 0.65 T, respectively (marked by arrows). It is there that there is also a slope change in the field-dependent MR. The presence and temperature-dependent behavior of these field-dependent MR slope changes is in qualitative agreement with the helimagnet transitions, from helical to conical and, finally, to ferromagnetic state. These results qualitatively agree with the MR signatures previously observed in bulk MnSi [53], but a difference is also noticed. The measured magnetic transitions from the helimagnetic state to conical helimagnetic state, and then to the induced ferromagnetic state, occur at slightly higher fields than in the bulk. More importantly, there is no skyrmion signals in the MR–B curves even as the temperature is close to T_C.

According to the above-mentioned mechanism for the formation of the highly stable skyrmion state in the introduction part, an effective

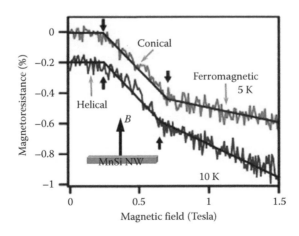

FIGURE 7.13 The magnetic field dependence of the MR as the magnetic field is applied perpendicular to the long axis of the NW. (From Higgins, J.M., et al., *Nano Lett.*, 10, 1605, 2010. With permission.)

uniaxial anisotropy may play an important role in the experimental observations [26]. In the MnSi NWs, the magnetostatic interaction would generate an effective uniaxial anisotropy with its easy axis along the NW axis. In this case, the MR measured configurations correspond to the magnetic field along the hard axis. Thus, the MR signals of skyrmions are hardly observed. By contrast, if a magnetic field is applied along the easy axis, skyrmion MR signals are expected to be observed easily.

Following the theoretical predication, Figure 7.14 shows typical experimental MR data as the magnetic field is applied along the easy axis, i.e., the long axis of the NW [46]. The sweeping data are recorded from the negative to positive saturation fields. Starting from $B_\parallel = 0$ to 7000 Oe, besides the kink at saturation field B_C, two anomalous kinks sequentially appear. Earlier studies on bulk MnSi single crystals have established that these two kinks in the *MR–B* isotherms correspond to the lower and upper critical fields driving the system into and out of the so-called A phase [53]. The dependence of the initial MR curves ($B_\parallel$ from 0 to 7000 Oe) on temperature is highly consistent with the *MR–B* isotherms of bulk MnSi single crystals under pressure, where a larger topological Hall effect is simultaneously observed [54]. These similar features in *M–B* curves between MnSi bulk and NWs provide clear proof of the skyrmion state in this 1D NW under an excitation of a moderate $B_\parallel$.

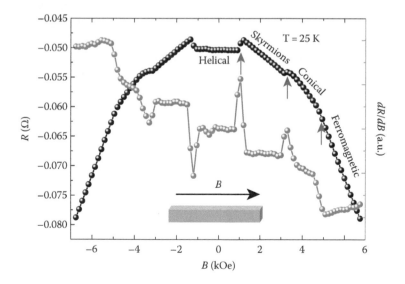

FIGURE 7.14　The magnetic-field dependence of the MR and its derivative on external magnetic field for a 410-nm NW at a representative temperature $T = 25$ K. For $T < T_C$ with increasing $H_\parallel$ from 0 to 7000 Oe, helimagnetic, skyrmions, conical phase, and field-polarized ferromagnetic state appear in turn. (From Du, H.F., et al., *Nano Lett.*, 14, 2026–2032, 2014. With permission.)

Based on the experimental MR data at various temperatures below T_C, values for critical magnetic fields are extracted. The helimagnetic phase is realized below the lower critical field. With the increase in magnetic field, the helimagnetic phase transforms into the skyrmion state via a first-order phase transition, manifested as very sharp peaks at the lower critical field due to the completely different topological properties between helimagnetic and skyrmion states [2]. Above the upper critical field, the system transforms into the conical phase from the skyrmion state. Beyond the saturated field at the shoulder, the system finally transitions into the field polarized ferromagnetic state. These results corroborate the well-established scenario in cubic helimagnets that helimagnetic phase, skyrmion state, and conical phase appear in turn with the increase in the external magnetic field [13].

The significantly extended T–B region for skyrmion state is observed in the NWs. Combining these magnetotransport data in $R(B_{\parallel})$, a magnetic phase diagram for the 410 nm NW was constructed as shown in Figure 7.15. The skyrmion state is clearly stabilized over a broad region

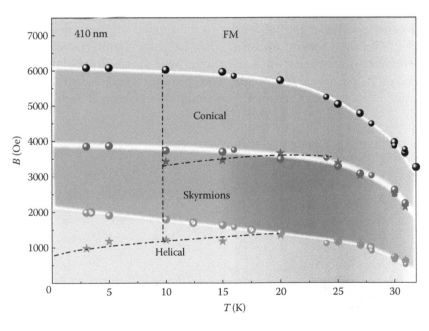

FIGURE 7.15 Magnetic phase diagram of a 410-nm NW under parallel magnetic-field $B_{\parallel}$ inferred from MR data. Skyrmions survive in a large T–B window down to the lowest temperature possible in the measurement (3 K) for increasing field strength. For decreasing field strength, the critical fields appear to decrease slightly below 20 K. Below 10 K, no obvious phase boundaries exist between the skyrmion state and the conical phase. (From Du, H.F., et al., *Nano Lett.*, 14, 2026–2032, 2014. With permission.)

Chapter 7

in the T–B plane, spanning from the T_C down to 3 K under zero field cooling (ZFC) and occupying almost one-third of the regions below Bc. This experimental observation of magnetic states is in contrast to the magnetic phase diagram of previously reported various shape single-crystal MnSi. In bulk single-crystal MnSi, the skyrmion phase can only exist as a metastable state in a small T–B plane from about 26 to 28.5 K and 1000 to 2500 Oe embedded in the conical phase region [55]. In thin plate [23] and nanostrip MnSi [19] carved from the corresponding bulk and NW, the skyrmion phase significantly extends its region from T_C down to the lowest measured temperature, 6 K. However, in both cases, the external field must be applied perpendicularly to the layer plane, and no conical phase appears in the magnetic phase due to the spatial confinement effect [22]. As the metastable behavior emerges under FC below 20 K, the dotted lines shown in Figure 7.15 are the phase boundaries for decreasing fields.

The highly stable skyrmion state in the 1D MnSi NW is further confirmed by using dynamic cantilever magnetometry [56]. The schematic of the experimental setup is shown in Figure 7.16a. A MnSi NW (light gray) is attached to the end of an Si cantilever (gray), whose long axis is either parallel or perpendicular to the applied magnetic field **B**. The cantilever oscillates in the x-direction. Laser light from the fiber interferometer is shown in white. By mounting a single MnSi NW on the end of an ultrasensitive Si cantilever and measuring the shifts in the cantilever's resonant frequency as a function of temperature, applied

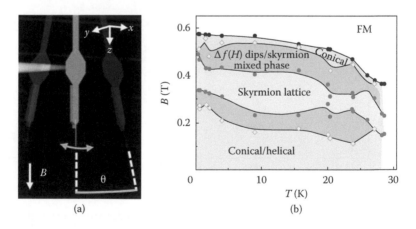

(a)   (b)

FIGURE 7.16   (a) Schematic of the experimental setup. (b) Phase diagram for MnSi NW parallel to the field. The magnetic phase diagram shows the boundaries between phases in T and B as determined from measurements of $M(B)$. (From Mehlin A., et al., *Nano Lett.*, 15, 4839–4844, 2015. With permission.)

magnetic field, and orientation, the nanostructure's magnetic phase diagram can be constructed. These shifts result from the magnetic torque produced by the NW's net magnetization M and an externally applied magnetic field B. Dynamic cantilever magnetometry is an ideal method for investigating the magnetization of individual nanostructures in defined magnetic field orientations.

From measurements at temperatures below T_C, the phase diagram is shown in Figure 7.16b. The skyrmion lattice phase measured for the NW extends from $T = 28$ K down to at least 0.4 K and stretches between $B \simeq 0.2$ and 0.5 T. This region is significantly larger than the small pocket near the critical temperature observed in bulk MnSi (from $T = 26$ to 28.5 K and $B = 0.1$ to 0.25 T) [55] and confirms the less direct magnetoresistance observations [46].

In this section, magnetotransport and dynamic cantilever magnetometry experiments show that skyrmion states are stable over a significantly larger T–B range in MnSi NWs under parallel magnetic field while the magnetic phase diagram of intact MnSi NWs under $B_\perp$ remains inconclusive using MR measurements. The mechanism to stabilize the skyrmions in parallel $B_\parallel$ is most likely the effects of shape-induced uniaxial anisotropy.

7.3.2 Electrical Probing of Individual Skyrmions in Helimagnetic NWs

Skyrmion states have been identified in B20 MnSi NWs, as discussed in Section 7.3.1. However, the widths of the NWs ($d \sim 400$ nm) in these studies are much larger than the skyrmion lattice constant ($Ld \sim 18$ nm) for MnSi, implying that the skyrmion state therein still condenses into the lattice as in bulk samples. An interesting question is whether skyrmions can survive in the ultra-thin NWs, especially, as their diameters decrease down to tens of nanometers comparable with their size.

In this section, the formation of *individual* skyrmions in thin MnSi NWs is introduced by MR measurements. A skyrmion cluster (SC), composed of sparsely distributed skyrmions rather than skyrmion lattices, is demonstrated in this confined system under an external magnetic field, B, aligned along the axis of the wire and confirmed by MC simulations. The *MR–B* curves exhibit striking jumps at specific fields that correspond to the creation or annihilation of individual skyrmions.

Figure 7.17 shows a typical curve of MR as a function of B for the 40 nm NW at $T = 14$ K. The origin of the MR is ascribed to the coupling

Chapter 7

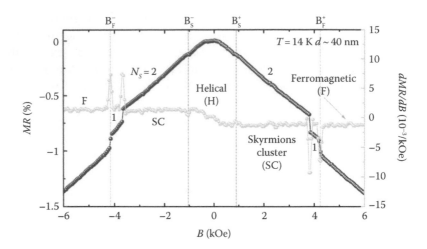

FIGURE 7.17 Variation of spin configurations with magnetic fields for a 40-nm NW. MR (dark gray dots) and dMR/dB (light gray dots) versus the magnetic field B. The measurements were carried out by increasing the field from −8 to 8 kOe. NS represents the number of skyrmions in the skyrmion cluster states. The transition fields, extracted from the dMR/dB data, at positive and negative sweeping magnetic-field branches are denoted by the superscripts "+" and "−," respectively. H, SC, C, and F stand for the helical, skyrmion clusters, conical, and ferromagnetic phases, respectively. (From Du, H.F., et al., *Nat. Commun.*, 6, 7637, 2015. With permission.)

of conduction electrons with chiral modulations [57,58]. Following the above-mentioned judgment that the magnetic scattering of the conduction electrons by the skyrmion phase would lead to two kinks in the MR–B isotherms, which correspond to the lower and upper critical fields driving the system into and out of the skyrmion phase. The data were recorded by increasing the field from −8 to 8 kOe. Following this field-sweeping sequence, $B_S^+ \cdot (B_S^-)$ and $B_F^+ \cdot (B_F^-)$ in the MR–B curve define the transition fields from helical to the skyrmion, and the skyrmion to ferromagnetic phase, respectively [31,59], or vice versa, where the superscript "+ (−)" denotes the positive (negative) branch of the sweeping field. In contrast to the continuous MR–B curves for thick wires (above 200 nm), this MR–B curve displays two discontinuous jumps with almost the same amplitudes in the skyrmion phase of $B_S < B < B_F$. This difference arises from the fact that the skyrmion lattice cannot be supported in the cross-section of such a narrow NW. Instead, a skyrmion cluster (SC) state with sparsely distributed skyrmions is present [35]. As the skyrmion is a topologically stable spin texture with an apparent "particle" character, a single skyrmion in the SC state cannot be created or destroyed by smooth variation of local moments from other phases. Thus, a field-driven cascading quantized transition is a natural result when the number of skyrmions, Ns, changes *one by one* under B. The maximum number, N_{max}, in a SC state can be easily estimated. The presence of a merohedral twin boundary splits the parallelogram cross-section into two triangles. Simple geometric analysis suggests that one skyrmion at most can exist in each triangle, resulting in a total number of $N_{max} = 2$. This number perfectly matches the number of discontinuous jumps (or drops) in MR.

This interpretation of the MR results are further supported by MC simulations based on the actual sample geometry, in which the 40 nm NW is divided into two merohedral lattices. Figure 7.18 shows the constructed model for the MC simulation. Concerning the real geometry, open and periodic boundary conditions were adopted in the cross-sectional plane of the NW and the long axis, respectively, and a boundary forming the same NW geometry as the experimental merohedral twin plane, which divides the NW into two equal parts with opposite chirality. Thus, the sign of the coefficient D is opposite in the two domains. According to the previous Lorentz TEM observations in another B20 compound FeGe, the inversion of the lattice chirality (handedness) of the B20 structure across certain grain boundaries would lead to the invariance of the sign of the spin–orbit interaction within FeGe [21]. It was also observed that the spins around the boundary are almost parallel with the grain boundary. This observation implies that the ferromagnetic interaction would play a dominant role in the narrow transition region where the inversion of the lattice chirality (handedness) occurs, yielding the nearly parallel spin arrangements in the thin region of crystal boundaries. Thus, the value of D is set to zero in the boundaries.

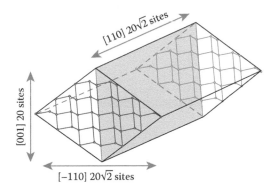

FIGURE 7.18   The orientation of the simple cubic lattice used in our MC simulations. The shaded areas mark the twin plane. (From Du, H.F., et al., *Nat. Commun.*, 6, 7637, 2015. With permission.)

Chapter 7

Snapshots of both 3D and the cross-sectional spin configurations are, respectively, shown in the upper and bottom part in Figures 7.19a through f, where the magnetic field B has the unit $B_S = J/\mu_B$, with J the exchange constant and μ_B the Bohr magnetron. At a low field, a distorted helical order is established with the propagation direction lying in the cross-section and perpendicular to the twin boundary (Figure 7.19a). This is different from the conventional bulk sample, where the propagation is along the [111] direction due to weak crystal anisotropy. This difference comes from the presence of the twin boundary, at which the DM interaction vanishes and only ferromagnetic exchange interaction survives. Therefore, ferromagnetic ordering is persistent along the twin boundary plane, and modulation along the NW is prohibited. When the field B/B_S is increased above a threshold, the SC appears (Figure 7.19c). Due to the geometric confinement, only one skyrmion is allowed in each merohedral domain and the skyrmions are aligned along the NW, forming two skyrmion tubes. At the NW boundary, spins align parallel to the boundary due to the DM interaction and missing spins near the boundary [35]. Consequently, the swirling direction at the boundary of each domain is opposite to that of the skyrmion within, and each skyrmion tries to sit at the center of each domain owing to the repulsion from the edge states [5–9].

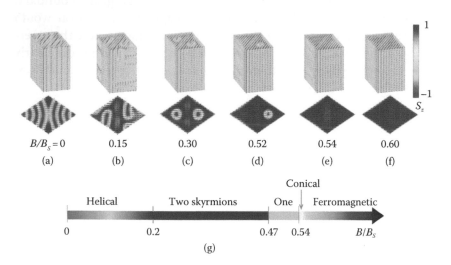

FIGURE 7.19 Calculated spin configurations for the typical states in the 40 nm NW. (a) Distorted helical; (b) intermediate state with meron in the interior of the NW; (c) two skyrmions; (d) one skyrmion; (e) distorted conical phase or 3D modulations; (f) field-polarized ferromagnetic state and the corresponding cross-sections of these states are also shown. (g) The phase diagram in B space, where $B_S = J/\mu_B$ is the unit of the magnetic field in the Monte Carlo simulations.

Notice that before entering the SC state, the system undergoes an intermediate state (Figure 7.19b), characterized by forming half-skyrmions in the interior of the NW. As a result, this state weakens the abrupt transition from the helical state to the skyrmion phase. With further increase in the field strength, one skyrmion disappears, leaving a mixed state of the 3D modulation in one domain and a skyrmion tube in the other domain (Figure 7.19d). Although in the simulations, the geometries of the two domains in the NW are identical except for different chiralities, the ferromagnetic exchange at the twin boundary gives rise to certain correlations between two domains. The result shows that the formation of 3D modulations in one domain can stabilize the remaining skyrmion in the other, so that two skyrmions do not disappear simultaneously. In real samples, the sizes of the two twins are not exactly identical, which may also lead to a difference in the corresponding critical fields. At an even larger field, the remaining skyrmions transform into the fully polarized ferromagnetic state (Figure 7.19f) via a distorted conical phase in a tiny magnetic field interval (Figure 7.19e), while swirling along the boundary persists (Figure 7.19f). The phase diagram for these states is shown in Figure 7.19g. This two-step destruction of the SC state corresponds to the two jumps observed in the $MR–B$ curve.

A systematic investigation on the transport properties of NWs with different diameters has also been performed. More discontinuous jumps are observed at a thicker NW with a diameter of 55 nm. A similar numerical calculation shows that only two skyrmions can exist in each triangle divided by the twin boundary and, therefore, $N_{max} = 4$. For an even thicker NW with a width of 80 nm, the isothermal MR curves display completely different transport behavior from the narrow NWs, but corroborate the well-established scenario for helimagnets [46]: the helical phase (H) is stable at low fields and then transits into the skyrmion state (S) at B_S via a discontinuous phase transition. At even higher fields the conical phase appears, and eventually turns into the ferromagnetic ordering above B_F. As the NW diameter ($d \sim 80$ nm) is much larger than the single skyrmion size ($LD \sim 18$ nm), closely packed skyrmions (CPSs) are hosted, leading to a skyrmion lattice, rather than a SC, and the continuous $MR–B$ curve between B_S and B_F. This is in contrast to the jumps in MR seen in narrow NWs due to the emergence of SCs. But when the NW is extremely narrow, the discontinuous jumps which originate from the creation or annihilation of an individual skyrmion should be absent. As a proof, a NW width of ~20 nm is examined. The MR follows a smooth curve with only two transition fields in the whole magnetic

Chapter 7

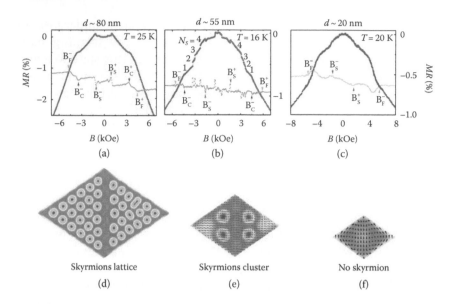

FIGURE 7.20 The NW diameter dependence of the MR behaviors and the skyrmion phase diagrams of temperature T and magnetic field B. MR as a function of B for three NWs with different diameters (dark gray dots: (a) 80 nm, (b) 55 nm, and (c) 20 nm). (d–f) The schematic skyrmions inferred from the MR–B data. The circular regions indicate the skyrmion states.

field region (Figure 7.20). This is expected for such narrow NWs because each small triangular lattice fails to host a single skyrmion (with a size of ~18 nm).

In summary, the presence of SC states in confined MnSi NWs with diameters comparable to single skyrmion domain size is demonstrated. The maximum number of skyrmions within this cluster is determined by the dimension of the NW cross-sections, and the skyrmion number can be controlled by external magnetic fields and revealed by quantized jumps in the MR curves. Indeed, the aforementioned theoretical results have demonstrated the possibility of SC states in ultra-thin nanodisk. Here, we observed the jumped MR signals in the MnSi NW with its diameter comparable to the featured skyrmion size. Actually, it is generally believed that skyrmions form in a tube shape in bulk or 2D films. The axis direction of the tube is along the magnetic field orientation, i.e., the spins within the skyrmions show the same arrangement along the tube or magnetic field orientation. As a result, the observed SC states in the NW is able to be simply regarded as the superposition of single-layer skyrmion states in the nanodisk. These results reveal new physics of the SC states in confined geometries and can guide the development of skyrmion-based memory devices in which the individual skyrmions could be utilized for multibit memory cells.

7.3.3   Edge–Mediated Skyrmion Chain and Its Dynamics in Nanostripes

Recently, many theoretical studies have demonstrated the possibility of hosting and manipulating individual skyrmions in the helimagnetic nanostripes [8–10]. In such schemes, the nanostripe is encoded via a precise control of a chain of individual skyrmions using spin-polarized current pulses. According to the theoretical results, skyrmion can be transferred from the domain-wall pair in junction geometry, and a strategy for the design of skyrmion-based race-track memory is also developed. By contrast, the experimental process is slow mainly due to the sample fabrication and resolution limit of measurement tools. Up to now, it is still unclear whether skyrmion can exist in such confined geometry and what is the stability if it can exist? In this section, the first experimental demonstration of individual skyrmions in nanostripes is introduced.

The B20 compound FeGe is chosen to investigate the effect of confined geometry on the formation and stability of skyrmions. FeGe is a typical helical magnet harboring a skyrmion phase near room temperature. Bulk FeGe samples exhibit a paramagnet to helimagnet transition at a Curie temperature $T_C \sim 280$ K under zero magnetic field [21]. The spin helix possesses a long range wavelength, $\lambda \sim 70$ nm, and a fixed wave vector, Q, along the high-symmetry crystal axis.

The FeGe nanostripes with various widths were fabricated from the bulk by a complex process. Figure 7.21a shows a typical TEM image of the nanostripe with a width of $w \sim 130$ nm, which is comparable to the featured skyrmion lattice constant, $a_{sk} \cdot \left(2\lambda / \sqrt{3} \sim 81 \, \text{nm}\right)$. The surrounding gray part is the amorphous PtCx, which can significantly reduce the effect of Fresnel fringes on the real magnetic structure around the edge and enable us to observe directly the magnetic structure in the ultra-narrow stripes. An external magnetic field, $\mathbf{B}$, is applied perpendicular to the stripe plane with its direction pointing upward (marked as black "$\odot$"). Figure 7.21b through h shows the evolution of the spin textures of this nanostripe with increasing B measured at 100 K, where the planar magnetic distributions were obtained by the magnetic transport-of-intensity equation (TIE) analyses of the high-resolution Lorentz TEM data [3]. The two FeGe/PtCx interfaces, marked by the white dotted lines, would induce weak artificial magnetic contrast in the PtCx region due to the Fresnel fringes. The magnetic state in the FeGe parts is marked with a small black triangle.

Chapter 7

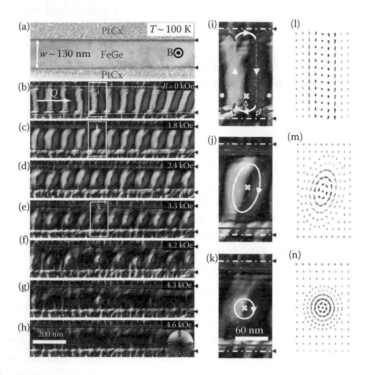

FIGURE 7.21 Variations of spin texture with magnetic field in a 130 nm FeGe nanostripe at the temperature T = 100 K. (a) TEM image of the FeGe nanostripe surrounded by an amorphous PtCx layer. (b–h) Magnetic-field dependence of the spin texture, represented by the lateral magnetization distribution, as obtained by transport-of-intensity equation (TIE) analysis of three Lorentz TEM images with defocus values of −144 μm, 0, and +144 μm. The color wheel represents the magnetization direction at every point. "Q" is the wave vector. (i–k) Enlarged regions marked by the white boxes in corresponding panels (b–e). The small white arrows represent the in-plane magnetization direction at each point. "•" and "×" stand for the upward and downward directions of spins, respectively. At B ~ 0 Oe, spin helices are terminated at the edge and form two types of half-disk domains by distinguishing the curling direction of in-plane magnetization around the edge (thick red lines, counterclockwise thick white lines, clockwise). The magnetic field–driven evolution of half-disk domains is illustrated by using small black arrows to point out the moving directions of the domains in (i). (l–n) The schematic spin arrangements shown in the corresponding panels (i–k). The white dot lines in (b) stands for the FeGe/PtCx interfaces, which are marked by small black triangular in the magnetic images (b–h) and (i–k). (From Du, H. F., et al., *Nat. Commun.*, 6, 8504, 2015. With permission.)

At low magnetic field, a spin helix is observed with a period of 70 nm (Figure 7.21b), consistent with the period in the bulk sample [21]. But, the wave vector, Q, is almost parallel to the long axis of the nanostripe. With increasing B, the magnetization undergoes a series of dynamical processes (Figure 7.21c and d), and eventually evolves to circular skyrmions at $B \sim 3.3$ kOe (Figure 7.21k and n). These skyrmions assemble into a perfect single chain due to the transverse constriction. This single skyrmion chain (SSC) in the interior of the nanostripe is accompanied by a chiral magnetization on the edge, as indicated by the red arrows in Figure 7.21k. Spins on the two edges tilt onto the plane and orient their planar components parallel to the boundary (schematic diagram in Figure 7.21n). That is because of an unbalanced torque acting on the edge spin originating from the DM interaction [8,9,35]. Reversal of the sample chirality would flip the edge spin orientations. Higher fields above $B \sim 4.2$ kOe melt the perfect SSC state into isolated skyrmions confined in the nanostripe (Figure 7.21f and g), and finally polarize all the interior magnetizations at about $B \sim 4.6$ kOe (Figure 7.21h), where magnetic contrast of the planar components disappears. However, the chiral edge magnetization is still persistent.

The temperature of ~ 100 K is far below the magnetic transition temperature T_C (~ 280 K) [21] and is even much lower than 200 K, the lower bound of the skyrmion phase reported in 2D FeGe plates with the same thickness of $t \sim 65$ nm to the 130 nm nanostripe. Very likely, the SSC phase extends through the whole low-temperature regime in this confined geometry. The comparable size between the nanostripe and single skyrmion allows us to identify that such a narrow stripe ($w \sim 130$ nm) can only accommodate one skyrmion in the transverse direction by simple geometric analyses. However, the wider extension of temperature for hosting the skyrmion phase in the nanostripe is in sharp contrast to the above-mentioned low stability of skyrmions in bulk or the thickness-dependent stability in 2D films [21]. Obviously, this observation suggests an important role played by geometric confinement on the formation of skyrmions.

To explore the physical origin of this effect, the magnetization process is closely examined. Figure 7.21b shows the spin helix at $B \sim 0$ Oe. Unlike the locking of wave vector, Q, along the high-symmetry crystal axis due to the weak crystal anisotropy reported so far, the spin helix in narrow stripes always orients its wave vector parallel to the edge of the nanostripe. In this case, spin helices are terminated at the edge, forming periodically modulated "half-disk" domains. Magnetic moments in these domains have fixed swirling directions (Figure 7.21i and l), which are indicated by white (clockwise) and red (counter clockwise) arrows, respectively. Moments at the disk center are out of plane with staggered polarity upward (marked as "•") and downward (marked as "×"). Owing to the chiral property of helical magnets, the rotation direction and polarity of disks are entangled so that only two types of "half disk" domains can occur. These two types of domains alternately appear along the edge since they are coupled to the periodically modulated helices inside the stripe.

When a weak external magnetic field is applied, the Zeeman energy enforces an increase in the proportion of the upward spins. According to the Lorentz TEM data, the increasing number of upward spins is accommodated by expanding the boundaries of the red "half-disk" domain (the moving directions are marked by small black arrows with white frame) as its polarity in the center is parallel to the magnetic field. Above a threshold value of magnetic field B ~1.8 kOe, the edge spins coalesce in the red parts, forming a uniform edge state (red straight arrows in Figure 7.21) [60], while the white parts are lifted away from the edge, forming half-skyrmions, i.e., merons [61,62]. Another meron is similarly formed at the opposite edge, and together they constitute a bimeron (Figure 7.21j and m), which can be regarded as an elongated skyrmion as it carries the same unit of topological charge as a skyrmion. With a further increase in magnetic field up to ~3.3 kOe, the elongated skyrmions are compressed and form a perfect SSC with the topological charge conserved.

The magnetization process in the 130 nm wide nanostripe provides a strong hint that the creation of skyrmions in confined geometries is closely related to the presence of half-disk domains, originating from spin helices propagating along the edge. Further systematic experiments give a common picture to create skyrmion in confined geometries at low temperature. The skyrmions originate from the sample shape–dependent orientation of the helical state. The helix with a distorted edge-state evolves into skyrmions, while those without distorted edge-state transfer into a ferromagnetic state directly. Taking the topological stability of skyrmions into account, it is commonly concluded that skyrmion cannot easily be destroyed and created from the single-twist magnetic structure including helical and conical phases. By contrast, the confined effect gives rise to the distorted edge states, which show "half-disk" arrangements, indicating their multi twist modulations. These results are displayed in Figure 7.22a.

Once the skyrmion chain is formed at the edge, another interesting observation is that this skyrmion chain tends to move collectively into the center of the stripe (Figure 7.22b) with the increase in magnetic fields, while keeping the number of skyrmions unchanged in a wide interval of magnetic fields. Such a fixed number of skyrmions further support the topological stability of skyrmions. The movement of skyrmions occurs possibly due to the repulsions between edge spins and skyrmions [62]. Consequently, a distorted skyrmion chain (DSC) centered in the nanostripe is formed. Finally, at even higher

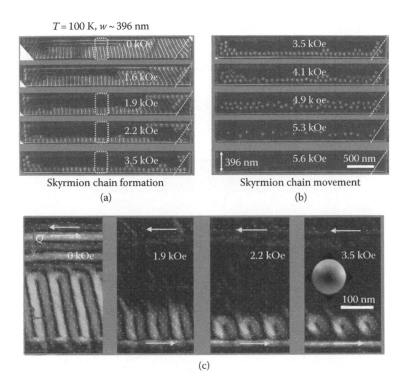

FIGURE 7.22 Variations of spin texture with magnetic field in 396 nm FeGe nanostripe at $T = 100\ K$. (a–b) Magnetic-field dependence of the spin texture; the crystal boundary in the right is marked by white dotted lines. (c) The corresponding enlarged region is marked by the dotted white rectangle in corresponding panels (a–b). The small white arrows stand for the in-plane magnetization direction at each point. The thick white arrows are plotted to guide the eyes. (From Du, H. F., et al., *Nature Commun.*, 6, 8504, 2015. With permission.)

magnetic fields, the topologically stable DSC is melted and skyrmions therein gradually disappear. A transition to the ferromagnetic spin textures occurs.

At elevated temperatures, thermal fluctuations would be dominant compared to the magnetization dynamics. It is well understood that a CPS crystal is favored by thermal fluctuations in bulk or 2D films. This crystallization of skyrmions is ascribed to interskyrmion interactions [13], which is an intrinsic property of skyrmions and should be independent of material details and sample sizes. Therefore, the CPS structure is expected in nanostripes when the temperature is high enough. This prediction is supported by Lorentz TEM images of nanostripes, which show that the skyrmions are closely packed (Figure 7.23).

Chapter 7

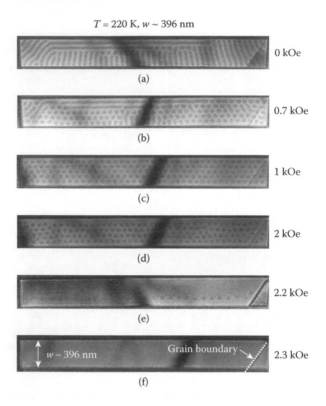

T = 220 K, *w* ~ 396 nm

(a)   0 kOe

(b)   0.7 kOe

(c)   1 kOe

(d)   2 kOe

(e)   2.2 kOe

w ~ 396 nm   Grain boundary   2.3 kOe

(f)

FIGURE 7.23 The magnetic-field dependence of spin textures in the 396 nm FeGe nanostripe at high temperature *T* ~ 220 K. (a) The helical state; (b) mixed state of skyrmion lattice and helical phase; (c) and (d) a packed skyrmion lattice; (e) isolated skyrmions; and (f) a field-polarized ferromagnetic state surrounded by edge vortex state. The images are acquired under over-focus conditions with the defocus value of 288 μm. The dotted lines indicate the grain boundary. (From Du, H. F., et al., *Nat. Commun.*, 6, 8504, 2015. With permission.)

In summary, the formation and evolution of skyrmions in the nanostripe depends on the temperature (Figure 7.24). In contrast to occupying a narrow temperature and magnetic field (*T–B*) window in bulk samples of FeGe, the skyrmion phase in such highly confined geometry is greatly extended, which is observed down to the lowest temperature accessible by our Lorentz TEM. Furthermore, by tracking the magnetization dynamics, the results reveal a generic edge-assisted mechanism of skyrmion formation in confined geometries. The edge spin distortion originating from the termination of spin helix at the boundary is the forerunner of skyrmions at low fields. A self-organized skyrmion chain and its subsequent collective motion are observed when the magnetic field is turned on. These results provide important

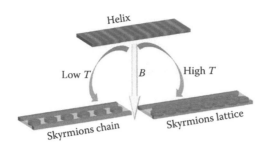

FIGURE 7.24 Schematic of field-driven evolution of spin textures in the nanostripes at different temperatures.

insight into the dynamical mechanism of the formation and evolution of the skyrmions in the confined geometry, and might, in the long term, shed light on applications.

7.4 Summary

Numerical calculations based on a fundamental model predicted a target skyrmion state and field-driven discrete one-by-one transition in SC states. In confined MnSi NWs, the SC state has been confirmed by MR measurements if the magnetic field is applied along the long axis of the NW. The maximum number of skyrmions within this cluster is determined by the dimension of the NW cross-sections, and the skyrmion number can be controlled by external magnetic fields and revealed by quantized jumps in the MR curves. Recently, lots of experiments have fabricated helimagnetic film with helix states by molecular-beam epitaxy (MBE) or magnetron sputtering, as in MnSi and $Fe_xCo_{1-x}Si$ and FeGe [24,25,63–65]. Meanwhile, single skyrmion writing with controllable and location-dependent properties is demonstrated by using spin-polarized scanning tunneling microscopy [66]. These advances provide a good basis for fabricating a nanoscale pattern of helimagnets.

References

1. Azzerboni, B., Asti, G., Pareti, L., and Ghidini. M. *Magnetic Nanostructures in Modern Technology* (Springer, Berlin, 237–306, 2008).
2. Mühlbauer, S., et al. Skyrmion lattice in a chiral magnet. *Science* **323**, 915 (2009).
3. Yu, X. Z., et al. Real-space observation of a two-dimensional skyrmion crystal. *Nature* **465**, 901 (2010).
4. Jonietz, F., et al. Spin transfer torques in MnSi at ultralow current densities. *Science* **330,** 1648 (2010).

Chapter 7

5. Nagaosa, N., and Tokura, Y. Topological properties and dynamics of magnetic skyrmions. *Nature Nanotech.* **8**, 899 (2013).

6. Neubauer, A., et al. Topological Hall effect in the A phase of MnSi. *Phys. Rev. Lett.* **102**, 186605 (2009).

7. Yu, X. Z., et al. Skyrmion flow near room temperature in an ultralow current density. *Nature Commun.* **3**, 988 (2012).

8. Fert, A., Cros, V., and Sampaio, J. Skyrmions on the track. *Nature Nanotechnology* **8**, 152–156 (2013).

9. Iwasaki, J., Mochizuki, M., and Nagaosa, N. Current-induced skyrmion dynamics in constricted geometries. *Nature Nanotech.* **8**, 742–747 (2013).

10. Sampaio, J., et al. Nucleation, stability and current-induced motion of isolated magnetic skyrmions in nanostructures. *Nature Nanotech.* **8**, 839–844 (2013).

11. Shibata, K., et al. Towards control of the size and helicity of skyrmions in helimagnetic alloys by spin–orbit coupling. *Nature Nanotech.* **8,** 723–728 (2013).

12. Brataas, A., Kent, A. D., and Ohno, H. Current-induced torques in magnetic materials. *Nature Mater.* **11**, 372–381 (2012), and reference therein.

13. Rößler, U. K., Leonov, A. A., and Bogdanov, A. N. Chiral skyrmionic matter in non-centrosymmetric magnets. *J. Phys.: Conf. Ser.* **303**, 012105 (2011).

14. Yu, X. Z., et al. Magnetic stripes and skyrmions with helicity reversals. *Proc. Natl. Acad. Sci. USA* **109**, 8856–8860 (2012).

15. Okubo, T., Chung, S., and Kawamura, H. Multiple-q states and the skyrmion lattice of the triangular-lattice Heisenberg antiferromagnet under magnetic fields. *Phys. Rev. Lett.* **108**, 017206 (2012).

16. Heinze, S., et al. Spontaneous atomic-scale magnetic skyrmion lattice in two dimensions. *Nature Phys.* **7**, 713 (2011).

17. Zhou, Y., and Ezawa, M. A reversible conversion between a skyrmion and a domain-wall pair in a junction geometry. *Nature Commun.* **5**, 4652 (2014).

18. Tomasello, R., et al. A strategy for the design of skyrmion racetrack memories. *Sci. Rep.* **4**, 6784 (2014).

19. Yu, X. Z., et al. Observation of the magnetic Skyrmion lattice in a MnSi nanowire by Lorentz TEM. *Nano Lett.* **13**, 3755–3759 (2013).

20. Rößler, U., Bogdanov, A., and Pfleiderer, C. Spontaneous skyrmion ground states in magnetic metals, *Nature (London)* **442**, 797 (2006).

21. Yu, X. Z., et al. Near room-temperature formation of a skyrmion crystal in thin-films of the helimagnet FeGe. *Nature Mater.* **10**, 106 (2011).

22. Rybakov, F. N., Borisov, A. B., and Bogdanov, A. N. Three-dimensional skyrmion states in thin films of cubic helimagnets. *Phys. Rev. B* **87**, 094424 (2013).

23. Tonomura, A., et al. Real-space observation of skyrmion lattice in helimagnet MnSi thin samples. *Nano Lett.* **12**, 1673 (2012).

24. Huang, S. X., and Chien C. L. Extended skyrmion phase in epitaxial FeGe(111) thin films. *Phys. Rev. Lett.* **108**, 267201 (2012).

25. Li, Y. F., et al. Robust formation of skyrmions and topological Hall effect anomaly in epitaxial thin films of MnSi. *Phys. Rev. Lett.* **110**, 117202 (2013).

26. Butenko, A. B., Leonov, A. A., Roessler, U. K., and Bogdanov A. N. Stabilization of skyrmion textures by uniaxial distortions in noncentrosymmetric cubic helimagnets. *Phys. Rev. B* **82**, 052403 (2010).

27. Wilson, M. N., et al. Extended elliptic skyrmion gratings in epitaxial MnSi thin films. *Phys. Rev. B* **86**, 144420 (2012).

28. Du, H. F., Ning, W., Tian, M. L., and Zhang, Y. H. Magnetic vortex with skyrmionic core in a thin nanodisk of chiral magnets. *EPL* **101**, 37001 (2013).

29. Bogdanov, A., and Hubert, A. The properties of isolated magnetic vortices. *Phys. Status Solidi B* **186**, 527 (1994).

30. Bogdanov A., and Hubert A. Thermodynamically stable magnetic vortex states in magnetic crystals. *J. Magn. Magn. Mater.* **138**, 255 (1994). Ibid. **195**, 182 (1999)
31. Yi, S. D., Onoda, S., Nagaosa, N., and Han, J. H. Skyrmions and anomalous Hall effect in a Dzyaloshinskii-Moriya spiral magnet. *Phys. Rev. B* **80**, 054416 (2009).
32. Uchida, M., et al. Topological spin textures in the helimagnet FeGe. *Phys. Rev. B* **77**, 184402 (2008).
33. Baudry, J., Pirkl, S., and Oswald, P. Looped finger transformation in frustrated cholesteric liquid crystals. *Phys. Rev. E* **59**, 5562 (1999).
34. Wachowiak, A., et al. Direct observation of internal spin structure of magnetic vortex cores. *Science* **298**, 577 (2002).
35. Du, H. F., et al. Field-driven evolution of chiral spin textures in a thin helimagnet nanodisk. *Phys. Rev. B* **87**, 014401 (2013).
36. Lee, S. H., Zhu, F. Q., Chien, C. L., and Marković, N. Effect of geometry on magnetic domain structure in Ni wires with perpendicular anisotropy: A magnetic force microscopy study. *Phys. Rev. B* **77**, 132408 (2008).
37. Clarke, D., Tretiakov, O. A., and Tchernyshyov, O. Stripes in thin ferromagnetic films with out-of-plane anisotropy. *Phys. Rev. B* **75**, 174433 (2007).
38. Ezawa, M. Compact merons and skyrmions in thin chiral magnetic films. *Phys. Rev. B* **83**, 100408(R) (2011).
39. Radhakrishnan, R., Gubbins, K. E., and Bartkowiak, M. S. Existence of a hexatic phase in porous media. *Phys. Rev. Lett.* **89**, 076101 (2002).
40. Leonov, A. O., Rößler, U. K., and Mostovoy, M. Target-skyrmions and skyrmion clusters in nanowires of chiral magnets. *EPJ Web Conferences* **75**, 05002 (2014).
41. Beg, M., et al. Ground state search, hysteretic behaviour, and reversal mechanism of skyrmionic textures in confined helimagnetic nanostructures *Sci. Rep.* **5**, 17137 (2015).
42. Sardella, E., and Brandt, E. H. Vortices in a mesoscopic superconducting disk of variable thickness. *Supercond. Sci. Technol.* **23**, 025015 (2010).
43. Higgins, J. M., Ding, R. H., DeGrave, J. P., and Jin, S. Signature of helimagnetic ordering in single-crystal MnSi nanowires. *Nano Lett.* **10**, 1605 (2010).
44. Seo, K., et al. Itinerant helimagnetic single-crystalline MnSi nanowires, *ACS Nano* **4**, 2569 (2010).
45. DeGrave, J. P., et al. Spin polarization measurement of homogeneously doped Fe1–xCoxSi nanowires by Andreev reflection spectroscopy, *Nano Lett.* **11**, 4431 (2011).
46. Du, H. F., et al. Highly stable skyrmion state in helimagnetic MnSi nanowires. *Nano Lett.* **14**, 2026–2032, (2014).
47. Du, H. F., et al. Electrical probing of field-driven cascading quantized transitions of skyrmion cluster states in MnSi nanowire. *Nature Commun.* **6**, 7637 (2015).
48. Du, H. F., et al. Edge-mediated skyrmion chain and its collective dynamics. *Nature Commun.* **6**, 8504 (2015).
49. Siwei Tang et al. Growth of skyrmionic MnSi nanowires on Si: Critical importance of the SiO2 layer. *Nano Res.* **7**, 1788–1796 (2014).
50. Szczech, J. R., and Jin, S. Epitaxially-hyperbranched FeSi nanowires exhibiting merohedral twinning. *J. Mater. Chem.* **20**, 1375 (2010).
51. Mena, F. P., et al. Heavy carriers and non-Drude optical conductivity in MnSi. *Phys. Rev. B* **67**, 241101(R) (2003).
52. Ritz, R., et al. Giant generic topological Hall resistivity of MnSi under pressure. *Phys. Rev. B* **87**, 134424 (2013).
53. Kadowaki, I. K., Okuda, K., and Date, M. Magnetization and magnetoresistance of MnSi. *J. Phys. Soc. Jpn.* **51**, 2433 (1982).

Chapter 7

54. Lee, M., et al. Unusual Hall effect anomaly in MnSi under pressure. *Phys. Rev. Lett.* **102**, 186601 (2009).
55. Bauer, A., and Pfleiderer, C. Magnetic phase diagram of MnSi inferred from magnetization and ac susceptibility. *Phys. Rev. B* **85**, 214418 (2012).
56. Mehlin A., et al. Stabilized skyrmion phase detected in MnSi nanowires by dynamic cantilever magnetometry. *Nano Lett.* **15**, 4839–4844 (2015).
57. Togawa, Y., et al. Interlayer magnetoresistance due to chiral soliton lattice formation in hexagonal chiral magnet $CrNb_3S_6$. *Phys. Rev. Lett.* **111**, 197204 (2013).
58. Kang, J., and Zang, J. D. Transport theory of metallic B2 helimagnets. *Phys. Rev. B* **91**, 134401 (2015).
59. Benjamin, J. C., et al. Large enhancement of emergent magnetic fields in MnSi with impurities and pressure. *Phys. Rev. B*, **88**, 214406 (2013).
60. Meynell, S. A., et al. Surface twist instabilities and skyrmion states in chiral ferromagnets. *Phys. Rev. B* **90**, 014406 (2014).
61. Lin, S. Z., Saxena, A., and Batista, C. D. Skyrmion fractionalization and merons in chiral magnets with easy-plane anisotropy. *Phys. Rev. B* **91**, 224407 (2015).
62. Zhang, X. C., et al. Skyrmion-skyrmion and skyrmion-edge repulsions in skyrmion-based memory. *Sci. Rep.* **5**, 7643 (2014).
63. Morley, S. A., Porte, N. A., and Marrows, C. H. Magnetism and magnetotransport in sputtered Co-doped FeSi films. *Phys. Status Solidi RRL* **5**, 429–431 (2011).
64. Wilson, M. N., et al. Extended elliptic skyrmion gratings in epitaxial MnSi thin films. *Phys. Rev. B* **86**, 114420 (2012).
65. Kanazawa, N., et al. Discretized topological Hall effect emerging from skyrmions in constricted geometry. *Phys. Rev. B* **91**, 041122 (2015).
66. Romming, N., et al. Writing and deleting single magnetic skyrmions. *Science* **341**, 636 (2013).

8. Magnetic Skyrmion Dynamics

Felix Büttner
Massachusetts Institute of Technology, Cambridge, Massachusetts

Mathias Kläui
Johannes Gutenberg-Universität Mainz, Mainz, Germany
Graduate School Materials Science in Mainz, Mainz, Germany

In this chapter, the dynamics of magnetic skyrmions is reviewed. Starting with a topological definition of what we call a magnetic skyrmion, we describe the topology and discuss the resulting general properties. To stabilize chiral skyrmions, we introduce a chiral exchange interaction and we present the spin canting that leads to a given handedness for chiral skyrmions. Based on the statics, we next describe the dynamics based on a one-dimensional model and then discuss the steady-state dynamics for instance in wire geometries as well as the gyrotropic relaxation eigenmodes. Finally, we present an experimental demonstration of both these types of dynamics and give a brief outlook on future challenges and opportunities in this field.

Chapter 8

8.1   Topological Definition of a Skyrmion

By definition, a vector field is called a skyrmion or said to be skyrmionic if it has a spherical topology. In the following, we will explain the concept of topology, illustrate how the mathematical definition is applied to a magnetic spin configuration, and discuss the implications of a spin structure being a skyrmion.

Topology is a mathematical concept to classify geometrical properties of continuous structures (where structures can be real-space objects, vector fields, momentum space functions, etc.). Two structures are considered equivalent if a continuous map from one to the other exists. There are many possible definitions of continuity inheriting from the large variety of possible complete sets of open subsets of the structure under consideration. Physical constraints, such as energy barriers or forbidden intermediate states, define allowed deformations and the set of prohibited or strongly suppressed transformations. They, therefore, specify meaningful topological distinctions. In physics, topologies are often classified according to homotopic transformations between spaces with Euclidian metric and a Euclidian definition of neighborhood and open sets. A homotopy is a continuous deformation, not necessarily bijective. In contrast, homeomorphisms—defining another possible topological classification—are bijective. For instance, a line and a point are homotopically equivalent, but they are not homeomorph.

To deduce implications of the topology on the physics of a system, the choice of topological classification has to be based on physical arguments. In magnetism, the following three arguments apply. First, in most magnetic systems, the magnetization profile varies on length scales much larger than crystal lattice constants due to the exchange mechanism. Hence, the associated vector field is well described in a continuum model, a prerequisite for a topological classification. Second, exchange interactions, which scale with the gradient of the magnetization, set a natural energy barrier for discontinuous deformations, justifying the concept of homotopy. And third, magnetostatic interactions (anisotropy and stray fields) stabilize the boundary of the structure. Hence, homotopy transformations are allowed, provided that they do not modify the boundary of the structure. This set of allowed transformations (all others are strongly suppressed) defines topological equivalence classes.

One example of a topologically nontrivial structure in real space is the skyrmion. Skyrmions, in general, refer to vector fields with a spherical topology, first identified by T. Skyrme in the field of nuclear physics.[61] Through topological arguments, Skyrme could show the existence of confined fermionic particle-like solutions of a nonlinear bosonic meson field theory. Specifically, he constructed a field operator that maps the three-dimensional (3D, or 3 + 1 to include the time dimension) domain space to the surface of a sphere.[61] Similarly, the macrospin in micromagnetism is a vector in $\mathbb{R}^3$ with a constant modulus, or, equivalently, a point on the two-dimensional (2D) surface of a sphere S^2. Of particular interest are skyrmionic spin structures in thin films (2 + 1 domain space), i.e., vector fields $\mathbb{R}^2 \rightarrow S^2$ that are homotopically equivalent to the identity map on a sphere id_{S^2}. In these structures, the domain space can be continuously deformed to a spherical shape such that the map to the spin space changes continuously, and the boundary of the domain space always maps to the same spin. This homotopy between such so-called magnetic skyrmion and the vector field of a sphere is visualized in Figure 8.1.

The homotopy between a vector field and the identity map on a sphere is described by the topologically invariant skyrmion number N. This counts the number of times the sphere is covered in the homotopical deformation (although if N is not equal to 1, the transformation is much more complicated than illustrated in Figure 8.1 and more difficult to visualize). In two dimensions, the skyrmion number N of a vector field $\mathbb{R}^2 \ni (x, y) \rightarrow m \in S^2$ can be calculated by $N = (8\pi)^{-1} \int dx dy\, n$ with the topological density[2] $n = 2(\partial_x \mathbf{m} \times \partial_y \mathbf{m}) \cdot \mathbf{m}$. Note that the sign in this formula is not consistently defined in the literature, and the definition used here is adapted from Belavin and Polyakov (the first published application of this formula in the field of magnetism that we are aware of).[2]

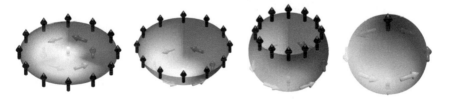

FIGURE 8.1 Homotopy between a skyrmion in a disk and the identity map of the sphere. The hue of the color represents the azimuthal angle of the spins, and the black/white level symbolizes the polar angle. The images show a continuous deformation of the skyrmion (left) to a spherical shape (right). All spins of the skyrmion boundary (black spins) are mapped to the north pole of the sphere, which is only possible because they all point in the same direction. The white spins of the inner domain are all mapped to the south pole. (From Büttner, F., *Topological mass of magnetic skyrmions probed by ultrafast dynamic imaging*. Dissertation, University of Mainz, Mainz, 2013. With permission.)

Chapter 8

For our definition and axially symmetric structures (or structures that can be deformed homotopically into axially symmetric shapes) with polarity p and winding number W, the skyrmion number can be calculated by the simplified formula $N = pW/2$.[39] This formula is useful to calculate the skyrmion number "by hand," because the winding number is strictly quantized and it is therefore easy to determine from a picture of the magnetization configuration. The winding number is most conveniently calculated by expressing the 3D macrospin in spherical angles (φ, θ). The winding number of a closed loop is given by the normalized difference of the azimuthal angle φ_i of the spin at an arbitrary point of the loop and the angle φ_f of the spin after going around the full loop once in positive orientation: $W = (\varphi_f - \varphi_i)/(2\pi)$. Naturally, the loop must not cross regions where φ is not well defined (singularities), such as points or areas with pure out-of-plane magnetization or points where two spins are pointing head to head. The winding number is always an integer, and it has nonzero value only if the loop encloses a singularity.[64] Often, only when $W = 1$ (and thus $N = \pm 1$), skyrmionic spin structures are called skyrmions. Note that the magnetization orientation $\mathbf{m}(r,\varphi)$ of circular $W = 1$ skyrmions can be described in polar real-space coordinates (r,φ) by $\mathbf{m}(r,\varphi) = (m_x, m_y, m_z) = (\sin(\theta)\cos(\varphi + \psi), \sin(\theta)\sin(\varphi + \psi), \cos(\theta))$, where $\theta = \theta\,(r)$ describes the cross-sectional domain wall profile and the constant ψ is the domain wall angle. Skyrmions with $\psi = 0$ (spins pointing outwards) and $\psi = \pi$ (spins pointing inwards) are called Néel skyrmions (sometimes in the literature the term hedgehog skyrmion is also used), whereas skyrmions with $\psi = \pi/2$ (spins rotating counterclockwise) and $\psi = 3\pi/2$ (spins rotating clockwise) are called Bloch skyrmions. The polarity is defined as $p = \dfrac{1}{2}(m_z(r = 0) - m_z(r = \infty))$.

The skyrmion number provides direct information about the domain structure. The only way to map a planar geometry continuously to a sphere is by contracting its boundary to one single point of the sphere (defining one of the poles). Therefore, all spins on the boundary of an $N = 1$ configuration must have the same orientation, forming a domain. Somewhere in the interior of the configuration, a connected area exists in which the spins point antiparallel to the outer domain (this inner domain can be arbitrarily small down to a single point). The transition between these two domains is a smooth domain wall winding around the inner domain to ensure that the whole sphere is represented. That is, the fact that a spin vector field has $N = 1$ implies the existence of the inner domain, the outer domain, and the domain wall, of which only the outer domain touches the boundary (thus confining the inner domain and the domain wall). The confined inner domain together with the domain wall has quasiparticle properties, which has led to the name skyrmion.

8.2 General Topological Properties of Skyrmions

The special topology of skyrmions has directly measurable implications; we will discuss four of the key ones here. First, skyrmions are protected against creation or annihilation by a topological energy barrier, as long as they do not move towards the edge of the sample. Due to this topological stability, there is an extended region in the phase diagram where skyrmions and the ferromagnetic state (uniform, parallel spin alignment) have a hysteretic coexistence.[52] In this regime, the number of skyrmions in a given area is a free parameter, which is very important when using skyrmions for data storage technologies. Still, it has been shown that there are ways to create and annihilate skyrmions artificially in this regime of the phase diagram. This can be achieved, for instance, by (1) locally injecting a spin-polarized current,[48,52] (2) locally heating the sample,[31] (3) local magnetic fields,[31] (4) sending a current through a wire with an appropriately shaped constriction,[24,31] (5) high-frequency bipolar excitations in combination with pinning sites,[70] or (6) moving two domain walls[76] or a stripe domain[26] from a constricted area to an extended area. In addition, skyrmions with a finite lifetime, so-called dynamic skyrmions, can be created uniformly in z-direction in a magnetized film by the combined action of the Oersted field and the Slowncewski spin torque of a local current in z-direction that is spin-polarized in z-direction.[75]

A second universal property of magnetic skyrmions is that electrons moving adiabatically across a magnetic skyrmion collect a Berry phase. This phase can be expressed through the Aharonov–Bohm effect caused by an "emergent" magnetic field of the skyrmion. This emergent magnetic field is proportional to the topological density n, and the Berry phase is proportional to the integrated enclosed flux, i.e., to the skyrmion number. The interference of different paths around the skyrmion leads to a deflection of the overall electron propagation direction, i.e., to a transverse current. This leads to a transverse voltage, the so-called topological Hall voltage.[42,53]

A third interesting property of magnetic skyrmions is their outstanding insensitivity to magnetic pinning, at least in the collective motion of densely packed skyrmions, so called skyrmion lattices. One scenario that is particularly interesting for technological applications, such as the racetrack memory proposed by Parkin et al.,[46] is the displacement of

Chapter 8

nontrivial magnetic textures by spin-polarized currents. Such nontrivial textures, of which skyrmions are one example, can potentially be used to encode information. The underlying phenomenon for the displacement of spin structures is the so-called spin-transfer torque of conduction electrons on localized spin moments, predicted by Berger[3] and Slonczewski.[62] The investigation of this effect is a subject of intense research nowadays, with a strong focus on the motion of straight domain walls.[5] However, extrinsic magnetic pinning of domain walls has been found to be significant, and mostly no motion has been observed for regular domain walls driven with current densities smaller than 10^{11} A/m^2. In contrast, for skyrmion lattices, this critical current density is five orders of magnitude smaller.[53] So far, no complete theoretical explanation of the particularly low pinning of skyrmions has been published, and it remains to be seen if these low critical current densities can be realized also for isolated skyrmions. However, the confinement of skyrmions intuitively leads to lower pinning compared to extended domain walls. This is firstly because, for topological reasons, skyrmions never touch the edge of the sample, reducing the sensitivity to edge roughness and thus to the main source of pinning. And second, a fully confined structure is flexible to deform and to move around obstacles (provided the obstacles are not attractive).[25,41]

A fourth phenomenon that is directly associated with the skyrmionic topology is a force that acts on a moving skyrmion, pointing perpendicular to its velocity vector. The spherical topology of a skyrmion leads to an intrinsic angular momentum in the quasiparticle equation of motion of a skyrmion,[45] expressed by a gyro term $\mathbf{G} \times \dot{\mathbf{R}}$, where $\mathbf{G} = (0,0,G)$ is the gyrocoupling vector and $\dot{\mathbf{R}}$ is the velocity of the center of mass $\mathbf{R}$ of the skyrmion.[67] The gyrocoupling strength $G = -4\pi NT M_s \gamma$ (where T is the material thickness, M_s its saturation magnetization, and $\gamma = 1.76 \times 10^{11}$ As/kg the gyromagnetic ratio) is proportional to the topological charge N of the skyrmion. Note that the intrinsic angular momentum is zero for skyrmions in antiferromagnets where $M_s = 0$. Intrinsic angular momentum is also found in structures with fractional skyrmion numbers, such as $N = \pm 1/2$ vortices,[23] but generally not in straight domain walls with $N = 0$.

8.3   Chiral Exchange Interactions

The spin structures present in a system generally result from the minimization of the relevant micromagnetic energy terms. In addition to the Heisenberg exchange that favors spin structures with random chirality, chiral skyrmions can be stabilized by chiral exchange interactions that can be found in systems with inversion asymmetry and strong spin–orbit coupling (SOC). The so-called Dzyaloshinskii–Moriya interaction (DMI) is a symmetry-breaking (antisymmetric) exchange interaction that is present in addition to the Heisenberg exchange. The DMI between two spins in a magnetic material is mediated via the SOC of an adjacent heavy-metal atom with strong spin–orbit interaction, as shown schematically in Figure 8.2. Taking into account SOC, Moriya found that the effective Hamiltonian for the interaction between two (macro-)spins $\mathbf{M}_1$ and $\mathbf{M}_2$ contains a term $\mathbf{D}_{12} \cdot (\mathbf{M}_1 \times \mathbf{M}_2)$.[38] As shown schematically in Figure 8.2, a DMI coupling can arise at the interface of a heavy metal layer and a ferromagnetic layer (FL) in a multilayer stack due to the broken inversion symmetry in z-direction and the strong SOC of the heavy metal. The resulting DMI vector, $\mathbf{D}_{12}$, points in the plane of the layers and perpendicular to the vector connecting $\mathbf{M}_1$ and $\mathbf{M}_2$. The magnitude and the sign of $\mathbf{D}_{12}$ are properties of the interface and the involved materials. The DMI leads to a favored chirality of a spin spiral state.[12,51] For sufficiently strong DMI, a spin spiral state is favored with the spiral axis in the plane.[47] For lower values of D, a skyrmion lattice is the ground state.[40] While DMI is not necessary to obtain skyrmion spin structures with a given topology as they can be stabilized by dipolar interactions for instance in thin films, the DMI modifies the spin structure.[30] In general, in thin film multi-layers, both dipolar interactions and DMI will be present as it has been shown that even for nominally symmetric stacks, DMI is present due to the different growth at the interfaces.[20]

Overall, strong DMI will increase the stability of spin structures and favor skyrmions with a certain topology ($W = 1$ winding number) and a fixed product ψp of domain wall angle ψ and polarity p, which, however, is not a topological quantity in a homotopic classification.

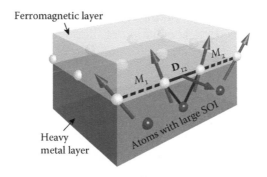

FIGURE 8.2 Illustration of the Dzyaloshinskii–Moriya interaction. In multilayer stacks with broken inversion symmetry, chiral coupling between two spins M_1 and M_2 is mediated by a heavy atom (dark gray) in one of the nonmagnetic layers (HL). The sign and strength of the resulting DMI (vector $\mathbf{D}_{12}$) are interface/materials' properties that lead to one favored chirality of the spin structure.

Chapter 8

8.4   Quasiparticle Equation of Motion

Here, we present the general quasiparticle equation of motion of a $W = 1$ rigid circular skyrmion, including the effects of spin-polarized electrical currents and uniform spin injection due to the spin Hall effect (SHE) in an adjacent heavy metal layer. An electrical current flowing through a ferromagnetic material gets polarized along the local magnetization direction. The spin-polarized current is parametrized by the spin drift velocity $\mathbf{u} = JP\mu_B/(eM_s)$, where J is the electron current density, P is the spin polarization of the material, $\mu_B = 9.27 \times 10^{-24}$ J/T is the Bohr magneton, and $e = 1.602 \times 10^{-19}$ C is the electron charge. At a gradient of the magnetization, the re-orientation of the conduction electron spins requires the transfer of angular momentum to the local magnetization. This transfer of angular momentum can be due to adiabatic or nonadiabatic spin-transfer torques, but the torque on the local magnetization is always proportional to the gradient of the magnetization and as such it is a function of position in the sample. In contrast, a heavy metal layer above or below the ferromagnetic layer becomes a source of a uniform (position-independent) spin current when it is transmitted by an electrical current $\mathbf{j}_{HM}$. The spin current flows perpendicular to $\mathbf{j}_{HM}$ and is polarized perpendicular to both $\mathbf{j}_{HM}$ and to the spin current direction. The injection of angular momentum leads to a torque on the magnetization, which can be described by a sum of the two terms: a contribution that acts like a uniform magnetic field along the polarization of the injected spins (field-like term) and one term that resembles the damping term of the Landau–Lifshitz–Gilbert equation (damping-like term). Here, we only consider the damping-like term because a uniform field does not cause motion if we assume the skyrmion to be rigid. Taking into account all these contributions, the resulting quasiparticle equation of motion for the center of mass position $\mathbf{R}$ reads[17,24,34,52,66,68]:

$$-M\ddot{\mathbf{R}} + \mathbf{G} \times (\dot{\mathbf{R}} - \mathbf{u}) - \tilde{D}(\alpha\dot{\mathbf{R}} - \beta\mathbf{u}) + 4\pi B\tilde{R}(\psi)\mathbf{j}_{HM} + \mathbf{F} = 0, \qquad (8.1)$$

where M is the effective mass of the skyrmion, $\tilde{D} = TM_s\gamma \int dxdy \left(\partial_x\mathbf{m}\right)^2$ is the dissipation constant, α is the Gilbert damping, β is the non-adiabaticity parameter, $B = \dfrac{\gamma^2\hbar\theta_{SH}}{2e}I$ is a coefficient that depends on the spin configuration (with the reduced Planck constant $\hbar = 6.63 \times 10^{-34}$ Js, the spin Hall angle θ_{SH}, and $I = \dfrac{1}{4}\displaystyle\int_0^\infty dr\left(\sin\theta\cos\theta + r\dfrac{d\theta}{dr}\right)$)

$$\tilde{R}(\psi) = \begin{pmatrix} \cos\psi & \sin\psi \\ -\sin\psi & \cos\psi \end{pmatrix}$$ is the rotation matrix corresponding to

the domain wall angle ψ, $\mathbf{j}_{HM}$ is the current density of an adjacent spin Hall heavy metal, and $\mathbf{F}$ is a driving force acting on the skyrmion for instance due to a gradient of the out-of-plane effective field. The center of mass of a skyrmion is determined from an average position of its spins (the spins enclosed by the loop of spins with zero out-of-plane moment) weighted by their out-of-plane component. Note that Equation 8.1 is written in a form that each term is given in units of force. Sometimes, another form is found in the literature, where the whole equation is divided by $TM_s\gamma$, thus giving each term the units of velocity.[24,52,68] In that form, the gyrocoupling constant and the dissipation constant reduce to $G' = -4\pi N$ (or sometimes $G' = 4\pi N$) and $\tilde{D}' = \int dxdy(\partial_x\mathbf{m})^2$, respectively.

The overall dynamics of skyrmions are well described by Equation 8.1. However, some details of the trajectories that are visible in micromagnetic simulations cannot be explained by this equation. Therefore, a number of modifications to Equation 8.1 were suggested. These include adding a gyrodamping[55] and making the parameters of the equation dependent on the excitation type (thermal, spin current, or field gradient) and the excitation frequency.[55] Yet, the magnitude and significance of such corrections still have to be determined from experiments. The few existing experiments are described well by Equation 8.1, and we therefore discuss the dynamics of skyrmions based on this equation in the following section.

8.5   Dynamics of Skyrmions

In this section, we discuss the dynamics of magnetic skyrmions in three distinct scenarios: (1) the steady-state motion, which skyrmions will enter if the excitation is constant or varies only very slowly, (2) the gyrotropic eigenmodes in a parabolic potential, and (3) the translational motion along a nanowire using pulsed forces.

8.5.1   Steady–State Motion

Due to the gyro term in Equation 8.1, the x and y position coordinates, R_x and R_y, are canonically conjugate variables,[25] similar to position and transverse spin angle for straight domain walls[36] (the transverse

Chapter 8

spin angle is the angle of the spins with respect to the plane formed by the set of all spins in the domains and in the domain wall at rest). Therefore, in the absence of confining potentials (and pinning, i.e., $\mathbf{F} = 0$), the steady-state motion of a skyrmion is a linear trajectory with a characteristic angle with respect to the direction of the driving force, the skyrmion Hall angle. The velocity depends linearly on the excitation strength, i.e., there is no intrinsic pinning, which is in stark contrast to the dynamics of nontopological domain walls. The steady-state velocity for a spin current in x-direction ($\mathbf{u} = (u,0,0)$) is given by[25]:

$$\dot{R}_x = \left(\frac{\beta}{\alpha} + \frac{\alpha - \beta}{\alpha^3 (\tilde{D}/G)^2 + \alpha} \right) u, \qquad (8.2)$$

$$\dot{R}_y = \frac{(\alpha - \beta)(\tilde{D}/G)}{\alpha^2 (\tilde{D}/G)^2 + 1} u. \qquad (8.3)$$

Often, $\alpha \ll 1$ and $\tilde{D}/G \leq 1$, which means that Equations 8.2 and 8.3 can be simplified to[25]:

$$\dot{R}_x \approx u, \qquad (8.4)$$

$$\dot{R}_y \approx (\alpha - \beta)(\tilde{D}/G)u. \qquad (8.5)$$

Hence, within this approximation, the velocity in the direction of the current flow $\dot{R}_x$ does not depend on α or β; only the skyrmion Hall angle $\tan(\xi) = \dot{R}_y / \dot{R}_x$ is influenced by these material parameters. Remarkably, the skyrmion Hall angle for current-driven skyrmion motion scales inversely with the topological charge N (remember $G \propto N$), whereas in cases of skyrmion motion driven by a finite force $F_x \neq 0$ (as, for instance, due to a magnetic field gradient), the spin Hall angle scales linearly with N[36]:

$$\tan(\xi) = -\frac{2N\Delta_0}{R\alpha}, \qquad (8.6)$$

where Δ_0 is the domain wall width parameter of the skyrmion domain wall and R is the skyrmion radius.

If the skyrmion is driven by a constant spin injection due to a spin Hall current $\mathbf{j}_{HM} = (j_{HM},0,0)$ instead of a spin current u, then the steady-state velocity depends on the chirality of the skyrmion. Essentially, the skyrmion is dragged in the direction where the spins in its domain wall are parallel to the injected spins.[68] The steady-state velocity for an $|N| = 1$ skyrmion in this case reads:

$$\dot{R}_x = \left[\frac{\alpha \tilde{D} B}{G^2 + \alpha^2 \tilde{D}^2}\cos(\psi) + \frac{GB}{G^2 + \alpha^2 \tilde{D}^2}\sin(\psi)\right] j_{\text{HM}}, \qquad (8.7)$$

$$\dot{R}_y = \left[-\frac{\alpha \tilde{D} B}{G^2 + \alpha^2 \tilde{D}^2}\sin(\psi) + \frac{GB}{G^2 + \alpha^2 \tilde{D}^2}\cos(\psi)\right] j_{\text{HM}}, \qquad (8.8)$$

indicating that a Néel skyrmion with low damping $\alpha \ll 1$ moves perpendicular to the spin Hall current, whereas a Bloch skyrmion moves in the direction of the injected spin Hall current.[68]

In case of a confining potential in y-direction and a spin current in x-direction (as experienced by a skyrmion in a nanowire), the current-driven skyrmion dynamics change drastically. Here, the skyrmion first moves on a diagonal line towards higher potential energy regions (towards the edge of the wire) before entering a steady-state longitudinal motion. The angle between the diagonal line of the initial motion and the x-axis is again the skyrmion Hall angle, and the velocity of the subsequent longitudinal motion of the skyrmion follows the same equations as that of a straight domain wall.[24]

8.5.2 Gyrotropic Eigenmodes

The dynamics of skyrmions is particularly rich in the case of non-steady-state motion, for instance due to pulsed spin currents or due to pinning ($\mathbf{F} \neq 0$). In those cases, inertia of the skyrmion becomes important. The simplest case is the motion of a skyrmion in a radially symmetric parabolic potential in the absence of currents or field gradients, i.e., for $\mathbf{F} = -K\mathbf{R}$ and $\mathbf{u} = 0$. Here, K is the potential stiffness. In reality, most potentials can be locally approximated by a parabola, which makes this analytically solvable approximation applicable to experimental situations. Therefore, we will discuss the solution of the equation of motion for this particular case in more detail.

As a reminder, the equation of motion for a skyrmion in a parabolic potential without current reads:

$$-M\ddot{\mathbf{R}} + \mathbf{G} \times \dot{\mathbf{R}} + \tilde{D}\alpha\dot{\mathbf{R}} - K\mathbf{R} = 0. \qquad (8.9)$$

By using the convenient complex parametrization $\bar{R} = R_x + iR_y$ of the location vector, we obtain:

$$-M\ddot{\bar{R}} + (\tilde{D}\alpha + iG)\dot{\bar{R}} - K\bar{R} = 0, \qquad (8.10)$$

Chapter 8

which is a simple harmonic oscillator with solution:

$$\bar{R}(t) = A\exp(i\bar{\omega}_1 t) + B\exp(i\bar{\omega}_2 t) \tag{8.11}$$

with:

$$\bar{\omega}_{1,2} = i\eta \pm \sqrt{\omega_0^2 - \eta^2}, \tag{8.12}$$

$$\eta = -\frac{1}{2M}(\tilde{D}\alpha + iG), \tag{8.13}$$

$$\omega_0 = \sqrt{K/M}. \tag{8.14}$$

The two terms of the sum in Equation 8.11 are called gyrotropic eigenmodes of the skyrmion. They describe a left-handed and a right-handed spiraling motion, respectively. The complex frequencies $\bar{\omega}_{1,2} = \omega_{1,2} + i/\tau_{1,2}$ combine the real frequencies of the spiraling motion $\omega_{1,2}$ and the damping $\tau_{1,2}$ in one number. When both modes are excited simultaneously, the resulting trajectories can have distinctly different shapes depending on the relative amplitude of the modes, see Figure 8.3. If the lower frequency mode has a smaller amplitude than the higher frequency mode, the trajectory looks like a deformed spiral, which rotates clockwise if N is positive, and counterclockwise if N is negative

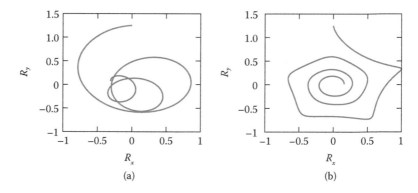

(a) (b)

FIGURE 8.3 Micromagnetically computed trajectories of a skyrmion in a parabolic potential. In both graphs, $\omega_1/\omega_2 = -4$ and N is negative. (a) The amplitude of the higher frequency mode ω_1 is initially four times larger than that of the lower frequency mode ω_2. The trajectory is a deformed spiral rotating counterclockwise. (b) The amplitude of the higher frequency mode is initially four times smaller than that of the lower frequency mode. The trajectory is a hypocycloid with $5 = -\omega_1/\omega_2 + 1$ cusps per turn, rotating clockwise.

(Figure 8.3a). If the lower frequency mode has a larger amplitude than the higher frequency mode; however, the trajectory becomes a hypocycloid and the rotation is reversed (counterclockwise if N is positive and clockwise if N is negative). The number of cusps is given then by $-\omega_1/\omega_2 + 1$ (Figure 8.3b). Note that the damping scales linearly with the frequency.[34] Therefore, a trajectory as in Figure 8.3a will at a later time transform into a motion that resembles Figure 8.3b (and thereby changing the sense of rotation). Even later, the amplitude of the lower frequency mode will be so small that the trajectory cannot be distinguished from that of a massless skyrmion within the experimental resolution.

Only $|N| = 1$ skyrmions show the previously discussed gyrotropic eigenmodes. Skyrmions of other topological charges show complex dynamics within their domain wall during the motion, which means that the spin structure no longer behaves like a rigid particle. Hence, Equation 8.1 is no longer able to describe the dynamics of the skyrmion as a whole on the time scale of the domain wall fluctuations.

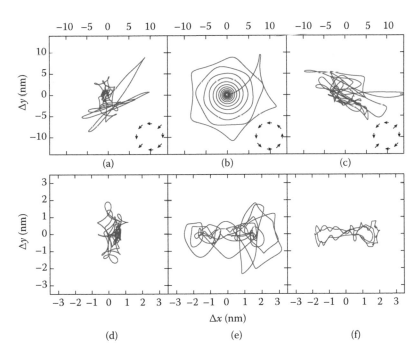

FIGURE 8.4 Trajectories of variable N skyrmions in a parabolic potential: (a) $N = 0$, (b) $N = 1$, (c) $N = 2$, (d) $N = 3$, (e) $N = 4$, and (f) $N = 5$. The red lines indicate the trajectory during the excitation with a magnetic field gradient. The blue lines indicate the relaxational motion. The insets in (a–c) schematically show the spin structures of the domain walls of an $N = 0$, $N = 1$, and $N = 2$ skyrmion, respectively. (From Felix, B, *Nat. Phys.*, 11, 225–228, 2015. With permission.)

Chapter 8

The resulting trajectories are visualized in Figure 8.4 for $N = 0,1, ..., 5$. For the dynamics on timescales much slower than the domain wall fluctuations, however, Equation 8.1 has been found to be very accurate even up to $N = 90$.[36]

The eigenfrequencies $\omega_{1,2}$ and the corresponding damping terms $\tau_{1,2}$ can be measured in an experiment, for instance by resonant excitations[43,57] or by imaging the trajectory.[11] Figure 8.5 shows the 2D relaxational trajectory of a skyrmion after displacement in a parabolic potential, measured with 3 nm spatial precision and 50 ps temporal resolution.

The experimental skyrmion trajectory in Figure 8.5 is in excellent agreement with the theoretical prediction of Equation 8.11 if both gyrotropic eigenmodes are considered. The existence of two gyrotropic eigenmodes is an unambiguous indicator for an $|N| = 1$ spin structure topology. Furthermore, the frequencies can be used to calculate the mass of the skyrmion:

$$\omega_1 + \omega_2 = \text{Re}(\bar{\omega}_1 + \bar{\omega}_2) \tag{8.15}$$

$$= \text{Re}(G/M - iD\alpha M) \tag{8.16}$$

$$= G/M. \tag{8.17}$$

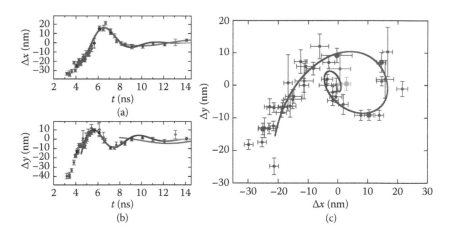

FIGURE 8.5 Gyrotropic trajectory of an $N = 1$ skyrmion, with (a) and (b) showing the temporal evolution of the x and y coordinate of the skyrmion, respectively, and (c) depicting the x-y trajectory. Data points plot positional changes of the skyrmion extracted from dynamic imaging of its relaxational motion after excitation in a parabolic potential. Colored lines represent a fit of the data points with the theoretical model of Equation 8.11. Gray lines depict the best fit with Equation 8.11 when setting the mass to zero. (From Büttner, F. *Nat. Phys.*, 11, 225–228, 2015. With permission.)

In the experiment, the frequencies were of the order of 1 GHz. The corresponding mass of the skyrmion has been found to exceed the inertia of its domain wall, the so-called Döring mass,[14] by at least a factor of 5.[11] In brief, the Döring mass of a domain wall originates from the fact that a domain wall can only move if its spins tilt out of their equilibrium position, which increases the magnetostatic energy. The increase in energy by tilting the spins is described by an effective anisotropy[9] $K_\perp$ and the Döring mass of a domain wall of width Δ_0 reads:

$$m_D = \frac{M_s^2(1+\alpha^2)}{K_\perp \gamma^2 \Delta_0}, \tag{8.18}$$

where $\gamma = 1.76 \times 10^{11}$ As/kg is the gyromagnetic ratio. Skyrmions, however, posses a mass that is significantly larger than expected from the domain wall theory, thus indicating further contributions to their inertia. It was suggested that the expansion and shrinking of the skyrmion, the so-called breathing mode, constitutes another source of inertia as this mode can store energy.[54] This source of inertia is only found in skyrmionic spin structures because the skyrmions are the only completely confined spin structures in magnetism, i.e., the only structures that can continuously grow and shrink. The large mass of the skyrmion has hence been called topological mass.[11]

8.5.3   Motion along Nanowires

The dynamics of skyrmions in fully confining potentials as discussed before is best suited for studying their quasiparticle properties as well as for applications in radio-frequency technology, where the skyrmions are excited at their resonance frequency.[57] Most other possible applications, however, rely on skyrmions being displaced to a new equilibrium position. In this complicated scenario, Equation 8.1 has to be solved numerically. Research in this field has just started and a more complete picture is still emerging. Here, we first demonstrate that inertia of a skyrmion does not impact how far the skyrmion is displaced but that the trajectory of the motion in between depends on the magnitude of the mass. Subsequently, we discuss the first experimental observation of the propagation of skyrmions.

A typical potential application of propagating skyrmions is in a magnetic shift register.[18] In this application, skyrmions are placed in a nanowire. A pulsed current or a pulsed magnetic field gradient are applied in the direction of the wire in order to move the skyrmions by a well-defined distance. Ideally, there is no other force along the

Chapter 8

direction of the wire (i.e., no pinning), and the potential perpendicular to the wire can be approximated by a parabola. The trajectories of a massive and a massless skyrmion in such an ideal scenario for the case of a pulsed force (and without current) are compared in Figure 8.6. The mass of the massive skyrmion, $M = 2 \times 10^{-21}$ kg, is the mass found in the experiment in Ref. [11]. Both skyrmions, the massive and the massless, travel the exact same distance. The massive skyrmion, however, needs more time to arrive at the new equilibrium position, and the transverse displacement during the motion is significantly smaller than for the massless skyrmion. Hence, a massless skyrmion is more likely to be annihilated when touching the edge of the wire.

One advantage of skyrmions is their low susceptibility to external fields as uniform fields do not induce skyrmion displacement. The dynamics due to field gradients as discussed before is thus difficult to implement for devices. Furthermore, field-induced dynamics exhibits poor scaling when shrinking the size of a potential device. An alternative approach is current-induced magnetization manipulation, which exhibits favorable scaling.[5] Previously, current-induced magnetization dynamics due to spin-transfer torque effects has been in the focus of research. Here, the transfer of spin angular momentum from conduction electrons to the magnetization is used to efficiently manipulate magnetization in multilayer stacks with potential applications in magnetic random access memories[27] or to move domain walls,[5] which can be the basis for racetrack memory devices.[46]

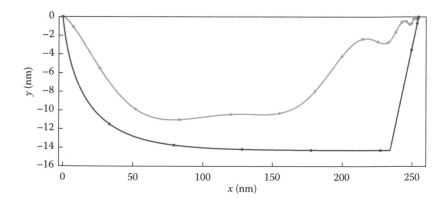

FIGURE 8.6 Trajectories of variable M skyrmions in wire geometry with a parabolic potential in y-direction. Light gray line: $M = 2 \times 10^{-21}$ kg. Dark gray line: $M = 0$. Both trajectories correspond to $G = 10^{-12}$ kg/s, $D\alpha = 7 \times 10^{-13}$ kg/s, and $K = 3.5 \times 10^{-3}$ N/m. The solid circles indicate time steps of 1 ns. (From Felix, B, *Nat. Phys.*, 11, 225–228, 2015. With permission.)

The adiabatic spin-transfer torque arises when an electron that traverses a spin structure continuously adjusts its spin orientation to match the local magnetization direction. For instance, if an electron passes adiabatically across a domain wall, it changes its spin direction by 180°. Due to conservation of angular momentum, a total of $1\hbar$ of angular momentum is transferred to the local spin structure in the process. Even when including nonadiabatic effects, the maximum angular momentum that an electron can transfer from its spin degree of freedom is $1\hbar$, and $1\hbar$/electron is hence a fundamental limit for the spin-transfer torque efficiency.

For the transfer of orbital angular momentum, however, there is no such fundamental limit and such a process can thus in principle be much more efficient. To transfer orbital angular momentum, one makes use of spin–orbit coupling in a heavy metal layer adjacent to a magnetic layer (Figure 8.8). Due to the SHE and the inverse spin galvanic effect (ISGE), the injected charge current j_{HM} leads to two torques acting on the magnetization. These two torques are distinct in their symmetry and are called field-like (FL) (sometimes also called reactive) and damping-like (DL) (sometimes also called dissipative or Slonczewski-like).[7] These torques result from two spin–orbit effects that occur in addition to spin-transfer torque:

1. When a charge current flows in a heavy metal layer, electrons experience a deflection that is perpendicular to their velocity and to their spin orientation. This so-called SHE leads to spin accumulations at all sides of the wire and each side has a different spin polarization, see Figure 8.7. Experimentally, a spin current flowing in z-direction (polarized in y-direction) is most interesting because it can be injected into a ferromagnet on top (or below) the HL, see Figure 8.8. The spin orientation in y-direction is called transverse polarization, to distinguish it from a longitudinal polarization (in the direction of current) and a perpendicular spin polarization (perpendicular to the interface). The transverse spin current that moves across the interface into the ferromagnet then interacts with the magnetization in the ferromagnetic layer via spin-transfer torque and thus manipulates the magnetization. Different microscopic origins have been proposed to explain this effect, including intrinsic band structure effects as well as extrinsic skew scattering and side jump effects.[60]

2. The ISGE (also termed Rashba–Edelstein effect) results from an electric field that originates from the symmetry breaking at the interface and then leads to an effective magnetic field for the

Chapter 8

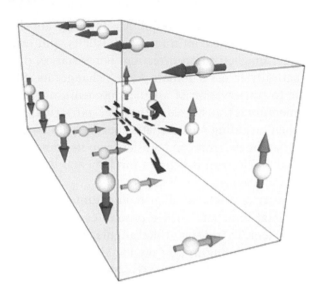

FIGURE 8.7 Illustration of the deflection of electrons in a heavy metal wire due to the spin Hall effect. Spheres indicate electrons and arrows their spin orientation. The dashed black arrows depict the electron trajectories. The electron flow is mostly along the wire axis, but due to the spin Hall effect, the electrons experience a deflection that depends on their spin orientation. In applications, typically a ferromagnetic layer would be placed on top or below the heavy metal layer and then experience an injection of spins that are polarized in a direction perpendicular to the current flow and perpendicular to the normal of the interface. (Reprinted from Pai, C.-F., *The spin Hall effect induced spin transfer torque in magnetic heterostructures.* PhD thesis, Cornell University, Ithaca, NY, 2015. With permission.)

moving electrons at the interface.[59] The sign of this field depends on the spin–orbit interaction and is independent of the Oersted field. It has a fixed direction at the interface within the sample, so that the spins moving at the interface experience a torque and a nonequilibrium spin density results. In contrast to the primarily bulk SHE, the ISGE is a pure interface effect only present in a multilayer configuration where the interface generates the required electric field. However, in such a multilayer configuration, the resulting spin density at the interface then acts by exchange on the magnetization in the ferromagnet and can thus manipulate it.

The damping-like torque and the field-like torque that can originate from the SHE or the ISGE are commonly called spin–orbit torques (SOTs). It is unclear and discussed controversially as to whether the ISGE alone can be responsible for the switching of the

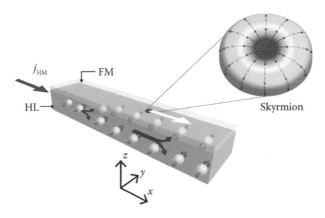

FIGURE 8.8 Schematic depiction of the multilayer stack suitable for spin–orbit torque-driven skyrmion motion. The multilayer consists of a heavy metal underlayer (HL) and a ferromagnetic layer (FL) where spin–orbit interaction effects, such as DMI and spin–orbit torques, occur. The injected charge current in the heavy metal layer j_{HM} along the x-direction in the HL splits due to the spin Hall effect, with spin-left electrons flowing up and spin-right electrons flowing down. Thus, a spin current in the z-direction with polarization in y-direction results. This spin current then acts on the magnetization in the FL. Furthermore, for electrons flowing at the interface, the inverse spin galvanic effect (sometimes also termed Rashba–Edelstein effect) leads to a nonequilibrium spin density and thus to an effective field at the interface that also acts on the magnetization in the FL. Both effects (spin Hall and inverse spin galvanic) lead to torques that can displace spin structures such as skyrmions (spin structure, see inset) along the wire with higher efficiency than the conventional approaches using spin-transfer torques.

perpendicular ferromagnetic layer, and the extent of the contribution from the SHE to the damping-like torque is not well known for metal systems. While originally it was predicted that the ISGE primarily leads to a field-like torque and the SHE to a damping-like torque, it is now generally accepted that both can lead to both torques.[21,33]

The current-driven motion of skyrmions in a thin transition metal ferromagnet at room temperature has recently been demonstrated experimentally.[70] In this experiment, Pt/Co/Ta and Pt/Co$_{60}$Fe$_{20}$B$_{20}$/MgO multilayer stacks with perpendicular magnetic anisotropy were studied with high-resolution magnetic transmission soft X-ray microscopy. Pt in contact with Co is known to generate strong DMI,[51] while Ta generates very weak DMI,[16] so that a large net DMI is anticipated in this asymmetric stack structure. Pt/Co$_{60}$Fe$_{20}$B$_{20}$/MgO multilayers are known for their low pinning due to the amorphous character of Co$_{60}$Fe$_{20}$B$_{20}$ and again the asymmetry in the stacking leads to a sizable DMI and SOT spin current.

Chapter 8

Recent simulations suggest that skyrmions in ultrathin films might be driven even more efficiently than in previous studies of bulk materials due to the increased efficiency of vertical spin-current injection that is only available in thin film multilayers. Pt/FM/Ta multilayer stacks with a ferromagnet (FM) are very favorable in that regard because Pt and Ta have spin Hall angles of opposite signs, meaning that the spin current injected from Pt upwards has the same polarization as the spin current injected by Ta downwards. Hence, the spin Hall currents generated at each interface work in concert to generate a large damping-like torque.[71] As the spin Hall effect direction of motion of a skyrmion depends on the skyrmion topology, which in turn is here dominated by the DMI, observations of current-induced displacement can serve to unambiguously verify the topology and chirality of the skyrmions in this system.

In the experiment, an external magnetic field B_z was applied to a 2-μm wide magnetic track to shrink the zero-field labyrinth domains into a few isolated skyrmions. The track was contacted by Au electrodes at either end for current injection, as shown in Figure 8.9a. Figure 8.9b shows a sequence of images of a train of four skyrmions stabilized by B_z. Each image was acquired after injecting 20 current pulses with a current-density amplitude of 2.2×10^{11} A/m^2 and a duration of 20 ns. The pulse polarity is indicated in the figure. Three of the four skyrmions move freely along the track and can be displaced forward and backward by current, while the leftmost skyrmion (which is highlighted by a white circle) remains immobile, evidently pinned by a defect. The propagation direction is along the current flow direction (against electron flow), and this same directionality was observed for skyrmions with oppositely oriented cores. Micromagnetic simulations show that the observed unidirectional spin-Hall-driven displacement is consistent with Néel skyrmions with left-handed chirality, confirming the topological nature of the skyrmions in this material.[70]

The average skyrmion velocity was measured versus current density in three different devices, shown in Figure 8.9c. The experiment yields a critical current density of $j_{crit} = 2.0 \times 10^{11}$ A/m^2, below which skyrmions remain largely pinned. Slightly above j_{crit} the skyrmions move at different average speeds in different regions of the track, suggesting a significant influence of local disorder on the dynamics. Interestingly, pinned skyrmions can be annihilated, as seen in the last image of Figure 8.9b, where only three skyrmions remain, and the leftmost skyrmion becomes pinned at the same location as was

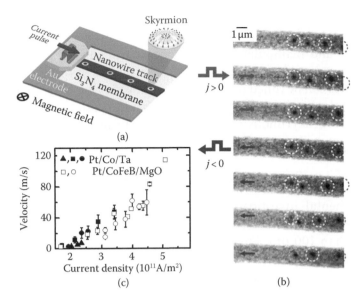

FIGURE 8.9   (a) Schematic of a magnetic track on $Si_3 N_4$ membrane with current contacts and skyrmions stabilized by a down-directed applied magnetic field. (b) Sequential STXM images showing skyrmion displacement after injecting 20 unipolar current pulses along the track, with an amplitude and polarity as indicated. Individual skyrmions are outlined in dotted circles for clarity. (c) Average velocity of skyrmions in Pt/Co/Ta (closed symbols) and Pt/CoFeB/MgO (open symbols) versus current density; error bars denote standard deviation of multiple measurements. (From Woo, S. et al., *Nature Materials*, 15, 501–506, 2016. With permission.)

the annihilated skyrmion. At higher current densities, faster skyrmion motion with velocities exceeding 100 m/s has been detected.

The very large value of the experimentally observed critical current density and the skyrmion velocities that are lower than those calculated for a defect-free sample are in sharp contrast to recent micromagnetic studies that predict high skyrmion mobility even in the presence of discrete defects.[18,24,52] These micromagnetic simulations suggest that a dispersion in the anisotropy energy due to interface disorder has little influence on the dynamics. However, a short length scale dispersion in the local DMI can cause skyrmion pinning and thus leads to a finite critical current and reduced velocities, which is in qualitative agreement with our experiments. This suggests that even higher velocities and lower critical currents might be achieved by engineering materials and enhancing interface quality. However, even with the observed velocities, fast switching as necessary for devices can be achieved already.

These results show that magnetic skyrmions and skyrmion lattices can be stabilized in common polycrystalline transition metal ferromagnets and manipulated at high speeds in confined geometries at room temperature. Since the magnetic properties of thin-film heterostructures can be tuned over a wide range by varying layer thicknesses, composition, and interface materials, it is now possible to engineer the properties of skyrmions and their dynamics using materials that can be readily integrated into spintronic devices.

8.6   Experimental Challenges

Experimental investigations of magnetic skyrmions are difficult because of their small size: typically, the diameter of a skyrmion is less than 100 nm. Furthermore, confirming the skyrmion topology requires either an image of its 3D spin structure close to atomic resolution or a measurement of the behavior of the skyrmion with unambiguous signatures of the actual skyrmion topology, as illustrated for the example of the skyrmion trajectory in Figure 8.4. Images of the 3D magnetization can be reconstructed from Lorentz transmission electron microscopy (TEM) data, as first demonstrated by Yu et al.[72] (see Figure 8.10). The technique is very powerful for static or quasi-static measurements of skyrmions due to its high spatial resolution. Ultrafast dynamic imaging, however, is challenging

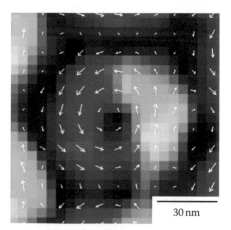

FIGURE 8.10   Real-space experimental image of a skyrmion. The image was acquired using Lorentz transmission electron microscopy. The color denotes the in-plane orientation of the magnetization, as indicated also by the arrows. Black areas are magnetized in the out-of-plane direction. (Reprinted by permission from Macmillan Publishers Ltd. [*Nature*] [Yu et al., 2010], copyright [2010]).

(1) because of the limited availability of stroboscopic TEM electron sources and (2) because the probing electrons of the TEM interact with the electric and magnetic fields that are normally used to excite dynamical behavior in magnetic systems. TEM pump-probe imaging of ferromagnetic domains has been available for decades, with nanometer spatial and sub-nanosecond temporal resolution.[4] However, thus far, no dynamic imaging of skyrmions using TEM has been reported.

Other ultra high spatial resolution imaging techniques with capabilities of detecting the spin configuration include scanning tunneling microscopy with a spin-polarized tip (SP-STM)[22] and nitrogen vacancy (NV) center magnetometry.[13,35] However, despite some advances in detecting spin waves with NV center magnetometry,[65,69] both techniques are presently not capable of pump-probe dynamic imaging.

There are several techniques suitable for pump-probe imaging of spin dynamics. The most common techniques are using light to probe the magnetic state, which is nonperturbative as long as heating of the sample is insignificant. Approaches using visible light, such as Kerr and Faraday microscopy, are widely accessible and have been successfully employed in studying bubble skyrmions for decades.[36] However, with some exceptions,[26] skyrmions in modern research are too small for visible light imaging. Therefore, X-ray imaging seems to be a rather appropriate method for imaging of skyrmion dynamics.

Generally, X-ray imaging techniques exploit the X-ray circular magnetic dichroism (XMCD)[56,63] to obtain X-ray absorption contrast for the spin component parallel to the propagation direction of the incident X-rays. A strong XMCD is observed when tuning the X-ray energy to the L-edge of the magnetic material, i.e., to wavelengths of a few nanometers. A weaker XMCD is found at the M-edge, i.e., at wavelengths of a few tens of nanometers. X-ray imaging techniques include transmission X-ray microscopy (TXM),[19] scanning transmission X-ray microscopy (STXM),[29] photo electron emission microscopy (PEEM),[32] and X-ray holography.[15] All of these techniques have been successfully employed in pump-probe imaging of spin dynamics and for imaging of skyrmions.[6,11,70]

Despite the many tools available for imaging of skyrmion dynamics, reports of such measurements are rare. The main challenge of such experiments is that pump-probe imaging requires one to repeat the experiment over and over again, and the dynamics of the specimen under investigation must be reproduced identically. Typically, a full movie with nanometer spatial and sub-nanosecond temporal

Chapter 8

resolution requires billions of repetitions. Almost any magnetic material grown today, in particular, systems that host magnetic skyrmions show some irregularities in the magnetic potential landscape, so-called pinning. In such rough potentials, tiny deviations in the initial conditions or thermal effects can lead to large changes in the subsequent dynamics. Advances on reducing magnetic pinning have been reported,[10] and individual samples have indeed proven suitable for pump-probe imaging.[11] However, further advances are pivotal for systematic investigations of skyrmion dynamics and for their application in devices.

8.7 Outlook

In this chapter, we have discussed the topological identity of skyrmions and how they manifest in solid-state magnetism. We have provided the quasiparticle equation of motion of magnetic skyrmions and reviewed theoretical and experimental studies revealing their dynamics in the three important cases of (1) quasi-static motion, (2) gyrotropic relaxation, and (3) propagation in a wire. However, skyrmion research is a rapidly developing field with new results appearing all the time. Some of the new research directions that have recently emerged include new materials systems beyond the well-established B20 bulk inversion asymmetry compounds and the multilayer systems with structural inversion asymmetry as many other materials are fundamentally inversion asymmetric. These include GaV_4S_8,[28,50] Heusler compounds,[37] and nitrides (such as $CoRh_{0.75}Fe_{0.25}Mo_3N$), and others. Furthermore, while here we focus on the dynamics of skyrmions in ferromagnets, other materials' systems are starting to attract attention. This includes skyrmions in synthetic antiferromagnets[74] and skyrmions in antiferromagnetic single-layer systems.[1,73] In Figure 8.6, the trajectories of skyrmions can be seen for different effective masses for an effective force acting on the skyrmion. Clearly, in addition to the motion in the direction of the force, a perpendicular motion is observed. This is often termed an effective Magnus force.[49] A key advantage of using antiferromagnetic or antiferromagnetically coupled layers with skyrmions is the opposite action of the Magnus force on both sub-lattices (both layers) leading to no transverse motion for the skyrmion and thus a reduction in the interaction of a skyrmion for instance with a wire edge. Finally, novel manipulation of skyrmions beyond field (gradients) and currents include optical as well as electric-field excitations as some multiferroic systems exhibit DMI and host skyrmions.[58]

Acknowledgments

We gratefully acknowledge discussions and input from many previous and present collaborators, students, and colleagues. We thank Roberto Zivieri for checking and commenting on the equations that describe the skyrmion dynamics. For the micromagnetic simulations, we thank the Department of Physics of the University of Hamburg for access to the PHYSnet-Computing Center. Much of this work was funded by the German Ministry for Education and Science (BMBF) through the projects MULTIMAG (13N9911) and MPSCATT (05K10KTB), the EU's 7th Framework Programme MAGWIRE (FP7-ICT-2009-5 257707), the European Research Council through the Starting Independent Researcher Grant MASPIC (ERC-2007-StG 208162), the Mainz Center for Complex Materials (COMATT), the Center for Innovative and Emerging Materials (CINEMA), the Graduate School of Excellence Materials Science in Mainz (MAINZ GSC266), and the Deutsche Forschungsgemeinschaft (DFG).

References

1. J. Barker and O. A. Tretiakov. Static and Dynamical Properties of Antiferromagnetic Skyrmions in the Presence of Applied Current and Temperature, *Physical Review Letters*, 116: 147203, 2016.
2. A. A. Belavin and A. M. Polyakov. Metastable states of two-dimensional isotropic ferromagnets. *JETP Letters*, 22(10):245, 1975.
3. L. Berger. Exchange interaction between ferromagnetic domain wall and electric current in very thin metallic films. *Journal of Applied Physics*, 55(6):1954–1956, 1984.
4. O. Bostanjoglo and Th. Rosin. Resonance oscillations of magnetic domain walls and Bloch lines observed by stroboscopic electron microscopy. *Physica Status Solidi (a)*, 57(2):561–568, 1980.
5. O. Boulle, G. Malinowski, and M. Kläui. Current-induced domain wall motion in nanoscale ferromagnetic elements. *Materials Science and Engineering: R: Reports*, 72(9):159–187, 2011.
6. O. Boulle, et al. Room-temperature chiral magnetic skyrmions in ultrathin magnetic nanostructures. *Nature Nanotechnology*, 11(5):449–454, 2016.
7. A. Brataas and Kjetil M. D. Hals. Spin-orbit torques in action. *Nature Nanotechnology*, 9(2):86–88, 2014.
8. F. Büttner. *Topological mass of magnetic skyrmions probed by ultrafast dynamic imaging*. Dissertation, University of Mainz, Mainz, 2013.
9. F. Büttner, B. Krüger, S. Eisebitt, and M. Kläui. Accurate calculation of the transverse anisotropy of a magnetic domain wall in perpendicularly magnetized multilayers. *Physical Review B*, 92(5):054408, 2015.
10. F. Büttner, et al. Magnetic states in low-pinning high-anisotropy material nanostructures suitable for dynamic imaging. *Physical Review B*, 87(13):134422, 2013.
11. F. Büttner, et al. Dynamics and inertia of skyrmionic spin structures. *Nature Physics*, 11:225–228, 2015.

Chapter 8

12. G. Chen, et al. Novel chiral magnetic domain wall structure in Fe/Ni/Cu(001) films. *Physical Review Letters*, 110(17):177204, 2013.

13. C. L. Degen. Scanning magnetic field microscope with a diamond single-spin sensor. *Applied Physics Letters*, 92(24):243111, 2008.

14. W. Döring. Über die Trägheit der Wände zwischen Weißchen Bezirken. *Z. Naturforsch. A*, 3:373, 1948.

15. S. Eisebitt, et al. Lensless imaging of magnetic nanostructures by X-ray spectro-holography. *Nature*, 432(7019):885–888, 2004.

16. S. Emori, et al. Spin Hall torque magnetometry of Dzyaloshinskii domain walls. *Physical Review B*, 90(18):184427, 2014.

17. K. Everschor, M. Garst, R. A. Duine, and A. Rosch. Current-induced rotational torques in the skyrmion lattice phase of chiral magnets. *Physical Review B*, 84(6):064401, 2011.

18. A. Fert, V. Cros, and J. Sampaio. Skyrmions on the track. *Nature Nanotechnology*, 8(3):152–156, 2013.

19. P. Fischer, et al. Magnetic imaging with full-field soft X-ray microscopies. *Journal of Electron Spectroscopy and Related Phenomena*, 189:196–205, 2013.

20. J. H. Franken, M. Herps, H. J. M. Swagten, and B. Koopmans. Tunable chiral spin texture in magnetic domain-walls. *Scientific Reports*, 4:5248, 2014.

21. P. M. Haney, et al. Current induced torques and interfacial spin-orbit coupling: Semiclassical modeling. *Physical Review B*, 87(17):174411, 2013.

22. S. Heinze, et al. Spontaneous atomic-scale magnetic skyrmion lattice in two dimensions. *Nature Physics*, 7(9):713–718, 2011.

23. D. L. Huber. Equation of motion of a spin vortex in a two-dimensional planar magnet. *Journal of Applied Physics*, 53(3):1899–1900, 1982.

24. J. Iwasaki, M. Mochizuki, and N. Nagaosa. Current-induced skyrmion dynamics in constricted geometries. *Nature Nanotechnology*, 8(10):742–747, 2013.

25. J. Iwasaki, M. Mochizuki, and N. Nagaosa. Universal current-velocity relation of skyrmion motion in chiral magnets. *Nature Communications*, 4:1463, 2013.

26. W. Jiang, et al. Blowing magnetic skyrmion bubbles. *Science*, 349(6245):283–286, 2015.

27. A. D. Kent and D. C. Worledge. A new spin on magnetic memories. *Nature Nanotechnology*, 10(3):187–191, 2015.

28. I. Kézsmárki, et al. Neel-type skyrmion lattice with confined orientation in the polar magnetic semiconductor GaV_4S_8. *Nature Materials*, 14(11):1116–1122, 2015.

29. A. L. D. Kilcoyne, et al. Interferometer-controlled scanning transmission X-ray microscopes at the advanced light source. *Journal of Synchrotron Radiation*, 10(2):125–136, 2003.

30. N. S. Kiselev, A. N. Bogdanov, R. Schäfer, and U. K. Rößler. Comment on "giant skyrmions stabilized by dipole-dipole interactions in thin ferromagnetic films". *Physical Review Letters*, 107(17):179701, 2011.

31. W. Koshibae, et al. Memory functions of magnetic skyrmions. *Japanese Journal of Applied Physics*, 54(5):053001, 2015.

32. H. Kronmüller and S. Parkin, editors. *Handbook of Magnetism and Advanced Magnetic Materials, volume 3: Novel Techniques for Characterizing and Preparing Samples*. Wiley, 2007.

33. K. S. Lee, et al. Angular dependence of spin-orbit spin-transfer torques. *Physical Review B*, 91(14):144401, 2015.

34. I. Makhfudz, B. Krüger, and O. Tchernyshyov. Inertia and chiral edge modes of a skyrmion magnetic bubble. *Physical Review Letters*, 109(21):217201, 2012.

35. P. Maletinsky, et al. A robust scanning diamond sensor for nanoscale imaging with single nitrogen-vacancy centres. *Nature Nanotechnology*, 7(5):320–324, 2012.
36. A. P. Malozemoff and J. C. Slonczewski. *Magnetic Domain Walls in Bubble Materials*. Academic Press, New York, 1979.
37. O. Meshcheriakova, et al. Large noncollinearity and spin reorientation in the novel Mn_2RhSn Heusler magnet. *Physical Review Letters*, 113(8):087203, 2014.
38. T. Moriya. Anisotropic superexchange interaction and weak ferromagnetism. *Physical Review*, 120(1):91–98, 1960.
39. C. Moutafis, S. Komineas, and J. A. C. Bland. Dynamics and switching processes for magnetic bubbles in nanoelements. *Physical Review B*, 79(22):224429, 2009.
40. S. Mühlbauer, et al. Skyrmion lattice in a chiral magnet. *Science*, 323(5916):915–919, 2009.
41. J. Müller and A. Rosch. Capturing of a magnetic skyrmion with a hole. *Physical Review B*, 91(5):054410, 2015.
42. A. Neubauer, et al. Topological Hall effect in the A phase of MnSi. *Physical Review Letters*, 102(18):186602, 2009.
43. Y. Onose, et al. Observation of magnetic excitations of skyrmion crystal in a helimagnetic insulator Cu_2OSeO_3. *Physical Review Letters*, 109(3):037603, 2012.
44. C. F. Pai. *The spin Hall effect induced spin transfer torque in magnetic heterostructures*. PhD thesis, Cornell University, Ithaca, NY, 2015.
45. N. Papanicolaou and T.N. Tomaras. Dynamics of magnetic vortices. *Nuclear Physics B*, 360(2–3):425–462, 1991.
46. Stuart S. P Parkin, M. Hayashi, and L. Thomas. Magnetic domain-wall racetrack memory. *Science*, 320(5873):190–194, 2008.
47. S. Rohart and A. Thiaville. Skyrmion confinement in ultrathin film nanostructures in the presence of Dzyaloshinskii-Moriya interaction. *Physical Review B*, 88(18):184422, 2013.
48. N. Romming, et al. Writing and deleting single magnetic skyrmions. *Science*, 341(6146):636–639, 2013.
49. A. Rosch. Skyrmions: Moving with the current. *Nature Nanotechnology*, 8(3):160–161, 2013.
50. E. Ruff, et al. Multiferroicity and skyrmions carrying electric polarization in GaV_4S_8. *Science Advances*, 1(10):e1500916, 2015.
51. K. S. Ryu, L. Thomas, S. H. Yang, and S. Parkin. Chiral spin torque at magnetic domain walls. *Nature Nanotechnology*, 8(7):527–533, 2013.
52. J. Sampaio, et al. Nucleation, stability and current-induced motion of isolated magnetic skyrmions in nanostructures. *Nature Nanotechnology*, 8:839–844, 2013.
53. T. Schulz, et al. Emergent electrodynamics of skyrmions in a chiral magnet. *Nature Physics*, 8(4):301–304, 2012.
54. C. Schütte and M. Garst. Magnon-skyrmion scattering in chiral magnets. *Physical Review B*, 90(9):094423, 2014.
55. C. Schütte, J. Iwasaki, A. Rosch, and N. Nagaosa. Inertia, diffusion, and dynamics of a driven skyrmion. *Physical Review B*, 90(17):174434, 2014.
56. G. Schütz, et al. Absorption of circularly polarized x rays in iron. *Physical Review Letters*, 58(7):737–740, 1987.
57. T. Schwarze, et al. Universal helimagnon and skyrmion excitations in metallic, semiconducting and insulating chiral magnets. *Nature Materials*, 14(5):478–483, 2015.
58. S. Seki, X. Z. Yu, S. Ishiwata, and Y. Tokura. Observation of skyrmions in a multiferroic material. *Science*, 336(6078):198–201, 2012.

Chapter 8

59. Ka Shen, G. Vignale, and R. Raimondi. Microscopic Theory of the inverse Edelstein effect. *Physical Review Letters*, 112(9):096601, 2014.

60. J. Sinova, et al. Spin Hall effects. *Reviews of Modern Physics*, 87(4):1213–1260, 2015.

61. T.H.R. Skyrme. A unified field theory of mesons and baryons. *Nuclear Physics*, 31:556–569, 1962.

62. J.C. Slonczewski. Current-driven excitation of magnetic multilayers. *Journal of Magnetism and Magnetic Materials*, 159(1–2):L1–L7, 1996.

63. J. Stöhr and H. C. Siegmann. *Magnetism - From Fundamentals to Nanoscale Dynamics*. Springer-Verlag, Berlin, Heidelberg, 2006.

64. O. Tchernyshyov and G. W. Chern. Fractional vortices and composite domain walls in flat nanomagnets. *Physical Review Letters*, 95(19):197204, 2005.

65. J.-P. Tetienne, et al. Nanoscale imaging and control of domain-wall hopping with a nitrogen-vacancy center microscope. *Science*, 344(6190):1366–1369, 2014.

66. A. Thiaville, Y. Nakatani, J. Miltat, and Y. Suzuki. Micromagnetic understanding of current-driven domain wall motion in patterned nanowires. *EPL (Europhysics Letters)*, 69(6):990, 2005.

67. A. A. Thiele. Steady-state motion of magnetic domains. *Physical Review Letters*, 30(6):230–233, 1973.

68. R. Tomasello, et al. A strategy for the design of skyrmion racetrack memories. *Scientific Reports*, 4:6784, 2014.

69. T. van der Sar, F. Casola, R. Walsworth, and A. Yacoby. Nanometre-scale probing of spin waves using single electron spins. *Nature Communications*, 6:7886, 2015.

70. S. Woo, et al. Observation of room-temperature magnetic skyrmions and their current-driven dynamics in ultrathin metallic ferromagnets. *Nature Materials*, 15:501–506, 2016.

71. S. Woo, et al. Enhanced spin-orbit torques in Pt/Co/Ta heterostructures. *Applied Physics Letters*, 105(21):212404, 2014.

72. X. Z. Yu, et al. Real-space observation of a two-dimensional skyrmion crystal. *Nature*, 465(7300):901–904, 2010.

73. X. Zhang, Y. Zhou, and M. Ezawa. Antiferromagnetic skyrmion: Stability, creation and manipulation. *Scientific Reports*, 6:24795, 2016.

74. X. Zhang, Y. Zhou, and M. Ezawa. Magnetic bilayer-skyrmions without skyrmion Hall effect. *Nature Communications*, 7:10293, 2016.

75. Y. Zhou, et al. Dynamically stabilized magnetic skyrmions. *Nature Communications*, 6:8193, 2015.

76. Y. Zhou and M. Ezawa. A reversible conversion between a skyrmion and a domain-wall pair in a junction geometry. *Nature Communications*, 5:4652, 2014.

9. Skyrmion Dynamics in Chiral Magnets

Shi-Zeng Lin
Los Alamos National Laboratory, Los Alamos, New Mexico

Jiadong Zang
University of New Hampshire, Durham, New Hampshire

Chapter 9

This chapter discusses the theoretical and experimental work on the dynamics of skyrmions in chiral magnets. A skyrmion is a stable particle-like emergent spin texture in magnets in the mesoscale. The experimental observations of skyrmions in B20 single crystal in 2009 have triggered enormous interest. Skyrmions have been found in a variety of systems, such as single crystals including metals, semiconductors, insulators, and artificial multilayer systems. Because of their efficient response to various external stimuli including electric current, magnetic field, thermal gradient, etc., skyrmions are deemed prime candidates for applications in the next generation spintronic devices. For this purpose, controlled creation, removal, and manipulation of skyrmions are crucial, and understanding their dynamics is essential. Recently, much progress has been made in understanding the interaction between conduction electrons and skyrmions in metal, the response of skyrmions to external drive, and the dynamics of skyrmions in insulators in the presence of a thermal gradient. In this chapter, we survey the recent progress in understanding the dynamics of skyrmion in chiral magnets. Vast information derived from these studies is shown to highlight unusual and unique dynamical properties of skyrmions in magnets, to reveal the challenges and opportunities in the research on skyrmion dynamics, and to demonstrate the huge potential for applications of skyrmions.

9.1   Introduction

In 1962, Tony Skyrme proposed a nonlinear field theory for the quantum chromodynamics [80,81]. This field theory supports a stable soliton solution that can be interpreted as baryons, which was the main motivation for Skyrme to construct and study this model. These solitons are called skyrmions in honor of Skyrme's pioneering work. The concept of skyrmion has profound consequences, not only in high-energy physics but also in condensed-matter physics. The skyrmion excitation has been studied in several condensed matter systems, such as crystal liquid [90], quantum Hall system [82], Bose–Einstein condensate [27], and magnets [8,10].

A skyrmion in magnets, or a magnetic skyrmion, is a spin texture with a characteristic size determining by the competing interactions. Basically, a skyrmion is an emergent mesoscale particle-like topological excitation in magnets. Mathematically, it belongs to the homotopy class $\pi_2(S^2)$; that is, the target manifold and base manifold of the fields are both defined on the surface of a sphere, and captured by the integer-valued topological charge defined as:

$$Q = \frac{1}{4\pi} \int d^2 r \mathbf{n} \cdot (\partial_x \mathbf{n} \times \partial_y \mathbf{n}) = \pm 1, \tag{9.1}$$

with a unit vector **n** representing the spin direction. It was found a long time ago that a magnetic skyrmion is a topological excitation in the nonlinear sigma model, but the size of the skyrmion is not determined because there is no characteristic length in the nonlinear sigma model. In 1989, Bogdanov and Yablonskii realized that skyrmions can be stabilized by introducing a term that involves first-order spatial derivative, which yields a characteristic length scale for the system [10]. In fact, this term is the Dzyaloshinskii–Moriya (DM) interaction [21,57,58], which is generally present in systems without inversion symmetry. These magnets without inversion symmetry are called chiral magnets or helimagnets. Under magnetic fields, they found a triangular lattice of skyrmion similar to the vortex lattice in type II superconductors. Later, Rößler et al. showed that skyrmion lattice can exist in chiral magnets as a ground state without external magnetic field when the magnetic moment is soft near the magnetic transition [68].

The field of magnetic skyrmions gained full momentum in 2009 when a triangular lattice of skyrmion was observed in the A phase of MnSi single crystal by the small-angle neutron scatter in Germany [59]. The mysterious A phase was known decades ago from other measurements [36,69], but the spin configuration in the A phase was not understood at that time probably because it is difficult to measure the long wavelength modulation of spin configuration by the neutron scattering. The skyrmion lattice is stabilized in a narrow temperature region from 28 to 30 K and magnetic field of the order to 200 mT in bulk MnSi single crystal. In 2010, the real-space configuration of a skyrmion and skyrmion lattice was observed by Lorentz transmission electron microscopy in thin films of $Fe_{0.5}Co_{0.5}Si$ in Japan [96]. The size of the skyrmion is about 30 nm. Remarkably, the skyrmion phase is stabilized in a wider temperature–magnetic field region in thin films than that in bulk. This has been demonstrated clearly in FeGe by reducing the film thickness [94]. The skyrmion is more stable in thin films because of the spin anisotropy [13,26,94,93] and/or the suppression of the competing conical phase by the boundary effect.

Shortly after the first observation of skyrmion in MnSi, a skyrmion lattice was found in many other crystals without inversion symmetry, such as metals [59,96], semiconductors [94], and even insulators [1,78]. The phase diagram in these materials are similar, indicating that they can be described by the same spin Hamiltonian in the long wavelength limit. From these findings, it seems that the skyrmion is ubiquitous in magnets, and can be regarded as a fundamentally new state of magnetic order. The skyrmion phase can be detected by various experimental measurements, such as transport measurements [20,44,61], elastic

Chapter 9

modulus measurements [62], thermodynamic measurements [4], spin polarized scanning tunneling microscopy [26,66,67], Lorentz transmission electron microscopy [19,96], and small-angle neutron scattering [1,59]. Among these skyrmion-hosting materials, the insulator Cu_2OSeO_3 is particularly interesting because of the magnetoelectric coupling in this compound [77,78]. Another advantage of insulating materials is the low dissipation of power. Recently, the skyrmion lattice is found to be stable even above room temperature in the β-Mn-type Co-Zn-Mn alloys [86]. Besides single crystals, skyrmions can also exist in artificial multilayer systems, where inversion symmetry is broken explicitly at the interface [23]. One example is Fe film deposited on the Ir substrate, where the skyrmion lattice is stabilized in the ground state without external magnetic field [26]. Skyrmions were observed in multilayer systems at room temperature [16,34]. These multilayer systems are extremely appealing for applications of skyrmion because they offer great opportunity to tune the skyrmion properties compared with that in single crystals. The skyrmion lattice can also be realized in magnets with inversion symmetry due to frustrated interaction [25,43,46,64] or dipolar interaction [97].

A skyrmion as a particle responds to various external stimuli, such as electric current [35,73,95], electric field gradient [88,89], thermal gradient [38,47,54], strain [79], magnetic field, and microwaves [55,56,63,65,76,99]. This makes skyrmions extremely promising for applications [23,60]. As skyrmions can be manipulated by electric current, skyrmions are believed to be promising candidates for application in spintronic devices. One novel feature of skyrmions is that they are highly mobile and can be depinned by a low current density [35,73,95]. The threshold current density to make skyrmion mobile in helimagnets is about 10^6 A/m^2, which is lower by 5–6 orders of magnitude in comparison with that for magnetic domain walls; thus, the Joule heating can be dramatically reduced. Meanwhile, the size of skyrmion is about tens of nanometers. For these unique properties, skyrmions are deemed to have huge potential in developing the next generation memory devices. A racetrack memory based on skyrmions was proposed in Ref. [23]. Several interesting proposals to utilize skyrmions in spintronic devices were considered in Refs. [51,100,70,30]. For this purpose, the understanding of skyrmion dynamics in the presence of current drive is crucial.

The first question is how the conduction electrons interact with the skyrmion. As will be discussed in Section 9.3, the spin of the conduction electrons are polarized by the magnetic moment associated with skyrmion due to the local exchange interaction. Conduction electrons then gain a Berry phase produced by the non-coplanar spin texture associated with skyrmion. Such a Berry phase is equivalent to an emergent magnetic field [11,98]. The emergent magnetic field is huge. For a skyrmion of size of 10 nm, the induced effective magnetic field is about 100 T. At the same time, the magnetic moment of the conduction electrons is transferred into the skyrmion and drive the skyrmion to move [5,45,84]. The skyrmion modifies the electric conductivity of the system in two different ways. First, the emergent magnetic field produced by the skyrmion affects the transport of conduction electrons and produces the topological Hall contribution to the conductivity [20,44,61]. Meanwhile, when skyrmions move, they also generate an emergent electric field and hence yield the so-called skyrmion flow conductivity because of the time-dependent emergent magnetic field, similar to Faraday's Law in electromagnetism [73].

The next question is how skyrmions move in the presence of a driving current. To answer this question, we need to quantify the rigidity of the skyrmion as a particle. A skyrmion is an extended object involving huge numbers of degrees of freedom. They can deform in response to external perturbations. The deformations are characterized by the internal modes of a skyrmion [48,74], which will be discussed in Section 9.4. In the high field region, the internal modes associated with the deformation of skyrmion are gapped and well separated from the Goldstone mode associated with the translational motion of the skyrmion. This allows one to treat the skyrmion as a rigid particle by neglecting the deformations, and to derive the equation of motion for skyrmions as particles using the Thiele collective coordinate approach [85]. The dynamics is non-Newtonian because the dynamics are governed by a Magnus force which is transverse to the velocity of skyrmions. On the basis of the derived equation of motion, we can understand why pinning for skyrmion is weak.

From the application's perspective, it is urgent to design the skyrmion creation and annihilation in a controlled way by an electric current. To this end, we study the nonlinear dynamics of chiral magnets with a current, which is the main topic in Section 9.5. In the ferromagnetic state, the electric current produces a Doppler shift in the magnon spectrum and destabilize the ferromagnetic state above a threshold current [49]. Then skyrmions are created and move as a result of the current. Meanwhile, the motion of skyrmion radiates magnon and generates an additional damping for the skyrmion. At high currents, skyrmions are no longer stable and the system evolves into a chiral liquid phase with strong fluctuation in skyrmion topological charge. This indicates an upper limit for the skyrmion velocity. We will construct the dynamical phase diagram of the chiral magnets under current drive.

Chapter 9

For skyrmions in insulators, the question of how to drive skyrmion naturally arises because there is no electric current in insulators. In Section 9.6, we will show that skyrmion can be driven by any magnon current induced by a thermal gradient [38,47]. The magnon current carries magnetic moment, which drives the skyrmion via magnonic spin transfer torque. This results in a very counterintuitive motion for skyrmion: it moves from the cold region to the hot region. In the presence of magnetoelectric coupling, the motion of skyrmion induces an oscillation in the electric polarization. If one embeds the insulator in a closed circuit, there is oscillation of electric current induced by the skyrmion motion.

The remainder of the chapter is organized as follows. In Section. 9.2, we will present a phenomenological Hamiltonian to describe the skyrmion physics. We will present a single skyrmion solution, the interaction between two skyrmions, and the interaction between a defect and a skyrmion. We then discuss the emergent electromagnetic fields for the conduction electrons generated by the skyrmion spin texture in Section 9.3. In Section 9.4, we will discuss the internal modes associated with a skyrmion. We proceed to derive the equation of motion for a skyrmion and discuss its implication for the pinning of skyrmions by defects in Section 9.5. In Section 9.5, we will study the dynamical creation and destruction of skyrmion by an electric current. In Section 9.6, we will discuss the motion of skyrmion driven by a thermal gradient and the induced electric polarization. Finally, the chapter is concluded by a brief discussion in Section 9.7.

9.2　Model

The description of skyrmion physics using a microscopic Hamiltonian approach seems to be a difficult task for the following reasons. First, the size of skyrmion is in the mesoscale, which poses a grand challenge for the first principle calculations using the atomic orbitals. Second, the magnetic moments that stabilize the skyrmion spin texture have an itinerant nature, and therefore it is questionable to apply the localized moment picture. Third, the magnetic moments in the skyrmion hosting materials are provided by transition metal ions where strong correlation is deemed to be important. Alternatively, the phenomenological free energy function is useful to describe skyrmion physics in a broad range of materials. This is because the length scale involved in skyrmion physics is usually much longer than the lattice parameter of the crystal. This allows one to consider the long wavelength limit of the corresponding microscopic Hamiltonian. In the long wavelength theory, only several terms are allowed by the symmetry. For the B20 crystal which has cubic structure, the effective low energy theory for magnetic moments was proposed by Bak and Jensen [2]. The free energy density in terms of the spin density $\mathbf{S(r)}$ has the following form:

$$\mathcal{F}(\mathbf{r}) = \frac{A_1}{2} \sum_{\mu} S_{\mu}^2 + D\mathbf{S} \cdot (\nabla \times S) + \frac{B_1}{2} \sum_{\mu} \left[(\nabla S_{\mu})^2 \right]$$

$$+ \frac{B_2}{2} \sum_{\mu} (\partial_{\mu} S_{\mu})^2 + C \left(\sum_{\mu} S_{\mu}^2 \right)^2 + A_2 \sum_{\mu} S_{\mu}^4 - \mathbf{H}_a \cdot \mathbf{S},$$

(9.2)

where $\mu = x, y, z$. Here, the dipolar interaction between moments are neglected because it is much weaker than the other energy scales. The B20 crystal does not have inversion symmetry; therefore, the DM interaction is allowed and is described by the term with a coefficient D. Here, we have assumed that the DM vector is along the bond direction, which is realized in the B20 crystal, if we discretize Equation 9.2 into square grids. The terms with coefficients A_1 and C are the standard Ginzburg–Landau expansion of the magnitude of the moments. The terms with coefficients B_2 and A_2 describe the anisotropy allowed by the cubic symmetry: the B_2 term describes the compass-like interaction between moments which determines the direction of the ordering wave vector, and the A_2 term is a single-ion anisotropy. Strictly speaking, Equation 9.2 is valid only close to the transition point where the moment $\mathbf{S}$ is small and the expansion up to the quartic order in $\mathbf{S}$ is sufficient. Here, we assume that Equation 9.2 is still valid in the low-temperature region and neglect the anisotropic terms, which are not essential for stabilization of the skyrmion lattice. Using a unit vector $\mathbf{n} = \mathbf{S}/|S|$ representing the direction of spin, we introduce a phenomenological Hamiltonian:

$$\mathcal{H} = \int d\mathbf{r}^3 \left[\frac{J_{\text{ex}}}{2} (\nabla \mathbf{n})^2 + D\mathbf{n} \cdot \nabla \times \mathbf{n} - \mathbf{H}_a \cdot \mathbf{n} \right],$$

(9.3)

to describe the skyrmion physics in chiral magnets. The competition between the DM interaction and exchange coupling gives rise to a characteristic length $\lambda = J_{\text{ex}}/D$, which is much bigger than the lattice constant a of the crystal. The DM interaction, here is crucial for stabilizing the skyrmion solution because Equation 9.3 does not have a characteristic length scale without the DM interaction, and according Derrick's theorem [18], there is no stable topological solution when $D = 0$.

Chapter 9

For thin films of chiral magnets, a uniaxial anisotropy is also allowed by symmetry and the corresponding Hamiltonian can be written as:

$$\mathcal{H} = d \int d\mathbf{r}^2 \left[\frac{J_{ex}}{2}(\nabla \mathbf{n})^2 + D\mathbf{n} \cdot \nabla \times \mathbf{n} - \mathbf{H}_a \cdot \mathbf{n} - \frac{A_z}{2} n_z^2 \right], \tag{9.4}$$

with $A_z > 0$ being an easy-axis anisotropy and $A_z < 0$ being an easy-plane anisotropy. Here, d is the thickness of the film and we have assumed that the spin texture is uniform in the direction perpendicular to the film. For convenience of simulations, one may discretize Equation 9.4 by defining spins on a square lattice with a lattice constant a:

$$\mathcal{H} = -\frac{d}{a} \sum_{\langle i,j \rangle} \left[J_{ex}\mathbf{n}_i \cdot \mathbf{n}_j + \mathbf{D}_{ij} \cdot (\mathbf{n}_i \times \mathbf{n}_j) \right] - \frac{d}{a} \sum_i \mathbf{H}_a \cdot \mathbf{n}_i - \frac{A_z d}{2a} \sum_i n_{i,z}^2. \tag{9.5}$$

The dynamics of spin is governed by the Landau–Lifshitz–Gilbert (LLG) equation [5,45,84]:

$$\partial_t \mathbf{n} = \frac{\hbar\gamma}{2e}(\mathbf{J} \cdot \nabla)\mathbf{n} - \gamma\mathbf{n} \times \mathbf{H}_{eff} + \alpha\partial_t \mathbf{n} \times \mathbf{n}, \tag{9.6}$$

with the effective magnetic field $\mathbf{H}_{eff} \equiv -\delta\mathcal{H}/\delta\mathbf{n}$. Here, α is the phenomenological Gilbert damping term and $\mathbf{J}$ is the electric current accounting for the adiabatic spin transfer torque. In metallic chiral magnets, the motion of skyrmions generates electric fields and causes dissipations in conduction electrons, which give an additional contribution to α. The effective theory in Equations 9.3, 9.4, and 9.6 has successfully captured the main features of the skyrmion physics, such as the phase diagram, collective excitation, and dynamics of skyrmion in experiments.

At zero temperature, $T = 0$, in two dimensions (2D), the Hamiltonian equation (Equation 9.3) stabilizes a spiral phase, triangular lattice of skyrmion, and ferromagnetic state upon increasing the magnetic field perpendicular to the plane. At $H_a = 0$, one spiral solution can be found exactly with $\mathbf{n} = (0, \cos(qx), \sin(qx))$ with a wave vector $q = D/J_{ex}$ along the x-direction. The direction of $\mathbf{q}$ is not fixed for the Hamiltonian in Equation 9.3 because it is isotropic in space. The compass-like interaction in Equation 9.2 can fix the direction of $\mathbf{q}$. Numerical calculations show that the spiral phase is stable for $H_a < 0.2\, D^2/J_{ex}$ [8]. A triangular lattice of skyrmion is stabilized in an intermediate magnetic field $0.2\, D^2/J_{ex} < H_a < 0.8\, D^2/J_{ex}$, and the ferromagnetic state is stabilized at high fields. The transition between these phases is of first order. A typical spin configuration in the skyrmion lattices are shown in Figure 9.1. When the magnetic field is tilted away from the normal of the plane, skyrmions are noncircular and the interaction between skyrmions is anisotropic [52].

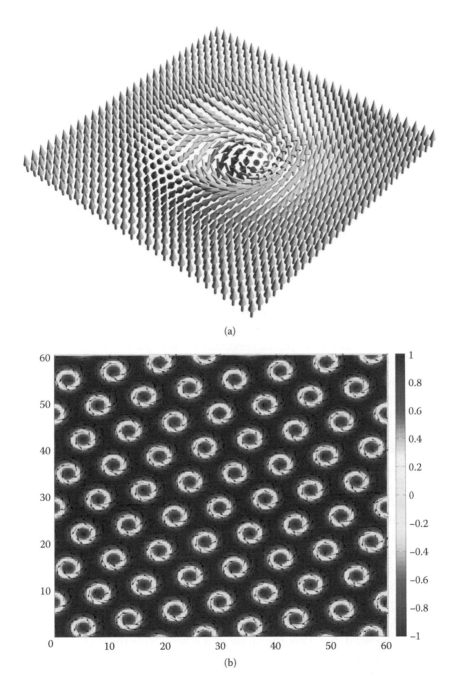

(a)

(b)

FIGURE 9.1 One typical spin configuration for (a) a skyrmion in the ferromagnetic background and (b) in the skyrmion lattice phase. The vectors denote the n_x and n_y components and the n_z component is represented by the color scale on the right.

Chapter 9

In three dimensions (3D), the skyrmion spin texture is uniform along the field direction and the skyrmions are string-like objects with a finite elastic energy. The skyrmion line lattice is stabilized only in a small temperature region close to the transition temperature in 3D [12]. The conical phase with the modulation wavevector along the field direction occupies the majority of the phase diagram. The spiral phase is stabilized in the low field region. The reason why the skyrmion line lattice is only stabilized near the transition temperature is as follows. The triangular skyrmion lattice can be regarded as a superposition of three magnetic spirals with the wavevector $\mathbf{q}_i$ being in the same plane but differing from each other by $2\pi/3$ rotation along the z-axis. The superposition of triple $\mathbf{q}$ spirals does not conserve the magnitude of the moments $|\mathbf{n}| \neq 1$. At low T, there must be induced modulation of moments with high order q's in order to make $|\mathbf{n}| = 1$, which costs energy in the DM and exchange interaction. While for the conical phase, the modulation of moments in the field direction reduces the DM and exchange interaction, and the canted moments along the field direction are favored by the Zeeman interaction. Therefore at low T, the conical phase has lower energy than the skyrmion lattice. Near the transition temperature, the moments are soft, $|\mathbf{n}| \neq 1$. In this case, the skyrmion line lattice has lower energy than that for the conical phase.

Because the transition between the skyrmion lattice phase and the ferromagnetic state is of the first order, a single skyrmion can coexist with the ferromagnetic state as a metastable state. Such a metastable state perhaps is also relevant for applications because manipulation on a single skyrmion is required for memory devices based on skyrmions. Below we construct a single skyrmion solution for the Hamiltonian Equation 9.3 in 2D with a perpendicular magnetic field. The single skyrmion solution was first found in Ref. [8]. We introduce the sphere coordinate for the spin unit vector $\mathbf{n} = (\sin\theta\cos\varphi, \sin\theta\sin\varphi, \cos\theta)$. The problem is centrosymmetric, which allows us to simplify the problem by introducing the polar coordinate in 2D (ρ, φ). In this case, θ depends on ρ only, $\theta(\rho)$, and φ depends on φ, $\varphi(\varphi)$. The DM interaction in Equation 9.3 favors a skyrmion solution with $\varphi = \varphi + \pi/2$. The phase shift $\pi/2$ depends on the direction of the DM vector. In Equation 9.3, the DM vector is along the bond direction for the corresponding discrete square spin lattice. For the DM vector perpendicular to the bond direction, the phase shift is zero. Note that the topological charge of skyrmion defined below does not depend on the phase shift.

Substituting the ansatz into Equation 9.3, we can obtain an equation for $\theta(\rho)$ by minimizing the energy:

$$-\rho\partial_\rho^2\theta - \partial_\rho\theta + \cos(2\theta) + \frac{\sin(2\theta)}{2\rho} + \frac{\beta}{2}\rho\sin(\theta) - 1 = 0, \qquad (9.7)$$

with the boundary condition $\theta(\rho = 0) = \pi$ and $\theta(\rho \to +\infty) = 0$. Here, $\beta = 2H_a J_{ex}/D^2$ and we have renormalized the distance ρ as $\rho \to \rho/(J_{ex}/D)$. Generally, Equation 9.7 cannot be solved exactly. The asymptotic behavior for $\theta(\rho)$ in the limit $\rho \to \infty$ and $\rho \to 0$ can be obtained. In the limit $\rho \to 0$, $\theta(\rho) = \pi - C_1\rho$ and $\rho \to \infty$, $\theta \sim K_0(\rho/\xi)$, with a healing length $\xi = \sqrt{2/\beta}$ and K_0 the modified Bessel functions. Here, $\theta(\rho)$ approaches 0 (the ferromagnetic background) exponentially because the magnetic field opens a gap in the magnon spectrum. For $\theta(\rho)$ at an arbitrary distance, we need a numerical solution, and the results for different β are presented in Figure 9.2. There is only one length scale involved in the single skyrmion solution, and the definition of the skyrmion core is arbitrary. One may choose the definition that the core region of skyrmion is $\theta(\rho < R_c) < \pi/2$.

We proceed to discuss the pairwise interaction between two skyrmions. The interaction is induced by the overlapping of spin texture associated with two skyrmions. When two skyrmions are well separated, the interaction is mediated by an exchange of massive magnons. In this linear region, the skyrmion interaction is pairwise. The interaction decays exponentially with a separation r_d and can be

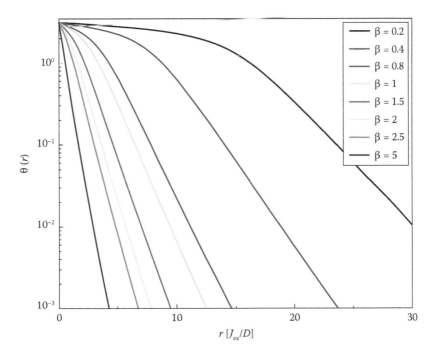

FIGURE 9.2 Result of $\theta(r)$ obtained from a numerical solution of Equation 9.7 for different values of the magnetic field β.

Chapter 9

described by $F_{ss} \sim K_1(r_d/\xi)$. When the separation between skyrmions is comparable with the size of skyrmion, the interaction is induced by the overlapping of the nonlinear cores. In this region, the interaction is nonlinear, and in principle, there is many-body interaction among skyrmions. To obtain the interaction between two skyrmions at an arbitrary distance, we calculate the energy of the system by fixing two skyrmions at a separation r_d by freezing the spins within a radius $r \leq J_{ex}/D$. The interaction energy as a function of r_d is shown in Figure 9.3. The interaction between two skyrmions are repulsive.

We next study the interaction between skyrmions and defects in inhomogeneous systems. We assume an inhomogeneous electron density, which gives rise to an inhomogeneous exchange interaction J_{ex} produced by the double-exchange mechanism. The defect is modeled by the following profile of J_{ex}:

$$J_{ex}(r) = J_0\left(1 + \sum_i J_d \exp\left[-|\mathbf{r} - \mathbf{r}_{d,i}|/\xi_d\right]\right), \tag{9.8}$$

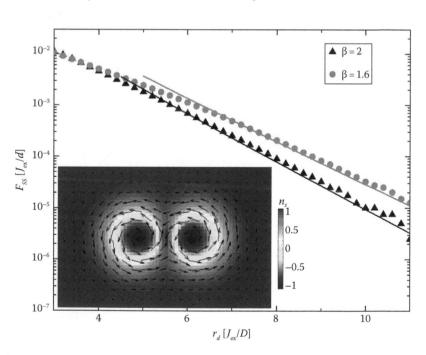

FIGURE 9.3 Interaction force between two skyrmions as a function of the separation r_d at two different magnetic fields. Lines are fitting to the expression $K_1(r_d/\xi)$, with $\xi = D/\sqrt{H_a J_{ex}}$, and symbols are obtained from a numerical solution of Equations 9.4 and 9.6. inset shows the stationary configuration of two skyrmions at $r_d = 3.6 J_{ex}/D$.

where ξ_d is the size of the defect and is comparable to the inter atomic separation. Here, $\mathbf{r}_{d,i}$ is the center of pinning potential and J_d characterizes the strength of the defects. Generally, skyrmions deform in order to adapt to the pinning potential. Such a deformation is weak for a weak pinning, and one can still use the rigid skyrmion approximation. In this case, the pinning energy is the self-energy of skyrmion by using $J_{ex}(r)$ in Equation 9.8. We employed two methods to calculate the pinning energy of a skyrmion. First, we obtain a stationary skyrmion structure from Equation 9.8 by using a uniform distribution of $J_{ex}(r)$, i.e., $J_d = 0$. Then, we calculate the self-energy of a skyrmion using $J_{ex}(r)$ in Equation 9.8. We also performed self-consistent calculation using $J_{ex}(r)$ in Equation 9.8 by holding the spin at the center of the skyrmion unchanged in order to pin the skyrmion at a desired position. Both results are in good agreement with each other (see Figure 9.4). The pinning for a large separation can be fitted by $F_d \sim J_d \exp(-r_d/\xi_d)$. The force per unit length is proportional to J_d, and it is repulsive for $J_d > 0$ and attractive for $J_d < 0$. The pinning force decreases exponentially with the decay length determined by ξ_d. The nonuniform region of

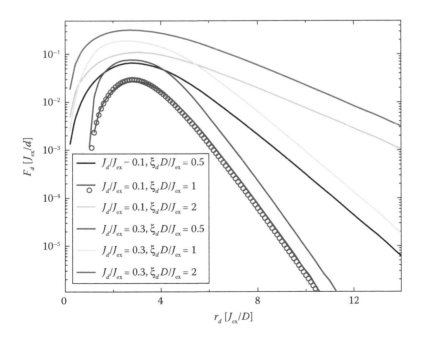

FIGURE 9.4 Interaction force between a skyrmion and a defect for different profiles of J_{ex}. Symbols are results obtained by assuming a rigid skyrmion structure, and lines are results obtained from full numerical calculations of Equations 9.4 and 9.6.

Chapter 9

electron density in typical solid is of the order of the inter-atomic size, $\xi_d \sim 0.1$ nm, which is much smaller than the typical size of skyrmion. The huge size mismatch between skyrmions and defects leads to a weak pinning of skyrmion by the defects.

9.3   Emergent Electromagnetism

An important issue of the skyrmion physics is to explore its dynamical behavior under various external stimuli, such as the electric current or temperature gradient. A good understanding of its dynamical behavior is crucial for the future skyrmion-based applications. From the physical point of view, we would like to see how the nontrivial real-space topology of the skyrmion brings fascinating phenomena associated with the topological nature of a skyrmion [98].

Let us first investigate the current-driven skyrmion motion. In order to do that, a full understanding of the interaction between conduction electrons and skyrmions becomes necessary. These electrons couple with the local magnetic moments by Hund's coupling:

$$S_{Hund} = \int dt d^3x J_H S \psi^\dagger \sigma \cdot \mathbf{n} \psi, \tag{9.9}$$

where ψ is the electron operator, and J_H is the coupling strength. The action of the full system is therefore given by:

$$S = S_{el} + S_{spin} + S_{Hund}, \tag{9.10}$$

where the free electron action is:

$$S_{el} = \int dt d^3x \left[i\hbar \psi^\dagger \dot{\psi} + \frac{\hbar^2}{2m} \psi^\dagger \nabla^2 \psi \right], \tag{9.11}$$

and the spin action is:

$$S_{spin} = \int dt d^3x (\mathcal{L}_B - \mathcal{H}), \tag{9.12}$$

with the Lagrangian for the Berry phase of spin:

$$\mathcal{L}_B = \frac{\hbar S}{a^3} (\cos\theta - 1)\dot{\varphi}. \tag{9.13}$$

Here m is the electron mass, S is the magnitude of the momentum and a is the lattice constant. Once J_H is large, one can apply the adiabatic approximation in which the electron spins align perfectly with the local moment. Then, ψ is projected into the state of the local magnetization by:

$$\psi = \chi|\mathbf{n}\rangle, \tag{9.14}$$

with $\sigma \cdot \mathbf{n}|\mathbf{n}\rangle = |\mathbf{n}\rangle$ and χ the orbital wave function. After this transformation, Hund's rule of coupling becomes $J_H S \chi^\dagger \chi$, which is simply a constant, while the action for the orbital wave function becomes:

$$S_{el} = \int dt d^3x \left[\chi^\dagger \left(i\hbar \frac{\partial}{\partial t} + e A_0 \right) \chi - \frac{1}{2m} \chi^\dagger \left(-i\hbar \nabla - \frac{e}{c} \mathbf{A} \right)^2 \chi \right], \tag{9.15}$$

where the emergent gauge potentials, A_0 and $\mathbf{A}$, are given by:

$$A_0 = -i\frac{\hbar}{2e}\langle \mathbf{n}|\dot{\mathbf{n}}\rangle, \tag{9.16}$$

$$\mathbf{A} = i\frac{\hbar}{2e}\langle \mathbf{n}|\nabla\mathbf{n}\rangle, \tag{9.17}$$

By choosing the spin coherent state:

$$|\mathbf{n}\rangle = \begin{pmatrix} \cos\dfrac{\theta}{2} \\ \sin\dfrac{\theta}{2} e^{i\varphi} \end{pmatrix}, \tag{9.18}$$

the emergent gauge fields have the following explicit form:

$$A_\mu = \frac{\hbar}{2e}(1 - \cos\theta)\partial_\mu \varphi. \tag{9.19}$$

Equation 9.15 shows explicitly that in the presence of a spatial/temporal nonuniform spin texture, the conduction electrons feel a fictitious electromagnetic field generated by a non-coplanar spin texture. The corresponding electric and magnetic fields are, respectively:

$$E_\mu = \frac{\hbar}{2e}\mathbf{n} \cdot (\partial_\mu \mathbf{n} \times \dot{\mathbf{n}}), \tag{9.20}$$

$$B_z = \frac{\hbar c}{2e}\mathbf{n} \cdot (\partial_x \mathbf{n} \times \partial_y \mathbf{n}). \tag{9.21}$$

Chapter 9

These fields are truly acting on the electrons and modifying the electron dynamics. The emergent magnetic field is of particular interest, which is proportional to the topological charge density of skyrmion:

$$q = \frac{1}{4\pi} \mathbf{n} \cdot \left(\partial_x \mathbf{n} \times \partial_y \mathbf{n} \right). \tag{9.22}$$

For a single skyrmion, the topological charge:

$$Q = \int q d^2 x = \pm 1, \tag{9.23}$$

thus implies the flux quantization of the fictitious magnetic field:

$$\Phi_0 = \int d^2 x \, B_z = \frac{hc}{e}. \tag{9.24}$$

For a skyrmion with nanometer size, the average fictitious magnetic field $\langle B \rangle = \Phi_0/\pi R^2$ can thus be elevated up to 1000 T, exceeding the highest magnetic field produced by any lab in the world. Here, R is the skyrmion radius. Such a high field would thereby generate a significant Hall signal, called the topological Hall effect due to its intimacy to skyrmion's nontrivial topology. Analogous with the classical Hall effect, the transverse electric field is $\mathbf{E} = \mathbf{v} \times \mathbf{B}/c$, and the Hall conductivity is:

$$\sigma_{xy}^{Top} / \sigma_{xx} \approx e \langle B \rangle \tau/m, \tag{9.25}$$

where τ is the relaxation time of the conduction electrons. This topological Hall effect has been extensively observed in B20 compounds like MnSi [61] and FeGe [28].

Assuming the magnetic moments vary very slowly at distances of the order of lattice constant a or k_F^{-1}, we can expand the electron action, Equation 9.15, in terms of the gauge field $\mathbf{A}$, which leads to minimal coupling between the vector potential and electric current:

$$S_{int} = -\int dt d^3 x \mathbf{J} \cdot \mathbf{A}, \tag{9.26}$$

where,

$$\mathbf{J} = \frac{i\hbar e}{m} \left[\chi \nabla \chi - (\nabla \chi^\dagger) \chi \right]$$

is the electric current. In the presence of a constant current $\mathbf{J}$, S_{int} is a function of the magnetic moment only. Adding this action to the spin action in Equation 9.12, one can thus derive by variation the Landau–Lifshitz equation for magnetization dynamics, Equation 9.6, except for the damping term.

However, for the time being, we are interested in the motion of a single skyrmion, say, as a whole. To this end, we use the collective coordinate approach. The spatial and time dependence of the magnetic moments $\mathbf{n}(\mathbf{r},t)$ is given by:

$$\mathbf{n}(\mathbf{r}, t) = \mathbf{n}_s(\mathbf{r} - \mathbf{u}(t)), \tag{9.27}$$

where $\mathbf{n}_s(\mathbf{r})$ is the skyrmion configuration in static, and the collective coordinate $\mathbf{u}$ describes the skyrmion position. Inserting it into S_{int}, we get:

$$
\begin{aligned}
S_{\text{int}} &= -\int dt d^3x J_i A_i \\
&= -\int dt d^3x J_i \frac{\partial A_i}{\partial \mathbf{n}} \frac{\partial \mathbf{n}}{\partial \mathbf{r}}(-\mathbf{u}) \\
&= \frac{\hbar}{e} \int dt d^3x J_i u_k \left(\mathbf{n}_0 \times \partial_i \mathbf{n}_0\right) \cdot \partial_k \mathbf{n}_0 \\
&= \frac{\hbar d}{e} \int dt d^2x J_i u_k \left(\mathbf{n}_0 \times \partial_i \mathbf{n}_0\right) \cdot \partial_k \mathbf{n}_0.
\end{aligned}
\tag{9.28}
$$

In the last equation, d is the sample thickness, and $\int d^3x$ is replaced by $d \int d^2x$, utilizing the fact that skyrmions along the field direction are simply stacked. In this hydrodynamic approach, the skyrmion position $\mathbf{u}$ varies much slower than the microscopic variable $\mathbf{n}$. Therefore, u_k can be approximated by $\int \frac{d^2x}{R^2} u_k$, where R is the skyrmion size. In the presence of a steady current $\mathbf{j}$, we have:

$$
\begin{aligned}
S_{\text{int}} &= \frac{d\hbar}{e} \int \frac{dt d^2x}{\xi^2} J_i u_k \int d^2x (\mathbf{n}_0 \times \partial_i \mathbf{n}_0) \cdot \partial_k \mathbf{n}_0 \\
&= \frac{d\hbar Q \varepsilon_{ik}}{e} \int \frac{dt d^2x}{\xi^2} J_i u_k \\
&= d \frac{\hbar Q}{e} \int \frac{dt d^2x}{\xi^2} \left(J_x u_y - J_y u_x\right).
\end{aligned}
\tag{9.29}
$$

Here, the definition of topological charge is used. Similarly, the free spin action is given by:

$$S_{\text{spin}} = \int dt d^3 x \mathcal{L}_B$$

$$= -\frac{eS}{a^3} \int dt d^3 x A_0$$

$$= -\frac{\hbar S}{a^3} \int dt d^3 x u_k (\mathbf{n}_0 \times \dot{\mathbf{n}}_0) \cdot \partial_k \mathbf{n}_0 \qquad (9.30)$$

$$= \frac{\hbar S}{a^3} \int dt d^3 x u_k \left(\mathbf{n}_0 \times \partial_j \mathbf{n}_0 \dot{u}_j\right) \cdot \partial_k \mathbf{n}_0$$

$$= d \frac{\hbar S}{a^3} \int \frac{dt d^2 x}{\xi^2} (\dot{u}_x u_y - \dot{u}_y u_x). \qquad (9.31)$$

Taken into account the spins of conduction electrons, the total spin S is replaced by $S + x/2$, the total spin average per lattice site, where x is the filling of conduction band. Denoting $\gamma = \dfrac{a^3}{\hbar(S+x/2)}$ and combining Equations 9.29 and 9.31 together and performing a variation with respect to $\mathbf{u}$, one can get the skyrmion's equation of motion:

$$\dot{\mathbf{u}} = -\frac{\hbar\gamma}{2e}\mathbf{J}. \qquad (9.32)$$

It shows a longitudinal motion of the skyrmion under an electric current. This current-driven skyrmion motion is the physical basis of future devices such as skyrmion racetrack memory.

The motion of skyrmion also produces an emergent electric field:

$$\mathbf{E} = -\frac{\hbar}{2e}\mathbf{n}_0 \cdot (\nabla \mathbf{n}_0 \times \partial_j \mathbf{n}_0)\dot{u}_j \qquad (9.33)$$

$$= -\mathbf{V}_{\|} \times \mathbf{B},$$

where $\mathbf{V}_{\|} = \dot{\mathbf{u}}$ is the skyrmion velocity. As explained above, a fictitious magnetic field is induced by the skyrmion, and it moves once skyrmion moves. The emergent electromagnetic fields follow the Faraday Law; that is, a moving magnetic field generates an electric field normal to its velocity. This field also generates the Hall motion of electrons, whose magnitude is given by

$$\Delta\sigma_{xy}/\sigma_{xx} = -\frac{x}{2S+x}\frac{e\langle B\rangle\tau}{m}. \qquad (9.34)$$

Notice the minus sign in Equation 9.34 in comparison to the topological Hall conductivity in Equation 9.25, which comes from the minus sign of electron charge. This new Hall effect differs by the factor of $-\dfrac{x}{2S + x}$ from the topological Hall effect. Its physical origin can be easily understood by noting that the total force acting on a single conduction electron is $\mathbf{F} = -e[(\mathbf{v} - \mathbf{V}_{\parallel}) \times \mathbf{B}]/c$; that is, the Lorentz force on electrons due to the internal magnetic field generated by the skyrmion lattice depends on the relative velocity of electrons and skyrmions. When the skyrmion lattice begins to slide above the threshold electric current J_c [3], the net topological Hall voltage will be suddenly reduced by the factor of $\dfrac{2S}{2S + x}$. In this way, the effect of the spin-motive force and the collective shift of skyrmions can be identified experimentally.

9.4 Internal Modes and Equation of Motion

A skyrmion as a topological excitation in spin systems is an extended object, which involves huge numbers of degrees of freedom. Therefore, a skyrmion can deform in the presence of external perturbations. These deformations with characteristic frequencies below the magnon gap are the internal modes of a skyrmion at rest. These internal modes determine the stability of a skyrmion and also tell the rigidity of a stable skyrmion. Mathematically, the internal modes are eigen vectors of a small perturbation around a stationary skyrmion. In this section, we will discuss the internal modes of a static skyrmion in the ferromagnetic background in 2D.

First let us consider the internal modes in a simple model as a reference. A soliton in a one-dimensional φ^4 model is defined by the following equation of motion:

$$\left(\partial_t^2 - \partial_x^2\right)\varphi + \frac{\partial V}{\partial \varphi} = 0, \tag{9.35}$$

with the potential $V = (\varphi^2 - 1)^2/8$. The potential V is minimized at $\varphi_m = \pm 1$; thus, it supports a soliton or kink solution running from $\varphi = -1$ at $x = -\infty$ to $\varphi = 1$ at $x = +\infty$ or vice versa. One of such kink solutions can be found exactly:

$$\varphi_K(x) = \tanh[(x - x_0)/2], \tag{9.36}$$

with x_0 being the center of the kink. Equation 9.35 is translationally invariant; thus, the system energy does not depend on x_0. This indicates the existence of a gapless mode associated with the translational motion of the kink. A small displacement of the field with $x_0 \ll 1$ can be expressed as:

$$\varphi_K(x) = \tanh\left(\frac{x}{2}\right) - \frac{x_0}{2}\mathrm{sech}^2\left(\frac{x}{2}\right). \tag{9.37}$$

We find the internal mode corresponding to the translational motion of the kink $\tilde{\varphi}_0 = \mathrm{sech}^2(x/2)$. To find other internal modes, we expand φ around the static kink solution $\varphi(x, t) = \varphi_K(x, t) + \tilde{\varphi}(x, t)$, and linearize Equation 9.35 with respect to $\tilde{\varphi}(x, t) = \int d\omega \exp(i\omega t)\, \tilde{\varphi}(x, \omega)$

$$-\partial_x^2 \tilde{\varphi} + \frac{\partial^2 V(\varphi_K)}{\partial \varphi^2}\tilde{\varphi} = \omega^2 \tilde{\varphi}. \tag{9.38}$$

Equation 9.38 is the Schrödinger equation describing a particle moving in the potential created by the kink $\partial^2 V(\varphi_K)/\partial\varphi^2$. The presence of a static kink breaks the translational invariance for $\tilde{\varphi}$ in Equation 9.38; therefore, the momentum is no longer a good quantum number. It is easy to verify that the translational mode $\tilde{\varphi}_0 = \mathrm{sech}^2(x/2)$ is a solution to Equation 9.38 with eigen frequency $\omega = 0$. The other internal mode with frequency below the plasma frequency $\omega_0 = 1$ is $\tilde{\varphi}_1 = \tanh(x/2)\mathrm{sech}(x/2)$ with the eigen frequency $\omega_1 = \sqrt{3}/2$.

The discussions of the internal modes in φ^4 model illustrate the basic properties of the internal modes, which are also shared by those in a skyrmion. The presence of a skyrmion in the ferromagnetic state introduces an centrosymmetric potential for the magnons, as sketched in Figure 9.5. Therefore, we can introduce an angular momentum as a

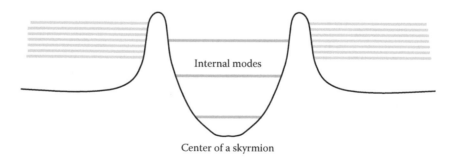

Internal modes

Center of a skyrmion

FIGURE 9.5 Schematic view of the effective magnon potential produced by the skyrmion. The potential is centrosymmetric and is drawn along the radial direction of the skyrmion.

quantum number for the magnon modes. This potential stabilizes several magnon modes that are localized inside the skyrmion, and these localized modes are the internal modes of a skyrmion. However, as the magnon energy increases, the magnons are no longer confined inside the skyrmion. These magnon modes extend into the ferromagnetic state and merge with the magnon continuum of the ferromagnetic state.

To describe the internal modes, we adopt the formulations in Ref. [74] when damping is absent. First, we need to find a stationary skyrmion in the ferromagnetic state, which can be obtained by solving Equation 9.7. The spin can be represented as $\mathbf{n}_s(\mathbf{r}) = (\sin\theta_s\cos\varphi_s, \sin\theta_s\sin\varphi_s, \cos\theta_s)$. The internal modes corresponding to the translational motion of skyrmion can be found immediately in analogy to the soliton case, $\tilde{\mathbf{n}}_0 = (\mathbf{l} \cdot \nabla)\mathbf{n}_s$ with eigen frequency $\omega = 0$ for the Hamiltonian with translational invariance. Here $\mathbf{l}$ is a unit vector representing the direction of skyrmion motion. We then introduce a local coordinate system evolving with the skyrmion spin texture, and with the three unit axes defined as:

$$\hat{\mathbf{x}}'(r) = (-\sin\varphi_s, \cos\varphi_s, 0), \tag{9.39}$$

$$\hat{\mathbf{y}}'(r) = (-\cos\theta_s\cos\varphi_s, -\cos\theta_s\sin\varphi_s, \sin\theta_s), \tag{9.40}$$

$$\hat{\mathbf{z}}'(r) = (\sin\theta_s\cos\varphi_s, \sin\theta_s\sin\varphi_s, \cos\theta_s). \tag{9.41}$$

For convenience, we also introduce the following chiral vector:

$$e_\pm = \frac{1}{\sqrt{2}}(\hat{\mathbf{x}}' \pm \hat{\mathbf{y}}'). \tag{9.42}$$

Then the spin including perturbations, ψ and ψ^*, in the skyrmion state can be written as:

$$\mathbf{n} = \mathbf{n}_s + \tilde{\mathbf{n}} = \hat{\mathbf{z}}\sqrt{1 - 2|\psi|^2} + e_+\psi + e_-\psi^*. \tag{9.43}$$

The problem is centrosymmetric; it is more convenient to introduce the polar coordinate $\mathbf{r} = (\rho, \varphi)$. We then substitute Equation 9.43 into Equation 9.4 and keep terms second order in ψ. The term zeroth order in ψ is the energy of a skyrmion, and the term first order in ψ vanishes for a stationary skyrmion solution. The spectrum of ψ is governed by the following Hamiltonian:

$$\mathcal{H}^{(2)} = \frac{1}{2}\hat{\psi}^\dagger \hat{H}^{(2)}\hat{\psi}, \tag{9.44}$$

Chapter 9

$$\hat{H}^{(2)} = -J_{ex}\sigma_0 \nabla^2 + 2\sigma_3 \left(J_{ex}\frac{\cos\theta_s}{\rho^2} - D\frac{\sin\theta_s}{\rho} \right) i\partial_\phi + \sigma_0 V_0 + \sigma_1 V_1, \quad (9.45)$$

$$V_0 = J_{ex}\frac{1+3\cos(2\theta_s)}{4\rho^2} - \frac{3D\sin(2\theta_s)}{2\rho} + H_a\cos\theta_s$$

$$-\frac{A_z}{2}\left(1-3\cos^2\theta_s\right) - D\partial_\rho\theta_s - \frac{J_{ex}}{2}\left(\partial_\rho\theta_s\right)^2, \quad (9.46)$$

$$V_1 = J_{ex}\frac{\sin^2\theta_s}{2\rho^2} + D\frac{\sin(2\theta_s)}{2\rho} - D\partial_\rho\theta_s - \frac{J_{ex}}{2}\left(\partial_\rho\theta_s\right)^2 + \frac{A_z}{2}\sin^2\theta_s, \quad (9.47)$$

where we have introduced the spinor operator:

$$\hat{\psi} = \begin{pmatrix} \psi \\ \psi^* \end{pmatrix} \quad \text{and} \quad \hat{\psi}^\dagger = \begin{pmatrix} \psi^* & \psi \end{pmatrix},$$

and σ_i with $i = 1,2,3$ are Pauli matrices and σ_0 is the unit matrix.

The presence of skyrmion gives rise to an emergent magnetic field acting on the magnon. This can be seen explicitly by introducing an emergent vector potential in the polar coordinate:

$$\mathbf{A} = \left(\frac{\cos\theta_s}{\rho} - \frac{D}{J_{ex}}\sin\theta_s \right) e_\varphi, \quad (9.48)$$

with $e\varphi$ being the unit vector in the azimuthal direction and $\nabla \cdot \mathbf{A} = 0$. Then, $\hat{H}^{(2)}$ can be written in a more compact form:

$$\hat{H}^{(2)} = J_{ex}\left(-i\nabla - \sigma_3\mathbf{A}\right)^2 + \sigma_0\left(V_0 - J_{ex}\mathbf{A}^2\right) + \sigma_1 V_1. \quad (9.49)$$

It is clear that the skyrmion generates an effective vector potential acting on the magnons, which gives rise to nontrivial response such as magnon Hall effect [54,74,29,71]. Such novel physics is absent in the discussions of the internal modes in the φ^4 model.

To correctly describe the dynamics of ψ, we also need to expand the Berry phase contribution of spin in Equation 9.13 to the second order in ψ:

$$\mathcal{L}_B^{(2)} = -i\frac{1}{2}\hat{\psi}^\dagger\sigma_3\hbar\partial_t\hat{\psi}. \quad (9.50)$$

Minimizing the action $S^{(2)} = \mathcal{L}_B^{(2)} - \mathcal{H}^{(2)}$ with respect to $\psi^\dagger$, we obtain the equation of motion for ψ:

$$-i\sigma_3\hbar\partial_t\hat{\psi} = \hat{H}^{(2)}\hat{\psi}. \qquad (9.51)$$

The interaction between the skyrmions and magnons is described by the bosonic Bogoliubov-de Gennes equation with an emergent electromagnetic field. The Hamiltonian $\hat{H}^{(2)}$ includes a potential and an effective vector potential $\mathbf{A}$ produced by the skyrmion. Introducing the frequency ω and angular momentum l,

$$\hat{\psi} = \exp(i\omega t + il\phi)\hat{\eta}_l(\rho). \qquad (9.52)$$

Equation 9.51 becomes a partial differential equation for $\hat{\eta}_l(\rho)$ in one dimension, which was analyzed in Ref. [74].

In the above formulation, the dissipation is neglected. The dissipation may be taken into account phenomenologically by using the replacement $\omega \leftarrow \omega + i\alpha\omega$. Moreover, the lattice of the spin system is neglected, which is valid in the large skyrmion size limit. In principle, the spin lattice breaks the translational invariance and introduces a gap for the translational motion of skyrmion. The magnitude of the gap depends on the ratio of the skyrmion size to the lattice parameter of the spin system. To treat the damping more naturally and to account for the effect of underlying spin lattice, we adopted the exact numerical diagonalization method as was employed in Ref. [48].

We consider a 2D square lattice of spin systems with an easy-axis anisotropy described by Equation 9.5. The dynamics of spin follows the LLG equation in Equation 9.6 without a current $\mathbf{J} = 0$. We use the local coordinate approach defined in Equations 9.39 through 9.41, in which the local z'-axis is parallel to the $\mathbf{n}_s$ direction, while the perturbations $\tilde{\mathbf{n}}$ lies in the local $x' - y'$ plane [91]. We then substitute $\mathbf{n}_s$ into Equation 9.5 and keep the contributions up to the first order in $\tilde{\mathbf{n}}$. The resulting equation can be expressed as a matrix equation in the frequency domain, and we obtain the lowest 30 eigenfrequencies and eigenmodes by diagonalizing the matrix with the Lanczos method. Details of this calculation are included in Ref. [48].

Several low eigenfrequencies as a function of magnetic fields are shown in Figure 9.6 for $A_z = 0$ and $D/J_{ex} = 0.1$. A magnon gap, $\omega_g = \gamma H_a$, in the magnon spectrum, is induced by magnetic field. To assign the internal modes, we calculate the profile of the spin configuration

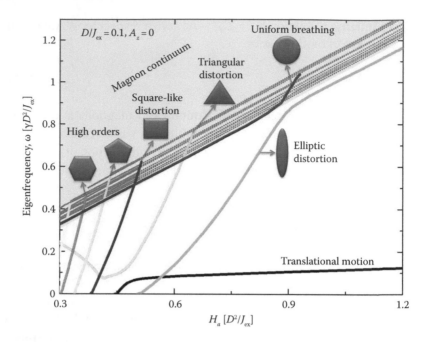

FIGURE 9.6 Magnetic field dependence of the eigenfrequencies of different modes in the absence of easy-axis anisotropy ($A_z = 0$). The modes are assigned to the lth order breathing mode and the translational mode. We show only several eigenfrequencies of the magnon continuum (shaded region) because only the lowest 30 eigenmodes have been obtained with the Lanczos method.

$\mathbf{n} = \mathbf{n}_s + F\tilde{\mathbf{n}}_v \sin(\omega t)$, where $\tilde{\mathbf{n}}_v$ is the eigenvector, F is the arbitrary amplitude, and ω is the eigenfrequency. Spin configurations for several internal modes at $t = 0$, $t = \pi/(2\omega)$, and $t = 3\pi/(2\omega)$ are shown in Figure 9.7. The translational mode exists in the whole magnetic field region. These modes can be characterized by an angular momentum l defined in Equation 9.52. The skyrmion gets deformed into lth order polygons when the internal mode with an angular momentum l is excited. Here, $l = 0$ corresponds to the uniform breathing mode.

As shown in Figure 9.6, the eigen modes become degenerate when levels cross at a special magnetic field. At a specific magnetic field H_l, the lth mode becomes gapless, meaning an instability of the skyrmion due to the increasing deformation associated with such a mode. Below the threshold field H_l, the eigenfrequency of the lth model becomes negative. The damping which is proportional to the eigen frequency also becomes negative. Therefore, the amplitude of the l mode increases

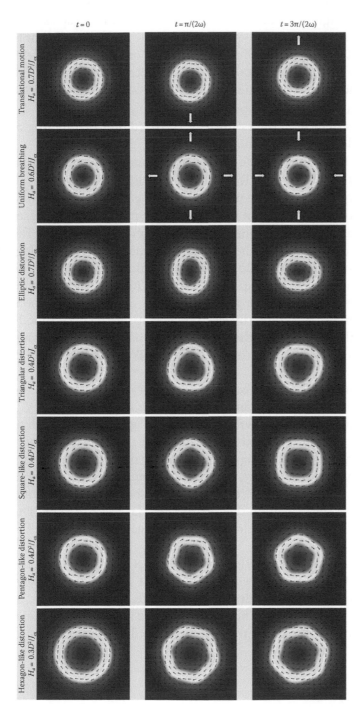

FIGURE 9.7 Deformation of the skyrmion associated with different internal modes. The spin configuration is obtained from the eigenvector $\mathbf{n}_\nu$ by using $\mathbf{n} = \mathbf{n}_s + F\mathbf{n}_\nu \sin(\omega t)$ with $F = 10$. The spin configuration is plotted at $t = 0$ (left column), $t = \pi/(2\omega)$ (middle column), and $t = 3\pi/(2\omega)$ (right column). The white arrows in the last column denote the direction of the translational motion and in the next to last column represent the direction of the uniform breathing. The vectors in the plots denote the n_x and n_y components, and the contour plot denotes the n_z component with red representing $n_z = 1$ and blue representing $n_z = -1$. Here, $D/J_{ex} = 0.1$ and $A_z = 0$.

exponentially with time and destabilizes the skyrmion solution. Upon reducing the magnetic field, the mode with $l = 2$ becomes gapless and therefore the skyrmion in the ferromagnetic state is locally stable only for $H_a > H_2$. To verify the stability of the skyrmion state at a low field, we also solve directly Equations 9.4 and 9.6 with a skyrmion in the ferromagnetic state as an initial condition. The skyrmion solution is unstable with respect to weak perturbations for field $H_a \lesssim 0.55 D^2/J_{ex}$, which is consistent with the estimate of H_2 based on the softening of the $l = 2$ mode. Such an instability was also found in Ref. [9].

There are three internal modes below the magnon continuum at fields above $H_a \gtrsim 0.55 D^2/J_{ex}$. The uniform breathing mode with $l = 0$ has eigen frequency above the magnon gap and gets buried inside the magnon continuum. The eigen frequency of the model $l = 2$ increases with field and approaches the magnon continuum for a large field. The lowest mode corresponds to the translation motion of skyrmion. It is gapped because the spin lattice used in the calculations breaks the translation invariance. The gap depends on the ratio of the skyrmion size-to-lattice constant of the spin system. For a smaller skyrmion, the gap increases. Therefore, the eigen frequency for the lowest mode increases with field because the skyrmion is smaller under a higher field. The gap is proportional to D^2 because the skyrmion size shrinks at a larger D.

We then investigate the effect of an easy-axis anisotropy A_z on the internal modes. In Figure 9.8a and b, the results for $A_z = 0.5 D^2/J_{ex}$ and $A_z = 1.0 D^2/J_{ex}$ are shown. The magnon gap is increased to $\omega_g = \gamma(H_a + A_z)$ by the easy-axis anisotropy, and the uniform breathing mode is now below and separated from the magnon continuum. With an easy-axis anisotropy, the eigenfrequencies of the internal modes increase because the effective potential for magnons becomes narrower when the skyrmion size shrinks for a large anisotropy. The number of internal modes below the magnon gap is also reduced at low fields.

Finally, we consider the effect of the Gilbert damping α on the internal modes. The eigenfrequencies acquire an imaginary part, and the lifetime of the modes becomes finite in the presence of damping: $\tau = 2\pi/\mathrm{Im}[\omega]$. For the magnons in the ferromagnetic state, the dependence of the eigen frequency on the damping constant α is given by:

$$\omega(\alpha) = \frac{1+\alpha i}{1+\alpha^2}\omega \quad (\alpha = 0). \tag{9.53}$$

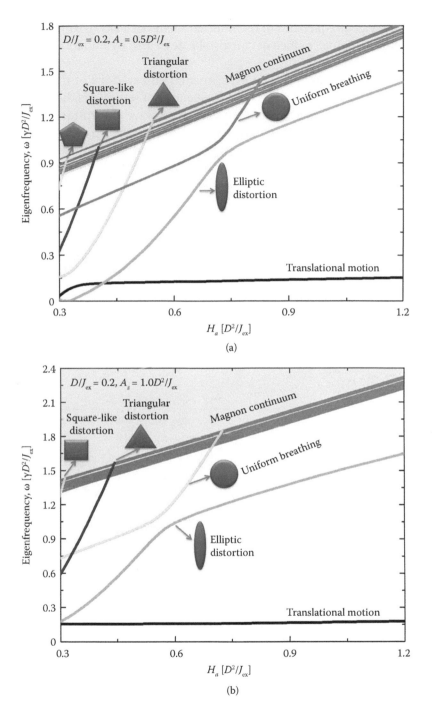

FIGURE 9.8 Same as Figure 9.6 but with $A_z = 0.5D^2/J_{ex}$ in (a) and $A_z = 1.0D^2/J_{ex}$ in (b). The eigenfrequencies for $D/J_{ex} = 0.1$ and 0.2 are almost identical in normalized units.

The eigenfrequencies of the internal modes of a skyrmion as a function of damping coefficient can also be described by Equation 9.53 (see Figure 9.9).

We next discuss possible experimental observations of the internal modes. The internal modes have frequencies of several gigahertz. Thus, they can be excited and observed with microwave absorption measurements. Because the wavelength of microwaves in the gigahertz region is much larger than the skyrmion size, only the uniform breathing mode can be significantly excited. The other modes with $l > 0$ can be measured with a local probe, such as a spin-polarized scanning tunneling microscope.

Here, we have considered the magnon modes in a single skyrmion. When skyrmions condense into lattice, the magnons couple with the periodic potential produced by the skyrmion lattice, which leads to the formation of magnon bands. The magnon bands in the limit of zero momentum $k = 0$ have been calculated numerically in Ref. [55] and observed by microwave resonance in Refs. [56,63,65,76]. There are three different modes: clockwise and counterclockwise rotation of skyrmions, and the breathing modes. The eigenfrequencies are of the order of several gigahertz.

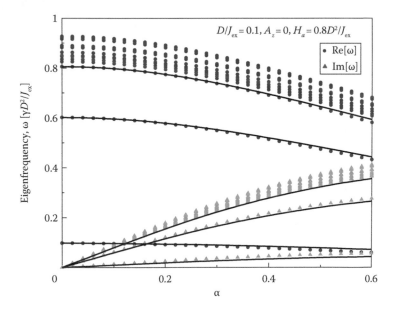

FIGURE 9.9   Dependence of the eigenfrequencies ω on the damping coefficient α. Symbols are from numerical diagonalization and lines are given by Equation 9.53. For clarity, not all lines are shown.

9.5 Dynamics of Skyrmion under Current Drive

9.5.1 Equation of Motion

From the discussions on the internal modes in the previous section, we see that internal modes associated with the skyrmion deformation are gapped in the high field region. For a skyrmion moving at a low velocity, we can neglect these internal modes and only account for the lowest mode, the translational motion of skyrmion. The translational mode is gapless in the continuum limit or in the limit that the skyrmion size is much bigger than the lattice parameter of the spin system. Note the translational mode is the only Goldstone of the Hamiltonian in Equation 9.3 because the Zeeman and DM terms remove all the continuous symmetries associated with the spin vector $\mathbf{n}$. In other words, we assume that the skyrmion is rigid and it behaves as a particle. We also consider a dilute concentration of skyrmions such that the overlap between skyrmions can be neglected.

We adopt Thiele's collective coordinate approach [85] to derive the equation of motion for skyrmion as a particle. For a rigid skyrmion traveling with velocity $\mathbf{v}$, the spins associated with a skyrmion can be written as $\mathbf{n}_s(\mathbf{r} - \mathbf{v}t)$. The dynamics of $\mathbf{n}_s$ is governed by the equation of motion:

$$\partial_t \mathbf{n}_s = \frac{\hbar\gamma}{2e}(\mathbf{J} \cdot \nabla)\mathbf{n}_s - \gamma\mathbf{n}_s \times \mathbf{H}_i + \alpha\mathbf{n}_s \times \partial_t \mathbf{n}_s, \tag{9.54}$$

where $\mathbf{H}_i = \mathbf{H}_s + \mathbf{H}_d$. Here, $\mathbf{H}_d$ is the effective field produced by defects and $\mathbf{H}_s$ is the field generated by other skyrmions. The effective magnetic field produced by $\mathbf{n}_s$ itself, $\mathbf{H}_0 \equiv -\delta/\delta\mathbf{n}_s$, does not enter into Equation 9.54 because $\mathbf{n}_s \times \mathbf{H}_0 = 0$ for a rigid skyrmion. We then integrate out the internal degree of skyrmion by multiplying both sides of Equation 54 by $\times \mathbf{n}_s$ (cross product) and then by $\partial_\mu \mathbf{n}_s$ (dot product). We then integrate around a skyrmion and use the definition of topological charge (Equation 9.1) and $\int dr^2 \partial_x \mathbf{n}_s \cdot \partial_y \mathbf{n}_s = 0$ for a rigid skyrmion. We obtain:

$$\alpha_s \mathbf{v} = \frac{\gamma}{4\pi}\left[\mathbf{F}_M + \mathbf{F}_L + \int dr^2 \mathbf{H}_\perp(\mathbf{r}' - \mathbf{r}) \cdot \nabla_r \mathbf{n}_s(\mathbf{r})\right], \tag{9.55}$$

where $\mathbf{H}_\perp$ is the field component perpendicular to $\mathbf{n}_s$ and α_s is the viscosity of the skyrmion. The self-energy density of the skyrmion in the presence of defects is:

$$E_s(\mathbf{r} - \mathbf{r}') = -\int dr''^2 \mathbf{n}_s(\mathbf{r} - \mathbf{r}'') \cdot \mathbf{H}_d(\mathbf{r}' - \mathbf{r}''),$$

with:

$$\mathbf{H}_d(\mathbf{r}) = J_{ex}(\mathbf{r})\nabla^2\mathbf{n}_s/2 - D(\mathbf{r})\nabla\times\mathbf{n}_s + \mathbf{H}_a.$$

The corresponding pinning force is then given by:

$$\mathbf{F}_d \equiv -\nabla E_s(\mathbf{r}-\mathbf{r}') = \int dr''^2\nabla\mathbf{n}_s(\mathbf{r}-\mathbf{r}'')\cdot\mathbf{H}_{d,\perp}(\mathbf{r}'-\mathbf{r}'').$$

The interaction potential between a skyrmion at $\mathbf{r}$ and an another skyrmion at $\mathbf{r}'$ is:

$$U_{ss}(\mathbf{r}'-\mathbf{r}) = -\int dr''^2\mathbf{n}_s(\mathbf{r}-\mathbf{r}'')\cdot\mathbf{H}_s(\mathbf{r}'-\mathbf{r}''),$$

and the corresponding force is:

$$\mathbf{F}_{ss} \equiv -\nabla U_{ss}(\mathbf{r}'-\mathbf{r}) = \int dr''^2\nabla_r\mathbf{n}_s(\mathbf{r}-\mathbf{r}'')\cdot\mathbf{H}_{s,\perp}(\mathbf{r}'-\mathbf{r}'').$$

Using the definition of $\mathbf{F}_d$ and $\mathbf{F}_{ss}$, we obtain the equation of motion for skyrmion from Equation 9.55:

$$\frac{4\pi\alpha_s}{\gamma}\mathbf{v}_i = \mathbf{F}_M + \mathbf{F}_L + \sum_j \mathbf{F}_d(\mathbf{r}_j-\mathbf{r}_i) + \sum_j \mathbf{F}_{ss}(\mathbf{r}_j-\mathbf{r}_i), \qquad (9.56)$$

Here, $\mathbf{F}_M$ is the Magnus force per unit length, which is perpendicular to the velocity.

$$\mathbf{F}_M = 4\pi\gamma^{-1}\hat{z}\times\mathbf{v}_i. \qquad (9.57)$$

$\mathbf{F}_L$ is the Lorentz force that arises from the emergent quantized magnetic flux $\Phi_0 = hc/e$ carried by the skyrmion in the presence of a finite current:

$$\mathbf{F}_L = 2\pi\hbar e^{-1}\hat{z}\times\mathbf{J}. \qquad (9.58)$$

Note that the forces in Equation 9.56 are defined as force per unit length. The term on the left-hand side of Equation 9.56 describes the damping of skyrmion motion. It has contribution from the underlying damping of the spin precession. In metallic compounds, there is additional damping due to the conduction electrons localized in skyrmions. The physical picture is the following. In the presence of an emergent electric field **E**, an additional electric current **J′** will be induced according to Ohm's law, **J′** = σ**E**, with σ the conductivity of the electrons, which should be included in the first term on the right-hand side of the LLG equation (Equation 9.54). Thus, the damping constant becomes $\alpha_s = \alpha\eta + \alpha_\sigma\eta'$ with:

$$\alpha_\sigma = 4\pi\left(\frac{\hbar}{2e\xi_s}\right)^2 \gamma\sigma, \tag{9.59}$$

$$\eta = \eta_\mu = \frac{1}{4\pi}\int_{\text{skyrmion}} d\mathbf{r}^2\left(\partial_\mu \mathbf{n}\right)^2, \tag{9.60}$$

$$\eta' = \frac{\xi_s^2}{16\pi^2}\int_{\text{skyrmion}} dr^2\left[\mathbf{n}\cdot(\partial_x \mathbf{n}\times\partial_y \mathbf{n})\right]^2, \tag{9.61}$$

where the integration in Equations 9.60 and 9.61 is performed around the skyrmions and $\xi_s \sim J_{ex}/D$ is the size of skyrmions. There exists another dissipation due to the radiation of magnon by the motion of skyrmions, which is not taken into account here. The fictitious electric field induced by the skyrmion motion thus paves a new dissipation path of the magnetic system. In contrast to the Gilbert damping, this new mechanism does not require relativistic effects and only involves Hund's coupling that conserves the total spin. The relaxation of the uniform magnetization, described by the Gilbert damping, is clearly impossible without the spin–orbit coupling, which breaks the conservation of the total spin. This argument, however, does not apply to inhomogeneous magnetic textures where the breaking of the rotational symmetry by non-collinear spin orders enables the relaxation without the spin–orbit coupling (note that α_σ vanishes as $\xi_s \to \infty$). Despite the nonrelativistic origin, α_σ depends on the DM coupling, as the latter determines the skyrmion size. In the case of MnSi, a = 2.9 Å, $\xi_s \approx 77$ Å, the electron density $n_e \approx 3.8 \times 10^{22}$ cm^{-3}, and the residual resistivity $\rho_r \approx 2\ \mu\Omega \cdot$ cm [42,61], we estimate $\alpha_\sigma \approx 0.1$, which is much larger than the intrinsic Gilbert damping.

According to Equation 9.56, a rigid skyrmion does not have an intrinsic mass. Skyrmions gain a dynamical mass when they deform and one needs to account for the higher internal modes to calculate the skyrmion mass. The mass of a skyrmion is studied in Refs. [14,53,75], and it is shown that the inertia term can affect the equation of motion. A similar equation of motion as that in Equation 9.56 was considered before in the context of vortices of Type II superconductors [7]. However, the Magnus force is negligibly small for superconducting vortices in most cases [7].

Chapter 9

The equation of motion for a skyrmion in a clean system driven by a constant current can be solved analytically using Equation 9.56. It has longitudinal velocity:

$$v_{\parallel} = -\frac{\gamma \hbar}{2e\left(1+\alpha_s^2\right)} J,\tag{9.62}$$

and transverse velocity:

$$v_{\perp} = \frac{\gamma \hbar \alpha_s}{2e\left(1+\alpha_s^2\right)} J.\tag{9.63}$$

Because $\alpha_s \ll 1$, skyrmions move almost antiparallel to the direction of electric current. There is a weak transverse motion due to the damping α_s. This transverse motion of the skyrmion is in close analogy with the Hall effect of the electron. We can define a Hall angle $\tan(\theta_H) = v_{\perp}/v_{\parallel} = -\alpha_s$. Because α_s is contributed from the dissipation due to conduction electrons and the Gilbert damping, the Hall angle is now $\theta_H = -\arctan(\alpha\eta + \alpha_{\sigma}\eta')$, which becomes phenomenal in metal due to a large α_{σ} and can be observed by real-space images of Lorentz force microscopy. In insulators, the Gilbert damping is usually weak $\alpha \approx 0.01$. The Hall angle is therefore tiny.

We have compared the analytical results in Equations 9.62 and 9.63 to those obtained directly by numerical simulations. The results for the Hall angle and velocity as a function of Gilbert damping α is shown in Figure 9.10. The velocity of a skyrmion center of mass is defined as $\mathbf{v} = \dot{\mathbf{r}}_c$ with the skyrmion center of mass $\mathbf{r}_c$:

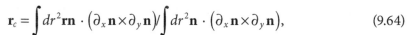

$$\mathbf{r}_c = \int dr^2 \mathbf{r} \mathbf{n} \cdot \left(\partial_x \mathbf{n} \times \partial_y \mathbf{n}\right) \Big/ \int dr^2 \mathbf{n} \cdot \left(\partial_x \mathbf{n} \times \partial_y \mathbf{n}\right),\tag{9.64}$$

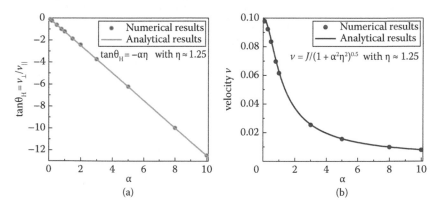

FIGURE 9.10 (a) The Hall angle $\tan\theta_H$ and (b) the velocity v vs. α for the motion of a single skyrmion in the ferromagnetic state obtained analytically (lines) and numerically (dots) at $J = 0.1$.

in the numerical calculations. In the simulations, we set $\alpha_\sigma = 0$; therefore, the skyrmion viscosity $\alpha_s = \alpha\eta$. The numerical and analytical results are in perfect agreement with each other as depicted in Figure 9.10.

The equation motion for a single skyrmion can also be derived heuristically. For simplicity, let us consider the case with $F_d = 0$ and $F_{ss} = 0$. In a stationary state, the skyrmion and conduction electrons move together at the same velocity so that there is no spin transfer torque. Effectively, the electrons and skyrmion form a composite object, and the situation is very similar to the vortex in superconductors where the localized quasiparticles play a role of electrons in the skyrmion [7]. One knows immediately the equation of motion for the skyrmion if one knows the equation of motion for conduction electrons, and in the following derivation, we focus on the conduction electrons inside the skyrmion. The electric current density inside the skyrmion is:

$$J_e = \sigma_\parallel \mathbf{E} + \sigma_\perp \hat{z} \times \mathbf{E}, \tag{9.65}$$

where $\sigma_\perp$ and $\sigma_\parallel$ are the Hall and longitudinal conductivity. Using the Drude model, we have for the conductivity:

$$\sigma_\parallel = \frac{e^2 n_e \tau_e}{m} \frac{1}{1+(\omega_c \tau_e)^2}, \quad \sigma_\perp = \frac{e^2 n_e \tau_e}{m} \frac{\omega_c \tau_e}{1+(\omega_c \tau_e)^2}, \tag{9.66}$$

where $n_e \sim 1/a^3$ is the electron density and τ_e is the electron relaxation time. Here, $\omega_c = eB/(mc)$ is the cyclotron frequency of electrons in the presence of an emergent magnetic field $B \approx \Phi_0/(\xi_s^2)$. Here, m_e is the electron mass. The emergent electric field is $\mathbf{E} = \mathbf{B} \times \mathbf{v}/c$. We obtain the equation motion for conduction electrons after substituting $\mathbf{E}$ into Equation 9.65 and taking the cross product $\times \Phi_0 \hat{z}/c$ ($\hat{z}$ is a unit vector perpendicular to the 2D plane) at both sides of Equation 9.65,

$$\pi n_e \frac{\omega_c \tau_e}{1+(\omega_c \tau_e)^2} \mathbf{v} = \pi n_e \frac{(\omega_c \tau_e)^2}{1+(\omega_c \tau_e)^2} \hat{z} \times \mathbf{v} + \hat{z} \times J_e \Phi_0/c, \tag{9.67}$$

which is also the equation of motion for the skyrmion. The emergent magnetic field is strong for a small skyrmion size, $\omega_c \tau \gg 1$, and therefore the Magnus force dominates over the dissipative force. Equation 9.67 has the same form as that was derived using Thiele's collective coordinate approach.

The corresponding action S_p for the equation of motion for skyrmion in Equation 9.56 can be written as:

$$\frac{S_p}{d} = S_{B,p} - U(\mathbf{r}) - \frac{4\pi\alpha_s}{\gamma} \int dt\, dt' \left[\frac{\mathbf{r}(t) - \mathbf{r}(t')}{t - t'}\right]^2, \tag{9.68}$$

Chapter 9

$$S_{B,p} = \frac{4\pi}{\gamma}\left[x\left(\frac{1}{2}\partial_t\, y - \frac{\hbar\gamma}{2e}J_y\right) - y\left(\frac{1}{2}\partial_t x - \frac{\hbar\gamma}{2e}J_x\right)\right], \tag{9.69}$$

where $U(\mathbf{r})$ is the potential per unit length induced by defects and other skyrmions. As a consequence of the Berry phase, the skyrmion coordinates x and y are conjugate with each other. According to Equation 9.69, the Magnus force $\mathbf{F}_M$ is originated from an effective transverse magnetic field, $B_z = 4\pi cd/(\gamma q_e)$, that couples to a moving particle with electric charge q_e. This emergent magnetic field is related to the skyrmion topological charge. Although the Magnus force does not produce work, it affects the skyrmion trajectory, which is one of the reasons why the pinning for skyrmion is weak as will be explained below.

So far we have focused on the adiabatic spin transfer torque described by the term, $\hbar\gamma(\mathbf{J} \cdot \nabla)\mathbf{n}/(2e)$, in Equation 9.54. In the presence of a non-adiabatic spin transfer torque, $-\zeta\hbar\gamma\mathbf{n} \times (\mathbf{J} \cdot \nabla)\mathbf{n}/(2e)$, there is an additional force for the skyrmion. The derivation of Equation 9.56 can be generalized straightforwardly by taking the nonadiabatic spin transfer torque into account. This yields additional force at the right-hand side of Equation 9.56:

$$\mathbf{F}_{non} = 2\pi\hbar\zeta\eta e^{-1}\mathbf{J}_B,$$

and the equation of motion for the skyrmion is:

$$\frac{4\pi\alpha_s}{\gamma}\mathbf{v}_i = \mathbf{F}_M + \mathbf{F}_L + \mathbf{F}_{non} + \sum_j \mathbf{F}_d(\mathbf{r}_j - \mathbf{r}_i) + \sum_j \mathbf{F}_{ss}\left(\mathbf{r}_j - \mathbf{r}_i\right). \tag{9.70}$$

In Ref. [31], the effects of the nonadiabatic spin transfer torque on skyrmion dynamics were studied.

In insulating chiral magnets, the damping contribution by the conduction electrons is absent and $\alpha_s = \alpha\eta$. Both the Lorentz force $\mathbf{F}_L$ and the force due to the nonadiabatic spin transfer torque $\mathbf{F}_{non}$ are also absent. In this case, the skyrmion can be driven by a magnetic field gradient or by a magnon current induced by temperature gradient.

Using the typical parameters for MnSi [98], $a \approx 2.9$ Å, $J_{ex} \approx 3$ meV$/a$, $D \approx 0.3$ meV$/a^2$, and $\alpha_s \approx 0.1$, we estimate the magnitude of the force in Equation 9.56. At a velocity $v = 1$ m/s, the Magnus force per unit length is $F_M \approx 5 \times 10^{-5}$ N/m. The dissipative force per unit length is $F_{diss} \equiv 4\pi\alpha_s v/(\gamma) \approx 5 \times 10^{-6}$ N/m. Thus, $F_M \gg F_{diss}$ and the dynamics of skyrmion is dominated by the Magnus force. The Lorentz force per unit length at a current density $J = 10^6$ A/m^2 is $F_L \approx 4 \times 10^{-9}$ N/m. According to the results in Figure 9.3, the repulsive force per unit length between skyrmions is $F_{ss} \approx 10^{-5}$ N/m at a separation $r_d = 10$ nm. To estimate the pinning, we need to know the distribution of defects in the system. Here, we estimate the pinning force using the depinning current measured in experiments by assuming that the skyrmion density is low. The measured depinning current for skyrmion is of the order of 10^6 A/m^2 [35,95,73], and we thus estimate the pinning force per unit length as $F_d \approx 4 \times 10^{-9}$ N/m.

Equation 9.56 allows us to explain why the pinning of skyrmions is weak. For a skyrmion, the Magnus force is dominant over the dissipative force $F_M \gg 4\pi\alpha_s v/\gamma$ for $\alpha_s \ll 1$. The velocity is almost perpendicular to the force. When a skyrmion travels around a pinning center or an obstacle, it is easily scattered with a velocity perpendicular to the force. Therefore, the skyrmion avoids passing through the pinning center as shown in Figure 9.11a and b, and the influence of the pinning centers or obstacles is minimized. In the region when the dissipative force is dominant for $\alpha \gg 1$ as realized in the vortex system in superconductors, the vortex has to pass through the pinning centers or obstacles, as shown in Figure 9.11c. Therefore, the pinning is strong for vortex in superconductors. Similar conclusions were reached by simulating directly the continuum model in Equations 9.3 and 9.6 in Ref. [31]. We remark that there are additional facts which minimize the pinning of skyrmions. Skyrmions condense into lattice and the effects of random defects are quickly smeared out on the scale of the skyrmion lattice. Moreover, the huge size disparity between skyrmions and defects also significantly suppresses the pinning of skyrmions as discussed in the previous section.

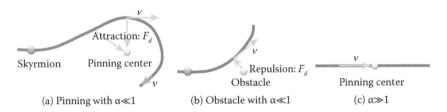

(a) Pinning with $\alpha \ll 1$ (b) Obstacle with $\alpha \ll 1$ (c) $\alpha \gg 1$

FIGURE 9.11   (a, b) Schematic view of a skyrmion passing through a pinning center (a) and obstacle (b) when the Magnus force is dominant. When the Magnus force is dominant over the dissipative force, the skyrmion is deflected by the pinning centers or obstacles. (c) Same as (a) and (b) except that the dissipative force is dominant. The skyrmion has to overcome the pinning site or obstacle by passing through it.

Chapter 9

9.5.2 Magnon Doppler Effect and Dynamical Creation of Skyrmion

In this section, we discuss the shift of the magnon frequency in the presence of conduction electrons, or the Magnon Doppler effect. Such an effect can be seen directly from Equation 9.6. For a magnon dispersion $\omega = \Omega(k)$, the adiabatic spin transfer torque term shifts the dispersion by the amount of $\mathbf{J} \cdot \mathbf{k}$ and the resulting magnon dispersion is:

$$\Omega(\mathbf{k}) = \frac{\hbar\gamma}{2e}\mathbf{J} \cdot \mathbf{k} + \frac{\gamma(1+i\alpha)}{\alpha^2+1}\left(H_a + J_{ex}\mathbf{k}^2\right). \tag{9.71}$$

The Magnon Doppler shift was predicted in Ref. [22] and was confirmed experimentally in Ref. [87], where a linear frequency shift as a function of current at a given magnon momentum k was observed.

An interesting observation is that the magnon gap vanishes for a large current. The critical current is given by [49]:

$$J_m = 4e\sqrt{H_a J_{ex}} \ / \left[\hbar\left(\alpha^2+1\right)\right], \tag{9.72}$$

at $k_x = \sqrt{H_a/J_{ex}}$ and $k_y = 0$ for a small α and for currents along the x direction. Once the magnon vanishes, the ferromagnetic state is instablized by the injected current. Such an instability also exists in conventional ferromagnets with inversion symmetry. The instability was not observed in Ref. [87] because the current is not enough to render the magnon gapless.

To verify the magnon instability due to the Doppler effect, we have performed numerical simulations of the ferromagnets without DM interaction [50]. The magnetization along the field direction, the emergent electric field **E**, and the scalar charity are shown in Figure 9.12. For $J > 1.6$, the ferromagnetic state is unstable and a chiral liquid phase is stabilized by current. It is a chiral liquid phase because the average chirality Q is zero while the average absolute value of chirality $P \equiv \int d\mathbf{r}^2 |q(\mathbf{r})|$ is nonzero, indicating a strong fluctuation of chirality in space and time. The emergent electric field parallel to the current $E_{||}$ is proportional to P and therefore increases with J. The component perpendicular to the current $E_\perp$ is proportional to Q and it is almost zero. Meanwhile the magnetization along the field direction M_z decreases linearly in the chiral liquid phase. A typical spin configuration of the chiral liquid is presented in Figure 9.13.

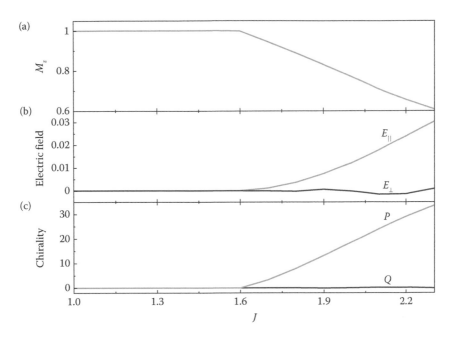

FIGURE 9.12   (a) Magnetization along the field direction M_z, (b) the electric field parallel to the current $E_{\parallel}$ and the electric field perpendicular to the current $E_{\perp}$, and (c) the total number of skyrmions $Q \equiv \int d\mathbf{r}^2 q(\mathbf{r})$ and the average of absolute scalar chirality $P \equiv \int d\mathbf{r}^2 |q(\mathbf{r})|$ as a function of the spin current J. Here, $\alpha = 0.1$, $D = 0$, and $H_a = 0.6\,J_{ex}$.

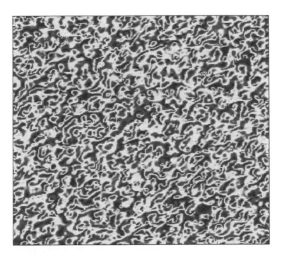

FIGURE 9.13   Typical spin configuration in the chiral liquid phase. Here, the color denotes n_z with the red for $n_z = 1$ and blue for $n_z = -1$.

Chapter 9

In the presence of the DM interaction, skyrmions are created dynamically by destroying the ferromagnetic state with current. As more and more energy is injected into the system, the skyrmion density keeps increasing until the system is fully occupied by skyrmions (see Figure 9.14c). As the emergent electric field is proportional to the skyrmion density, it also increases in a series of steps (Figure 9.14b). We also compare the analytical expression for J_m in Equation 9.72 to that obtained from the simulations. For a small damping relevant for most materials $\alpha \gg 1$, they agree with each other very well (see Figure 9.15). For a strong damping, J_m obtained by numerical

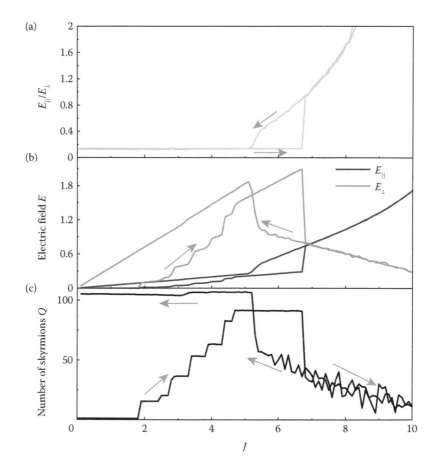

FIGURE 9.14   (a) Magnetization along the field direction M_z, (b) the electric field parallel to the current $E_\parallel$ and the electric field perpendicular to the current $E_\perp$, and (c) the total number of skyrmions $Q \equiv \int d\mathbf{r}^2 q(\mathbf{r})$ and the average of absolute scalar chirality $P \equiv \int d\mathbf{r}^2 |q(\mathbf{r})|$ as a function of the spin current J. The arrows in (a) and (b) indicate the direction of current sweep. Here, $\alpha = 0.1$ and $H_a = 0.6\ D^2/J_{ex}$.

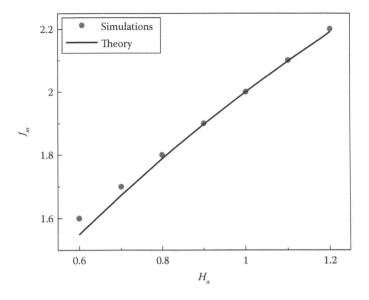

FIGURE 9.15 Comparison between J_m obtained by simulations and that in Equation 9.72.

simulations are higher than the one predicted by the linear analysis. This is because the energy pumped into the spin wave dissipates quickly for a strong damping, and the instability transition shifts to a higher current.

The reason why skyrmions are created after the instability is as follows. The skyrmion produces an emergent vector potential **A** (see Equation 9.17) generated by non-coplanar spin configurations. The current couples to the vector potential via the Lagrangian term $\mathcal{L}_{JA} = \mathbf{J} \cdot \mathbf{A}$ (see Equation 9.26). Therefore, the state with a nonzero **A** can lower the energy. In the presence of the DM interaction, the skyrmion state thus is induced by the current.

The Magnon Doppler effect can be used to create skyrmions electrically in a controlled way. We have proposed a setup as schematically shown in Figure 9.16a to manipulate skyrmion in a chiral magnetic nanodisk [51]. After injecting a current pulse, the ferromagnetic state is unstable if the current is strong enough to trigger the instability according to Equation 9.72. Initially, several skyrmions are created (see Figure 9.16e). Because of the repulsion between skyrmions, the skyrmions near the boundary are expelled from the disk. Meanwhile the skyrmion experience a geometry confinement produced by the nanodisk. As a result, a skyrmion is stabilized at the center of the disk after switching off the current pulse.

Chapter 9

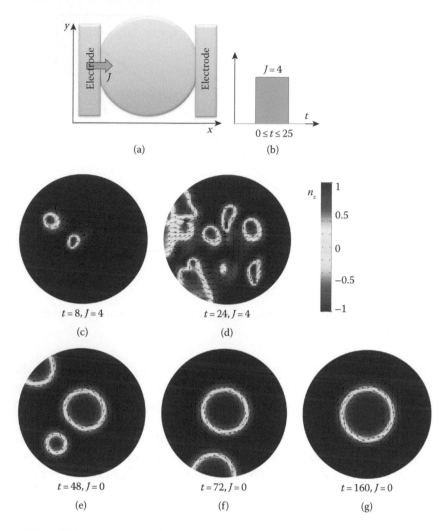

FIGURE 9.16   (a) Proposed setup to create skyrmions in a chiral magnetic nanodisk using a current pulse. (b) The shape of current pulse used in the simulations. (c–g) Magnetization dynamics after applying a current pulse in (b). In the end, a skyrmion is created at the center of the disk.

9.5.3   Distortion and Destruction of Skyrmions at High Currents

In this section, we study nonlinear dynamics of a skyrmion in the ferromagnetic state. A skyrmion moves fast with a high current and they start to distort in the presence of damping. This amounts to the excitations of internal modes and extended modes of skyrmion by the skyrmion motion. The Thiele collective coordinate approach, by assuming a rigid skyrmion, is questionable. Meanwhile, the motion of skyrmion disturbs the ferromagnetic background state and radiates spin waves. Such effects are more prominent when a skyrmion is decelerated or accelerated by impurities in inhomogeneous systems. For a skyrmion moves at a constant velocity, the radiation of spin wave is maximal at the resonance condition:

$$\Omega(\mathbf{k}) = \mathbf{v} \cdot \mathbf{k}. \tag{9.73}$$

It requires a threshold skyrmion velocity and therefore needs a threshold current above which the resonance condition (Equation 9.73) can be fulfilled. The skyrmion velocity is $\mathbf{v} \approx -\mathbf{J}$ for a weak damping $\alpha \ll 1$. The threshold current is $J_r = \gamma\sqrt{H_a J_{ex}}/(\alpha^2 + 1)$, which corresponds to the excitation of the magnon modes with wave numbers $k_x = \sqrt{H_a/J_{ex}}$ and $k_y = 0$. One typical configuration for n_x when the resonance condition is satisfied is displayed in Figure 9.17. Because additional energy is dissipated through the radiation of magnon into the ferromagnetic background, this causes additional damping to the skyrmion motion.

The skyrmion develops a tail at high velocities, which is opposite to the velocity of the skyrmion (see Figure 9.18). The distortion is more prominent for a strong damping. For a high enough velocity, the distortion is so strong that the skyrmion becomes unstable. This can be

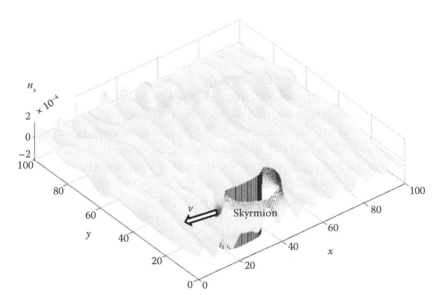

FIGURE 9.17 Spatial structure of the n_x component of the spin wave radiated by the skyrmion motion. The skyrmion location is labelled in the figure, and the arrow denotes the direction of skyrmion motion. Here, $\alpha = 0.1$, $J = 1.4$, and $H_a = 0.6$.

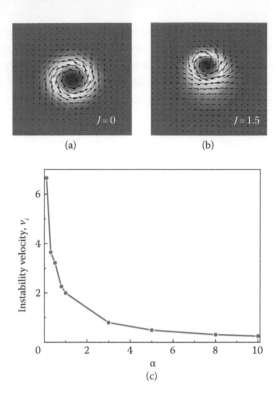

FIGURE 9.18   (a) Skyrmion at rest. (b) Skyrmion at $J = 1.5$ where it is distorted due to the damping. Here $\alpha = 10.0$. (c) The instability velocity v_i where skyrmions are destroyed as a function of α.

understood as follows. The skyrmion consists of spins which have a characteristic relaxation time. When the skyrmion moves fast, the underlying spins are unable to follow the skyrmion motion. As a result, the skyrmion breaks up. When the skyrmion is destroyed, the relation $E_\perp/E_\parallel = -\alpha\eta$ breaks down (displayed in Figure 9.14a). This means an upper limit for the skyrmion velocity. The skyrmions appear again at a lower current than that for skyrmion destruction when the current is reduced. Therefore, the instability transition of skyrmion is hysteretic. The number of skyrmions does not change when the current decreases down to $J = 0$. The instability velocity v_i where the skyrmions are destroyed depends on the damping constant α as $v_i \sim 1/\sqrt{1+\alpha^2\eta^2}$, as shown in Figure 9.18c. The instability of topological excitations at high velocities has been demonstrated in other systems, such as the Walker breakdown of domain walls [72] and the Larkin-Ovchinnikov instability of the vortex lattice in super-conductors [41].

We then study the resulting state after the destruction of skyrmions by high currents. According to the numerical simulations, the average skyrmion number $Q \rightarrow 0$ while $P \equiv \int d\mathbf{r}^2 |q(\mathbf{r})|$ increases and then saturates as a function of increasing current, similar to those in Figure 9.12c. This indicates that the disordered state when the skyrmions are destroyed is also a chiral liquid, discussed in the previous section. In the chiral liquid phase, there are strong spatial and temporal fluctuations of the scalar chirality $q(\mathbf{r})$ (or the emergent magnetic field) that changes sign. As a consequence, the mean value of the chirality is much smaller than the amplitude of fluctuations. The average chirality also fluctuates as a function of current as indicated by the sawtooth-like curve for Q shown in Figure 9.14c for the high current region. The emergent electric field component parallel to current $E_\parallel$ does not depend on the sign of chirality. Therefore, $E_\parallel$ increases with current and it exhibits a smooth behavior in the chiral liquid phase, consistent with the results shown in Figure 9.14b. Contrarily, the emergent electric field component perpendicular to current $E_\perp$ changes sign when the chirality is reversed. Therefore, $E_\perp$ fluctuates strongly and decreases with J in the chiral liquid phase, in good agreement with the numerical simulation displayed in Figure 9.14b.

9.5.4 Dynamics Starting from the Ground State Configurations

In equilibrium, the ground-state configurations of chiral magnetic thin films at low temperatures are spiral phase, skyrmion lattice, and ferromagnetic state. Here, we investigate the dynamics starting from these ground-state configurations. The case of the ferromagnetic state is discussed in the previous section. Upon increasing the current, it transits into the skyrmion flow phase and finally the chiral liquid phase. The current-voltage curve starting from the spiral state is shown in Figure 9.19. At low currents, the spiral is pinned because of the DM interaction. This intrinsic pinning occurs in clean systems without defects [31], similar to the case of magnetic domain walls [83].

The spiral structure becomes unstable above a threshold current and again skyrmion lattice is created. According to the numerical simulations, it is a two-step process. A small number of skyrmions ($Q = 5$) is first generated in the background of magnetic spiral when the current is higher than $J = 0.34$. Above $J \approx 0.6$, the system transits into the skyrmion lattice completely and the skyrmion lattice moves as a whole in response to the external current. At high currents, the skyrmion lattice is destroyed and the system evolves into the chiral liquid phase discussed above. There is strong hysteresis upon decreasing the current.

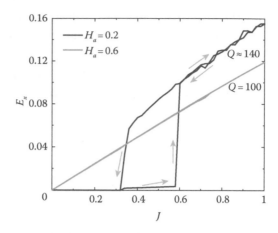

FIGURE 9.19 Dependence of the longitudinal emergent electric field E_x on current at different fields $H_a = 0.2$ and $H_a = 0.6$. For $H_a = 0.2$, the number of skyrmions Q is not conserved at high currents and only an averaged value of Q is given. Arrows denote the direction of current sweep.

If we start with the skyrmion lattice ground state, the system evolves into the skyrmion flow phase in the presence of a current. At high currents, finally the system evolves into the chiral liquid. Based on the numerical results, a $H_a - J$ phase diagram is constructed (see Figure 9.20a). The skyrmion flow phase is stabilized in the intermediate current region while the chiral liquid is stabilized at high currents. There is a strong hysteresis when the current is changed across the phase boundary, which indicates that the transition between different phases is of the first order. In Figure 9.20a, we show the phase boundary obtained by increasing current. For completeness, the $D - J$ phase diagram is sketched schematically in Figure 9.20b starting from the ferromagnetic state at high fields. When $D = 0$, the system evolves from the ferromagnetic state directly into the chiral liquid. While for a nonzero D, the skyrmion flow phase is stabilized by the DM interaction in the intermediate current region.

Using typical parameters for MnSi, we estimate the instability occurs at a current of order of 10^{12} A/m^2, which is experimentally accessible. The fastest possible velocity for skyrmions is of the order of 100 m/s. The instability current depends on the magnon gap, and one can diminish the instability current by using materials with a smaller magnon gap.

Finally, we compare the dynamic creation and destruction of vortices in type II superconductors by current to the case with skyrmions. In superconductors, vortices can be induced by a current

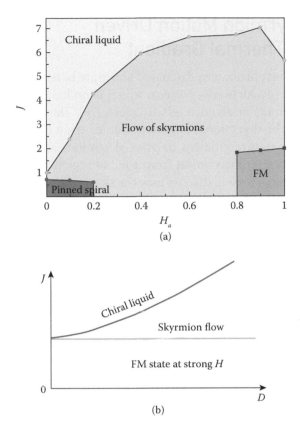

FIGURE 9.20   (a) Dynamical phase diagram for the model defined in Equations 9.3 and 9.6 with a nonzero D starting from the ground state configurations. (b) Dynamical phase diagram for the model defined in Equations 9.3 and 9.6 with $D = 0$ starting from a ferromagnetic state.

if the magnetic field produced by the current is bigger than the lower critical field Hc_1. In this case, vortex and antivortex pairs are created. At high currents, the motion of a normal core of vortex generates an electric field and shifts the quasiparticle distribution inside the normal core. As a result, quasiparticles are pumped out of the normal core and the size of normal core shrinks [41]. At a critical current, the vortex becomes unstable and the system jumps into the normal state. The dynamical creation and destruction of skyrmions discussed here has completely different origins. We also remark that the random defects have weak effects on the dynamical phase transition of skyrmions. This is because the random pinning potential produced by defects is quickly averaged out by the fast moving skyrmions, similar to the case of vortices [6,39].

Chapter 9

9.6 Skyrmion Motion Driven by Thermal Gradient

All fascinating phenomena discussed above are based on one fact—that the electron conduction is present, which is no longer true in insulating skyrmion materials such as Cu_2OSeO_3 [78]. The question is how to manipulate the skyrmion motion without itinerant electrons. One may apply a temperature gradient to drive skyrmions as one can expect a diffusion motion of skyrmion from the hot region to the cold region. In the following, we will discuss how the skyrmion moves under a temperature gradient [38,47].

9.6.1 Stochastic LLG Equation

In order to investigate the spin dynamics at finite temperature, we need to formulate the magnetization dynamics in the presence of thermal fluctuations, which has been extensively discussed in previous material [24]. In this section, we will derive the stochastic LLG equation using the Caldeira–Leggett approach [15]. The idea is to couple each magnetic moment to a fictitious reservoir consisting of harmonic oscillators φ_i, which are 3D vector fields interacting with the magnetic moments $\mathbf{n}$ in the simplest form:

$$\mathcal{H}_{int} = -\sum_i c_i \, \phi_i \cdot \mathbf{n}, \tag{9.74}$$

with c_i being the interaction strength. The Hamiltonian of the whole system is thus given by:

$$\mathcal{H} = \mathcal{H}_S + \mathcal{H}_{HO} + \mathcal{H}_{int}, \tag{9.75}$$

where $\mathcal{H}_S$ is the spin Hamiltonian, and $\mathcal{H}_{HO}$ is the Hamiltonian for the bath:

$$\mathcal{H}_{HO} = \frac{1}{2} \sum_i \left(m_i \, \dot{\phi}_i^2 + m_i \omega_i^2 \, \phi_i^2 \right), \tag{9.76}$$

with m_i and ω_i being the mass and frequency of each harmonic oscillator, respectively. One can readily derive the equation of motion for these oscillators, given by

$$m_i \, \ddot{\phi}_i + m_i \omega_i^2 \phi_i = c_i \mathbf{n}. \tag{9.77}$$

Its solution can be written in an integral form as:

$$\phi_i(t) = \phi_i(0)\cos\omega_i t + \dot{\phi}_i(0)\frac{1}{\omega_i}\sin\omega_i t + \int_0^t \frac{c_i}{m_i\omega_i}\sin[\omega_i(t-\tau)]\mathbf{n}(\tau)d\tau, \quad (9.78)$$

where $\phi_i(0)$ and $\dot{\phi}_i(0)$ are initial values. By varying the total Hamiltonian and Berry phase, contribution for $\mathbf{n}$ with respect to the moment $\mathbf{n}$, we thus obtain the Landau–Lifshitz equation:

$$\dot{\mathbf{n}} = -\gamma\mathbf{n}\times\left(-\frac{\partial\mathcal{H}}{\partial\mathbf{n}} + \sum_i c_i\phi_i\right) = -\gamma\mathbf{n}\times\left(-\frac{\partial H}{\partial\mathbf{n}}\right)$$

$$-\gamma\mathbf{n}\times\sum_i c_i\left[\phi_i(0)\cos\omega_i t + \dot{\phi}_i(0)\frac{1}{\omega_i}\sin\omega_i t \right. \qquad (9.79)$$

$$\left. + \int_0^t \frac{c_i}{m_i\omega_i}\sin[\omega_i(t-\tau)]\mathbf{n}(\tau)d\tau\right],$$

where γ is the gyromagnetic ratio, and the term $\sum_i c_i\phi_i$ serves as a new field action on spin $\mathbf{n}$. The last term can be manipulated via the integral by part:

$$\int_0^t \frac{c_i}{m_i\omega_i}\sin[\omega_i(t-\tau)]\mathbf{n}(\tau)d\tau = \int_0^t \frac{c_i}{m_i\omega_i^2}\frac{d}{d\tau}\cos[\omega_i(t-\tau)]\mathbf{n}(\tau)d\tau$$

$$(9.80)$$

$$= -\int_0^t \frac{c_i}{m_i\omega_i^2}\cos[\omega_i(t-\tau)]\dot{\mathbf{n}}(\tau)d\tau + \frac{c_i}{m_i\omega_i^2}[\mathbf{n}(t) - \mathbf{n}(0)\cos\omega_i t].$$

We denote $\mathbf{H}_{\text{eff}}$ as an effective magnetic field defined by $\mathbf{H}_{\text{eff}} = -\partial\mathcal{H}/\partial\mathbf{n}$. Consequently, we have:

$$\dot{\mathbf{n}} = -\gamma\mathbf{n}\times\mathbf{H}_{\text{eff}} - \gamma\mathbf{n}\times\mathbf{L} + \int_0^t \alpha(t-\tau)\dot{\mathbf{n}}(\tau)d\tau, \qquad (9.81)$$

where:

$$\mathbf{L} = \sum_i\left[c_i\phi_i(0) - \frac{c_i^2}{m_i\omega_i^2}\mathbf{n}(0)\right]\cos\omega_i t + c_i\dot{\phi}_i(0)\frac{1}{\omega_i}\sin\omega_i t,$$

$$(9.82)$$

$$\alpha(t-\tau) = \gamma\sum_i \frac{c_i^2}{m_i\omega_i^2}\cos\omega_i(t-\tau).$$

The last term in Equation 9.81 describes the dissipation of the system because it involves the first-order time derivative. We assume that the heat bath is in thermal equilibrium and the field φ obeys the Boltzmann distribution at T. As a result, we have:

$$
\langle \phi_j(0) \rangle = \frac{\displaystyle\int_{-\infty}^{+\infty} \phi_j \exp\left[-\sum_i \left(m_i \dot{\phi}_i^2 + m_i \omega_i^2 \phi_i^2 - c_i \phi_i \cdot \mathbf{n}(0) \right) \right]}{\displaystyle\int_{-\infty}^{+\infty} \exp\left[-\sum_i \left(m_i \dot{\phi}_i^2 + m_i \omega_i^2 \phi_i^2 - c_i \phi_i \cdot \mathbf{n}(0) \right) \right]}
\tag{9.83}
$$

$$
= \mathbf{n}(0) c_j / m_j \omega_j^2,
$$

and similarly:

$$
\langle \phi_i(0) \cdot \phi_j(0) \rangle = \frac{k_B T}{m_i \omega_i^2} \delta_{ij} + \frac{c_i}{m_i \omega_i^2} \frac{c_j}{m_j \omega_j^2} (1 - \delta_{ij}),
\tag{9.84}
$$

$$
\langle \dot{\phi}_i(0) \cdot \dot{\phi}_j(0) \rangle = \frac{k_B T}{m_i}.
\tag{9.85}
$$

Therefore, the field $\mathbf{L}$ satisfies the stochastic relation:

$$
\langle \mathbf{L}(t) \rangle = 0,
\tag{9.86}
$$

$$
\langle \mathbf{L}(t) \cdot \mathbf{L}(t') \rangle = k_B T \sum_i \frac{c_i^2}{m_i \omega_i^2} \cos \omega_i (t - t') = \frac{1}{\gamma} \alpha(t - t').
\tag{9.87}
$$

These two equations are the well-known fluctuation–dissipation relation required by a thermal equilibrium condition where thermal fluctuations carried by $\mathbf{L}$ are balanced by the dissipation carried by $\alpha(t - t')$. The Landau–Lifshitz equation now becomes:

$$
\dot{\mathbf{n}}(t) = -\gamma \mathbf{n}(t) \times \left(\mathbf{H}_{\text{eff}} + \mathbf{L}(t) \right) + \gamma \mathbf{n}(t) \times \int_0^t \langle \mathbf{L}(t) \cdot \mathbf{L}(\tau) \rangle \dot{\mathbf{n}}(\tau) d\tau.
\tag{9.88}
$$

In the simplest case, we assume that the stochastic field $\mathbf{L}$ is white noise with the correlation:

$$
\langle \mathbf{L}(t) \cdot \mathbf{L}(t') \rangle = k_B T \sum_i \frac{c_i^2}{m_i \omega_i^2} \cos \omega_i (t - t') = \frac{k_B T}{\gamma} \alpha \delta(t - t').
\tag{9.89}
$$

Finally, we obtain the stochastic Landau–Lifshitz–Gilbert equation in the presence of thermal fluctuations:

$$\dot{\mathbf{n}} = -\mathbf{n} \times (\mathbf{H}_{eff} + \mathbf{L}) + \alpha \mathbf{n} \times \dot{\mathbf{n}}, \tag{9.90}$$

which generalizes the deterministic LLG equation in Equation 9.6. We will use this equation to simulate skyrmion dynamics at finite temperatures.

9.6.2   Skyrmion Motions

In simulations, we generate a Gaussian distribution of $\mathbf{L}$, and the stochastic LLG equation (Equation 9.90) is integrated using the deterministic Heun scheme [24]. Here, a uniform but small temperature gradient is applied. The thermal fluctuation of each spin is about $k_B T$. As long as it is larger than the temperature difference between neighboring sites, the local equilibrium can be established and this stochastic LLG approach is justified. In this case, $\langle \mathbf{L}(\mathbf{r},t)\mathbf{L}(\mathbf{r},t) \rangle$ proportional to T is a linear function of the position. In what follows, the temperature gradient is applied along the x direction.

We employed the discretized version of the model Hamiltonian in Equation 9.5 by setting $A_z = 0$. The magnetic field $\mathbf{H}_a = H_a \hat{z}$ is perpendicular to the film. Here, H_a relates the real magnetic field h by $H_a = g\mu_B h$ with g the g-factor of spin. In the simulations, the Heisenberg exchange $J_{ex}/k_B = 50K$, and the strength of DM interaction $D = 0.5 J_{ex}$. Note in reality, D is an order of magnitude smaller. The advantage of using a large D in simulations is to reduce the skyrmion radius and minimize the calculation time. The lattice spacing a is 5 Å, and the system size in simulations is $150\,a \times 50\,a$, which is much larger than the skyrmion radius (about $5\,a$). Therefore, the finite size effect is negligible. Gilbert damping α is set to $\alpha = 0.1$ to accelerate the relaxation to local equilibrium in simulations.

In simulations, we initialize a skyrmion in the sample and the single skyrmion starts to move under the effect of the stochastic field. Although the instant velocity appears to be random, the overall velocity is nonzero. Figure 9.21a shows a typical simulation result of the time dependence of the center of mass of skyrmion according to Equation 9.64. At short time scales, the skyrmion oscillates around an average position because of thermal fluctuations. In the long run, the skyrmion drifts directionally. The mean velocity is derived by averaging over 1000 simulated events. Its relation with the temperature gradient is shown in Figure 9.21b. The longitudinal velocity is proportional to

Chapter 9

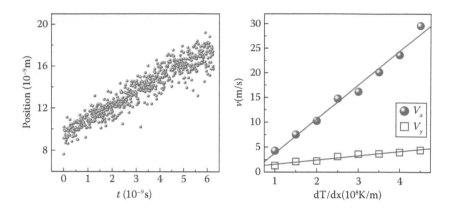

FIGURE 9.21   (a) A typical simulation results showing the skyrmion's instant longitudinal positions as a function of time. Although it fluctuates under finite temperature, a forward average velocity is observed. (b) A linear relation between the longitudinal velocity and the temperature gradient (top line) is displayed. A nonvanishing v_y (bottom line) indicates the Skyrmion Hall effect.

the temperature gradient. Meanwhile, the transverse velocity is nonzero and is proportional to the temperature gradient. The magnitude of the transverse velocity is one order of magnitude smaller than that of the longitudinal one. The transverse motion of skyrmions under a temperature gradient is the skyrmion's analogy of the Nernst effect.

A counterintuitive observation is that the skyrmion moves from the low temperature region to the high temperature one, as shown in Figure 9.22b. It is generally known that under a temperature gradient, particles like electrons should move to the cold terminal, due to the low density of particles at the cold end. This directional Brownian motion gives rise to various phenomena such as the Seebeck effect. Our results are entirely different from this conventional diffusion, suggesting a new mechanism of the skyrmion motion under a thermal gradient. This effect even holds also for the entire skyrmion crystal in which the whole skyrmion lattice is driven by the temperature gradient. As shown in Figure 9.22d, the whole crystal shifts towards high temperatures in a similar way. However, the skyrmion crystal distorts slightly because of thermal noise during the drifting process.

9.6.3   Magnon–Pulling Theory

To understand the counterintuitive motion of skyrmion under a thermal gradient, we use a magnon-pulling theory [40]. Here we consider magnon excitation in the ferromagnetic state with the

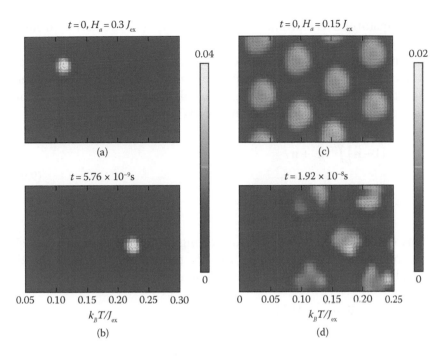

FIGURE 9.22   Snapshots of skyrmion motions. Color bars stand for the topological charge density q. (a) At the critical magnetic field of $H_a = 0.3J_{ex}$, a skyrmion is generated. (b) Under a temperature gradient, it moves from the low to high temperature region. (c, d) Skyrmion crystal moves in a similar way under a lower magnetic field $H_a = 0.15J_{ex}$.

spin fully polarized in the z-axis. The magnon is described by the spin components n_x and n_y, which are small. The magnon creation operator is $a^\dagger = (n_x - in_y)/\sqrt{2}$, and the magnon number operator is $\rho = a^\dagger a = \frac{1}{2}\left(n_x^2 + n_y^2\right)$. The spin component along the equilibrium direction is: $n_z = \sqrt{1 - \left(n_x^2 + n_y^2\right)} \approx 1 - \rho$. This result shows that each magnon carries spin one antiparallel to the equilibrium direction. Therefore, once there is a skyrmion under a temperature gradient, magnon as a low-lying excitation responds much more actively than the skyrmion itself because the density of the state of magnon is higher. As a typical quasiparticle, the magnon diffuses from the hot to the cold end to carry entropy. Because of the antiparallel alignment of the spin, the magnon current provides a negative transfer torque on the skyrmion. As a consequence of the conservation of the total angular momentum, the skyrmion moves in the opposite way to the magnon diffusion.

Chapter 9

To quantitatively formulate this physical picture and derive an equation of motion describing the effect of magnons on the equilibrium magnetization configurations, one has to separate the fast magnon dynamics from the collective slow dynamics of the skyrmion itself. To this end, let's decompose each local magnetization $\mathbf{n}$ into slow modes $\mathbf{n}_s$ and orthogonally fast modes $\mathbf{n}_f$:

$$\mathbf{n} = \left(1 - \mathbf{n}_f^2\right)^{1/2} \mathbf{n}_s + \mathbf{n}_f$$

$$\approx \mathbf{n}_s + \mathbf{n}_f - \frac{1}{2}\left(\mathbf{n}_f^2\right)\mathbf{n}_s + O\left(\mathbf{n}_f^3\right), \tag{9.91}$$

where $\mathbf{n}_s$ describes the equilibrium configuration, while $\mathbf{n}_f$ is responsible for the magnonic dynamics. The effective field $\mathbf{H}_{\mathrm{eff}}$ in the Stochastic Landau–Lifshitz–Gilbert equation is therefore:

$$\mathbf{H}_{\mathrm{eff}} = J_{\mathrm{ex}}\nabla^2\mathbf{n} - 2D\nabla\times\mathbf{n} + \mathbf{H}_a$$

$$= J_{\mathrm{ex}}\nabla_a\left[\nabla_a\mathbf{n}_s - \mathbf{n}_f(\nabla_a\mathbf{n}_f)\mathbf{n}_s - \frac{1}{2}\mathbf{n}_f^2\nabla_a\mathbf{n}_s + \nabla_a\mathbf{n}_f\right]$$

$$- 2D\left[\nabla\times\mathbf{n}_s + \nabla\times\mathbf{n}_f - \mathbf{n}_f(\nabla\mathbf{n}_f)\times\mathbf{n}_s - \frac{1}{2}\mathbf{n}_f^2\nabla\times\mathbf{n}_s\right] \tag{9.92}$$

$$+ \mathbf{H}_a + o\left(\mathbf{n}_f^3\right).$$

As $\mathbf{n}_f$ has a characteristic length scale of lattice constant a, compared to a length scale of $(J_{\mathrm{ex}}/D)a$ for $\mathbf{n}_s$, $\nabla_a\mathbf{n}_f \gg \nabla_a\mathbf{n}_s$. The torque are then reduced to:

$$-\mathbf{n}\times\mathbf{H}_{\mathrm{eff}} \approx -J_{\mathrm{ex}}\mathbf{n}_s\times\nabla^2\mathbf{n}_f - J_{\mathrm{ex}}\mathbf{n}_f\times\nabla^2\mathbf{n}_f.$$

Here, the DM contributions of $\mathbf{n}_f$ and $\mathbf{n}_s$ can be completely neglected because they involve the first-order spatial derivative, which is much smaller than the Heisenberg term. Contribution from the Zeeman term is negligible in a similar way. Besides, as $\mathbf{n}_f$ is a fast mode behaving as a function of sine or cosine in time, the terms linear in $\mathbf{n}_f$ vanish after averaging over time. Therefore,

$$-\mathbf{n}\times\mathbf{H}_{\mathrm{eff}} \approx -J_{\mathrm{ex}}\mathbf{n}_f\times\nabla^2\mathbf{n}_f = -J_{\mathrm{ex}}\nabla\left(\mathbf{n}_f\times\nabla\mathbf{n}_f\right). \tag{9.93}$$

Because $\mathbf{n}_f \perp \mathbf{n}_s$, we write $\mathbf{n}_f = \mathbf{n}_s \times \mathbf{d}_s$, where $\mathbf{d}_s$ is an arbitrary vector perpendicular to $\mathbf{n}_s$. Substituting it into the expression of the torque, we then obtain:

$$
\begin{aligned}
\mathbf{n}_f \times \nabla \mathbf{n}_f &= (\mathbf{n}_s \times \mathbf{d}_s) \times \nabla(\mathbf{n}_s \times \mathbf{d}_s) \\
&= (\mathbf{n}_s \times \mathbf{d}_s) \times (\nabla \mathbf{n}_s \times \mathbf{d}_s + \mathbf{n}_s \times \nabla \mathbf{d}_s) \\
&= -\mathbf{d}_s \nabla \mathbf{n}_s \cdot (\mathbf{n}_s \times \mathbf{d}_s) + \mathbf{n}_s \nabla \mathbf{d}_s \cdot (\mathbf{n}_s \times \mathbf{d}_s) \quad (9.94) \\
&\approx \mathbf{n}_s \nabla \mathbf{n} \cdot (\mathbf{n}_s \times \mathbf{d}_s) \\
&= \mathbf{n}_s \mathbf{j},
\end{aligned}
$$

where $\mathbf{j} = \nabla \mathbf{n} \cdot (\mathbf{n}_s \times \mathbf{n})$ is the magnon current. Here, the magnon current is a vector because we have projected the magnetic moment of the magnon current in the local magnetization direction. Here, we again use $\nabla_a \mathbf{n}_f \gg \nabla_a \mathbf{n}_s$. Consequently, we have:

$$
\nabla(\mathbf{n}_f \times \nabla \mathbf{n}_f) = (\partial_\mu \mathbf{n}_s) j_\mu + \mathbf{n}_s \partial_\mu j_\mu. \quad (9.95)
$$

We assume magnon current is steady so that $\nabla \cdot \mathbf{j} = 0$. Then, the torque is given by:

$$
-\mathbf{n} \times \mathbf{H}_{\text{eff}} = -J_{\text{ex}} j_\mu (\partial_\mu \mathbf{n}_s). \quad (9.96)
$$

The stochastic part of the spin transfer torque is:

$$
-\gamma \mathbf{n} \times \mathbf{L} = -\gamma(\mathbf{n}_s + \mathbf{n}_f) \times \mathbf{L} \approx -\gamma \mathbf{n}_s \times \mathbf{L}. \quad (9.97)
$$

The term $-\gamma \mathbf{n}_f \times \mathbf{L}$ is neglected as it is a linear function of $\mathbf{n}_f$ only, whose time average vanishes. Similarly, the Gilbert damping term becomes:

$$
\alpha \mathbf{n} \times \dot{\mathbf{n}} \approx \alpha \mathbf{n}_s \times \dot{\mathbf{n}}_s + \alpha \mathbf{n}_f \times \dot{\mathbf{n}}_f. \quad (9.98)
$$

As discussed above, $\mathbf{n}_f$ describes the cyclotron rotations with respect to $\mathbf{n}_s$, which is basically a sine or cosine function in time. The time derivative turns a sine function to cosine and vice versa. Therefore, the time average of $\alpha \mathbf{n}_f \times \dot{\mathbf{n}}_f$ vanishes so that:

$$
\alpha \mathbf{n} \times \dot{\mathbf{n}} \approx \alpha \mathbf{n}_s \times \dot{\mathbf{n}}_s. \quad (9.99)
$$

As a result, the LLG equation for $\mathbf{n}_s$ is given by:

$$
\dot{\mathbf{n}}_s = -\gamma J_{\text{ex}} j_\mu \partial_\mu \mathbf{n}_s - \gamma \mathbf{n}_s \times \mathbf{L} + \alpha \mathbf{n}_s \times \dot{\mathbf{n}}_s. \quad (9.100)
$$

Chapter 9

The magnon part of this equation of motion shows similarity with the magnetization dynamics in the presence of a charge current (cf. Equation 9.6). This similarity indicates that the magnon current $\mathbf{j}$ couples to the emergent gauge field $\mathbf{A}$ in the same way. The magnon naturally obeys the adiabatic approximation to the leading order as the spin of a magnon is antiparallel to the equilibrium direction of $\mathbf{n}_s$. This leads to the same equation as the case with an electric current. The only difference is the sign in front of $j_\mu \partial_\mu \mathbf{n}_s$ in Equations 9.100 and 9.6. This is because the magnetization $\mathbf{n}_s$ is parallel with the spin of itinerant electron, while antiparallel with the spin of magnon.

Let's again apply the collective coordinate $\mathbf{n}_s(r,t)=\mathbf{n}_s^0(\mathbf{r}-\mathbf{u}(t))$, where $\mathbf{n}_s^0$ is the ground configuration, and $\mathbf{u}(t)$ describes the position of the skyrmion. Inserting it into Equation 9.100 and integrating over the ground configuration, one finally gets the equation motion for the collective coordinates:

$$Q\varepsilon^{ij}\dot{u}_j(t)=Q\gamma J_{ex}\varepsilon^{ij}j_j+2\alpha\eta\dot{u}_i(t)+\frac{\gamma}{4\pi}\int d^2r\,\partial_i\mathbf{n}_s^0\cdot\mathbf{L}(\mathbf{r}+\mathbf{u},t),\quad(9.101)$$

where $\eta=\dfrac{1}{8\pi}\displaystyle\int d^2r\,\partial_i\mathbf{n}_s^0\times\partial_i\mathbf{n}_s^0$ is the shape factor. We define a collective stochastic force l_i acting on the skyrmion as a whole by $l_i(\mathbf{u},t)=\displaystyle\int d^2r\,\partial_i\mathbf{n}_s^0\cdot\mathbf{L}(\mathbf{r}+\mathbf{u},t)$, whose average then satisfies

$$\langle l_i(\mathbf{u},t)\rangle=0,\tag{9.102}$$

$$\langle l_i(\mathbf{u},t)l_j(\mathbf{u}',t')\rangle=8\pi\eta a^2\frac{\alpha k_BT}{\gamma}\delta_{ij}\delta(\mathbf{u}-\mathbf{u}')\delta(t-t').\tag{9.103}$$

The collective equation of motion resembles the standard Langevin equation. Let $P(\mathbf{r},t)$ be the probability to find the skyrmion at position $\mathbf{r}$ and time t. It thus satisfies the Fokker-Planck equation [24]:

$$\langle l_i(\mathbf{u},t)\rangle=0,$$

$$\frac{\partial P}{\partial t}=-\left[\gamma J_{ex}j_x-2\left(\frac{\gamma}{4\pi Q}\right)^2(\partial_x\xi')\right]\partial_xP-\gamma J_{ex}j_x\cdot 2\alpha\eta\partial_yP\tag{9.104}$$

$$+\left(\frac{\gamma}{4\pi Q}\right)^2\xi'\left(\partial_x^2+\partial_y^2\right)P.$$

At the current stage, we are only interested in the lowest order traveling wave solution of the Fokker-Planck equation, namely $P(\mathbf{r},t) = P(\mathbf{r} - \mathbf{v}t)$. The nonlinear last term describes broadening the wave package and can be neglected. Finally, we obtain the average velocity of the skyrmion in both the longitudinal and transverse directions:

$$v_x = \gamma J_{ex} j_x - \frac{\gamma}{\pi Q^2} \alpha \eta a^2 k_B \frac{dT}{dx} \equiv v_x^M - v^B,$$ (9.105)

$$v_y = 2\alpha \eta v_x^M.$$ (9.106)

The contributions from the magnon and the Brownian motion are separable and are denoted, respectively, by $v_{x,y}^M$ and v^B. Equation 9.105 shows explicitly that their effects are completely opposite: the skyrmion is pushed by the Brownian motion towards the cold terminal, while it is pulled back to the hot end by the magnon. On the other hand, as the temperature gradient is exerted along the x-direction, the Brownian motion along the y-direction vanishes on the average. Only the magnon transfer torque contributes to the transverse velocity, which is a factor α smaller than the longitudinal one, in good agreement with the numerical simulation in Figure 9.21b. This Hall effect of the skyrmion motion is closely related to the topology of the skyrmion texture captured by the nonzero topological charge Q. Generally speaking, a directional transverse motion requires the breaking of time reversal symmetry. Here, it is the dissipative damping α that breaks time reversal symmetry. Therefore, one expects that the transverse velocity is proportional to the Gilbert damping α.

In the previous discussion, we have considered a weak thermal noise where the skyrmion number does not change in the time scale of numerical calculations. Here we consider a high temperature case where skyrmions are created and destroyed dynamically by thermal fluctuations in the hot region. In the presence of temperature gradients, the thermally generated skyrmions diffuse into the cold region and remain there. Skyrmions are pushed to the cold region by newly generated skyrmions in the hot region because of the repulsion between them. Therefore, in the high temperature case, we recover the usual diffusion of skyrmions [47].

9.6.4   Electric Polarization

In chiral insulating magnets, electric polarization is allowed because of the breaking of the inversion symmetry. There exists magnetoelectric coupling; that is, the magnetization can induce electric polarization and vice versa. Here, we study the dynamics of electric polarization when a skyrmion in the ferromagnetic state is driven by a monochromatic magnon current [47]. The magnon current is generated by applying an ac magnetic field along the x-direction H_x, $_{ac}(x = L_x) = A\sin(\omega_m t)$ on one edge of the sample in addition to a dc magnetic field. The magnon current can be excited only if the driving frequency ω_m is bigger than the magnon gap: $\omega_m > \gamma H_a/(1 + \alpha^2)$. We consider two mechanisms for generation of electric polarization. Electric polarization can be induced by the so called d–p hybridization mechanism, which arises from the coupling between a transition metal ion and ligand (such as oxygen) ion with a single magnetic moment [32,33,92]. It was demonstrated experimentally that the electric polarization in Cu_2OSeO_3 is induced by the d–p hybridization mechanism [77,78]. In this case, the electric polarization depends on the direction of external magnetic field. We use a new coordinate system with the x-axis along the $[\bar{1}01]$ direction and the z-axis parallel to $\mathbf{H}_a$. Here, we study the case when $\mathbf{H}_a \parallel [110]$. The electric polarization in this configuration is given by:

$$\mathbf{P}_{dp} = \frac{P_0}{2}\left(-2n_x n_y, n_z^2 - n_x^2, 2n_y n_z\right), \tag{9.107}$$

with the magnetoelectric coupling constant $P_0 \approx 50$ $\mu C/m^2$ for Cu_2OSeO_3 [77].

In many multiferroic materials, the electric polarization was found to be induced by the so-called inverse DM mechanism [17,37]. The electric polarization is given by:

$$\mathbf{P}_{IDM} = \mathcal{P}_0\left[\hat{e}_x \times (n \times \partial_x \mathbf{n}) + \hat{e}_y \times (n \times \partial_y \mathbf{n})\right]. \tag{9.108}$$

When a skyrmion moves, the change of magnetization induces change in the electric polarization. When the sample is embedded in a circuit, the time-dependent polarization generates an ac electric current in the circuit. The electric current is given by $\mathbf{J}_e = \partial_t \mathbf{P}$. This electric current can be expressed in terms of the magnon current, for the inverse DM mechanism:

$$J_{e,IDM}(\omega) = \text{Im}\left[\frac{\mathcal{P}_0\omega}{\gamma}\left[\hat{e}_x \times \mathbf{j}_x(\omega) + \hat{e}_y \times \mathbf{j}_y(\omega)\right]\right]. \tag{9.109}$$

The ac electric current in the circuit induced by the motion of a skyrmion driven by magnon current is shown in Figure 9.23. The electric current oscillates at the same frequency as the magnon current,

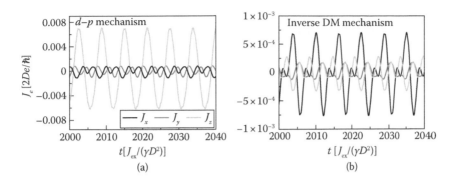

FIGURE 9.23 The ac current induced by the motion of a skyrmion in the ferromagnetic state driven by a monochromatic magnon current, according to (a) the d–p hybridization and (b) the inverse DM mechanism. The magnon current is generated by applying an ac magnetic field along the x-direction $H_{x,ac}(x = L_x) = A \sin(\omega_m t)$ on one edge with $A = 1.0 D^2/J_{ex}$, $\omega_m = 1.0 \gamma D^2/J_{ex}$. Here, $\alpha = 0.03$.

and there is no dc component. The amplitude of the oscillation is about 5×10^3 A/m^2 for typical parameters. For the inverse DM mechanism, the electric polarization vanishes for collinear spin textures ($\mathbf{P}_{IDM} = 0$) and skyrmions are then necessary to induce electric polarization. In the d–p hybridization mechanism, both the collinear background and skyrmion contribute to the electric polarization. This is the reason why the magnitude of the ac current for the d–p hybridization mechanism is much stronger than that for the inverse DM mechanism (see Figure 9.23).

9.7 Summary and Outlook

In summary, we have pedagogically discussed the dynamics of skyrmions in metallic and insulating systems. In helimagnet metals, an elegant structure of emergent electromagnetism, and various nontrivial effects such as magnon Doppler effects and dynamical distortions of skyrmion are studied. In the insulating materials, a temperature gradient drives the motion of skyrmions and induces electric polarization in the presence of magnetoelectric coupling. A counter-temperature gradient motion is observed and explained by the magnon pulling mechanism.

An important question is how to observe these theoretical predictions. To trace the skyrmion dynamics, a direct way is to apply time-resolved real-space imaging, such as the Lorentz transmission electron microscopy. However, in most transmission electron microscopy facilities, the temporal resolution is only on the scale of second, which might not be adequate in monitoring the motion of a single skyrmion. Up to now, the simplest current-driven skyrmion in nanostructured samples

Chapter 9

has not been observed yet. More advanced techniques and experimental designs are thus required to capture the real-time motion of skyrmions.

A major motivation of the skyrmion studies is its application in future memory devices. One can put skyrmions on the racetrack and use them as the digit carrier; "1" for skyrmion and "0" for ferromagnet. Our analysis in principle demonstrates the feasibility of current-driven skyrmion motions with low dissipation. The small sizes of skyrmions also indicate the high storage capacity of such devices. However, a practical question to be asked is how can we detect skyrmions by electric measurement rather than microscopy and neutron scattering. As introduced in this chapter, the topological Hall is a routine way, but the signal is tiny and highly entangled with the anomalous Hall effect. Simple longitudinal resistance signals being compatible with spintronic devices are highly desired.

The future direction of both theoretical and experimental efforts is to search for materials hosting very small skyrmion size. In this case, quantum effects become important and the classical approaches introduced in this chapter are questionable. Especially, the electrons exhibit strong quantum fluctuations, whose effect on the spin dynamics cannot be simply captured by the adiabatic spin transfer torque in the Landau–Liftshitz–Gilbert equation. A complete quantum transport analysis of the electrons and their interaction with local moments is needed. The technique of nonequilibrium Green's function could be employed, but it is a challenge to realize large-scale simulations.

Acknowledgments

The authors thank Cristian D. Batista, Charles Reichhardt, and Avadh Saxena, N. Nagaosa, Y. Tokura, M. Mostovoy, J. H. Han, M. Mochizuki, C. Chien and O. Tchernyshyov for helpful discussions and fruitful collaborations on this topic. Computer resources for numerical calculations were supported by the Institutional Computing Program at LANL. The work at LANL was carried out under the auspices of the NNSA of the US DOE at LANL under Contract No. DE-AC52-06NA25396, and was supported by the LANL LDRD-DR Program. The work at New Hampshire was supported by US NSF-ECCS 1408168.

References

1. T. Adams, et al. Long-wavelength helimagnetic order and skyrmion lattice phase in Cu_2OSeO_3. *Phys. Rev. Lett.*, 108(23):237204, 2012.
2. P. Bak and M. H. Jensen. Theory of helical magnetic structures and phase transitions in MnSi and FeGe. *J. Phys. C: Solid State Phys.*, 13(31):L881, 1980.

3. S. E. Barnes and S. Maekawa. Current-spin coupling for ferromagnetic domain walls in fine wires. *Phys. Rev. Lett.*, 95(10):107204, 2005.

4. A. Bauer, M. Garst, and C. Pfleiderer. Specific heat of the skyrmion lattice phase and field-induced tricritical point in MnSi. *Phys. Rev. Lett.*, 110:177207, 2013.

5. Ya. B. Bazaliy, B. A. Jones, and S.-C. Zhang. Modification of the Landau-Lifshitz equation in the presence of a spin-polarized current in colossal- and giant-magnetoresistive materials. *Phys. Rev. B*, 57:R3213, 1998.

6. R. Besseling, N. Kokubo, and P. H. Kes. Dynamic melting of confined vortex matter. *Phys. Rev. Lett.*, 91:177002, 2003.

7. G. Blatter, et al. Vortices in high-temperature superconductors. *Rev. Mod. Phys.*, 66:1125–1388, 1994.

8. A. Bogdanov and A. Hubert. Thermodynamically stable magnetic vortex states in magnetic crystals. *J. Magn. Magn. Mater.*, 138(3):255–269, 1994.

9. A. Bogdanov and A. Hubert. The stability of vortex-like structures in uniaxial ferromagnets. *J. Magn. Magn. Mater.*, 195(1):182–192, 1999.

10. A. N. Bogdanov and D. A. Yablonskii. Thermodynamically stable "vortices" in magnetically ordered crystals: The mixed state of magnets. *Sov. Phys. JETP*, 68:101, 1989.

11. P. Bruno, V. K. Dugaev, and M. Taillefumier. Topological hall effect and berry phase in magnetic nanostructures. *Phys. Rev. Lett.*, 93:096806, 2004.

12. S. Buhrandt and L. Fritz. Skyrmion lattice phase in three-dimensional chiral magnets from Monte Carlo simulations. *Phys. Rev. B*, 88:195137, 2013.

13. A. B. Butenko, A. A. Leonov, U. K. Rößler, and A. N. Bogdanov. Stabilization of skyrmion textures by uniaxial distortions in noncentrosymmetric cubic helimagnets. *Phys. Rev. B*, 82:052403, 2010.

14. F. Bttner, et al. Dynamics and inertia of skyrmionic spin structures. *Nat. Phys.*, 11(3):225–228, 2015.

15. A. O. Caldeira and A. J Leggett. Quantum tunnelling in a dissipative system. *Ann. Phys.*, 149(2):374–456, 1983.

16. G. Chen, A. Mascaraque, A. T. N'Diaye, and A. K. Schmid. Room temperature skyrmion ground state stabilized through interlayer exchange coupling. *Appl. Phys. Lett.*, 106(24):242404, 2015.

17. S.-W. Cheong and M. Mostovoy. Multiferroics: A magnetic twist for ferroelectricity. *Nat. Mater.*, 6(1):13–20, 2007.

18. G. H. Derrick. Comments on nonlinear wave equations as models for elementary particles. *J. Math. Phys.*, 5(9):1252, 1964.

19. H. Du, et al. Edge-mediated skyrmion chain and its collective dynamics in a confined geometry. *Nat.Commun.*, 6:8504, 2015.

20. H. Du, et al. Highly stable skyrmion state in helimagnetic MnSi nanowires. *Nano Lett.*, 14(4):2026–2032, 2014.

21. I. Dzyaloshinsky. A thermodynamic theory of weak ferromagnetism of antiferromagnetics. *J. Phys. Chem. Solids*, 4:241, 1958.

22. J. Fernández-Rossier, M. Braun, A. S. Núñez, and A. H. MacDonald. Influence of a uniform current on collective magnetization dynamics in a ferromagnetic metal. *Phys. Rev. B*, 69:174412, 2004.

23. A. Fert, V. Cros, and J. Sampaio. Skyrmions on the track. *Nat. Nanotechnol.*, 8:152–156, 2013.

24. J. L. Garca-Palacios and F. J. Lzaro. Langevin-dynamics study of the dynamical properties of small magnetic particles. *Phys. Rev. B*, 58(22):14937–14958, 1998.

25. S. Hayami, S. Z. Lin, and C. D. Batista. Bubble and skyrmion crystals in frustrated magnets with easy-axis anisotropy. *Phys. Rev. B*, 93:184413, (2016).

Chapter 9

26. S. Heinze, et al. Spontaneous atomic-scale magnetic skyrmion lattice in two dimensions. *Nat. Phys.*, 7(9):713, 2011.
27. T.-L. Ho. Spinor bose condensates in optical traps. *Phys. Rev. Lett.*, 81:742–745, 1998.
28. S. X. Huang and C. L. Chien. Extended skyrmion phase in epitaxial FeGe (111) thin films. *Phys. Rev. Lett.*, 108:267201, 2012.
29. J. Iwasaki, A. J. Beekman, and N. Nagaosa. Theory of magnon-skyrmion scattering in chiral magnets. *Phys. Rev. B*, 89:064412, 2014.
30. J. Iwasaki, M. Mochizuki, and N. Nagaosa. Current-induced skyrmion dynamics in constricted geometries. *Nat. Nanotechnol.*, 8(10):742–747, 2013.
31. J. Iwasaki, M. Mochizuki, and N. Nagaosa. Universal current-velocity relation of skyrmion motion in chiral magnets. *Nat. Commun.*, 4:1463, 2013.
32. C. Jia, S. Onoda, N. Nagaosa, and J. H. Han. Bond electronic polarization induced by spin. *Phys. Rev. B*, 74:224444, 2006.
33. C. Jia, S. Onoda, N. Nagaosa, and J. H. Han. Microscopic theory of spin-polarization coupling in multiferroic transition metal oxides. *Phys. Rev. B*, 76:144424, 2007.
34. W. Jiang, et al. Blowing magnetic skyrmion bubbles. *Science*, 349(6245):283–286, 2015.
35. F. Jonietz, et al. Spin transfer torques in MnSi at ultralow current densities. *Science*, 330(6011):1648, 2010.
36. K. Kadowaki, K. Okuda, and M. Date. Magnetization and magnetoresistance of MnSi. *J. Phys. Soc. Jpn.*, 51(8):2433–2438, 1982.
37. H. Katsura, N. Nagaosa, and A. V. Balatsky. Spin current and magnetoelectric effect in noncollinear magnets. *Phys. Rev. Lett.*, 95:057205, 2005.
38. L. Kong and J. Zang. Dynamics of an insulating skyrmion under a temperature gradient. *Phys. Rev. Lett.*, 111:067203, 2013.
39. A. E. Koshelev and V. M. Vinokur. Dynamic melting of the vortex lattice. *Phys. Rev. Lett.*, 73:3580–3583, 1994.
40. A. A. Kovalev and Y. Tserkovnyak. Thermomagnonic spin transfer and Peltier effects in insulating magnets. *EPL (Europhys. Lett.)*, 97(6):67002, 2012.
41. A. I. Larkin and Y. N. Ovchinnikov. Nonlinear conductivity of superconductors in the mixed state. *Sov. Phys. JETP*, 41:960, 1976.
42. M. Lee, et al. Unusual hall effect anomaly in MnSi under pressure. *Phys. Rev. Lett.*, 102:186601, 2009.
43. A. O. Leonov and M. Mostovoy. Multiply periodic states and isolated skyrmions in an anisotropic frustrated magnet. *Nat. Commun.*, 6:8275, 2015.
44. Y. Li, et al. Robust formation of skyrmions and topological hall effect anomaly in epitaxial thin films of MnSi. *Phys. Rev. Lett.*, 110:117202, 2013.
45. Z. Li and S. Zhang. Domain-wall dynamics and spin-wave excitations with spin-transfer torques. *Phys. Rev. Lett.*, 92:207203, 2004.
46. S. Z. Lin and S. Hayami. Ginzburg-Landau theory for skyrmions in inversion-symmetric magnets with competing interactions. *Phys. Rev. B 93,064430* (2016).
47. S. Z. Lin, C. D. Batista, C. Reichhardt, and A. Saxena. AC current generation in chiral magnetic insulators and skyrmion motion induced by the spin Seebeck effect. *Phys. Rev. Lett.*, 112:187203, 2014.
48. S. Z. Lin, C. D. Batista, and A. Saxena. Internal modes of a skyrmion in the ferromagnetic state of chiral magnets. *Phys. Rev. B*, 89(2):024415, 2014.
49. S. Z. Lin, C. Reichhardt, C. D. Batista, and A. Saxena. Driven skyrmions and dynamical transitions in chiral magnets. *Phys. Rev. Lett.*, 110(20):207202, 2013.
50. S. Z. Lin, C. Reichhardt, C. D. Batista, and A. Saxena. Dynamics of skyrmions in chiral magnets: Dynamic phase transitions and equation of motion. *J. Appl. Phys.*, 115(17):17D109, 2014.

51. S. Z. Lin, C. Reichhardt, and A. Saxena. Manipulation of skyrmions in nanodisks with a current pulse and skyrmion rectifier. *Appl. Phys. Lett.*, 102(22):222405, 2013.

52. S. Z. Lin and A. Saxena. Noncircular skyrmion and its anisotropic response in thin films of chiral magnets under a tilted magnetic field. *Phys. Rev. B*, 92:180401, 2015.

53. I. Makhfudz, B. Krüger, and O. Tchernyshyov. Inertia and chiral edge modes of a skyrmion magnetic bubble. *Phys. Rev. Lett.*, 109:217201, 2012.

54. M. Mochizuki, et al. Thermally driven ratchet motion of a skyrmion microcrystal and topological magnon Hall effect. *Nat. Mater.*, 13:241, 2014.

55. M. Mochizuki. Spin-wave modes and their intense excitation effects in skyrmion crystals. *Phys. Rev. Lett.*, 108:017601, 2012.

56. M. Mochizuki and S. Seki. Magnetoelectric resonances and predicted microwave diode effect of the skyrmion crystal in a multiferroic chiral-lattice magnet. *Phys. Rev. B*, 87(13):134403, 2013.

57. T. Moriya. Anisotropic superexchange interaction and weak ferromagnetism. *Phys. Rev.*, 120:91, 1960.

58. T. Moriya. New mechanism of anisotropic superexchange interaction. *Phys. Rev. Lett.*, 4:228–230, Mar 1960.

59. S. Mühlbauer, et al. Skyrmion lattice in a chiral magnet. *Science*, 323(5916):915, 2009.

60. N. Nagaosa and Y. Tokura. Topological properties and dynamics of magnetic skyrmions. *Nat. Nanotechnol.*, 8(12):899–911, 2013.

61. A. Neubauer, et al. Topological Hall effect in the A phase of MnSi. *Phys. Rev. Lett.*, 102:186602, 2009.

62. Y. Nii, et al. Elastic stiffness of a skyrmion crystal. *Phys. Rev. Lett.*, 113:267203, 2014.

63. Y. Okamura, et al. Microwave magnetoelectric effect via skyrmion resonance modes in a helimagnetic multiferroic. *Nat. Commun.*, 4, 2013.

64. T. Okubo, S. Chung, and H. Kawamura. Multiple-q states and the skyrmion lattice of the triangular-lattice Heisenberg antiferromagnet under magnetic fields. *Phys. Rev. Lett.*, 108:017206, 2012.

65. Y. Onose, et al. Observation of magnetic excitations of skyrmion crystal in a helimagnetic insulator Cu_2OSeO_3. *Phys. Rev. Lett.*, 109:037603, 2012.

66. N. Romming, et al. Writing and deleting single magnetic skyrmions. *Science*, 341(6146):636–639, 2013.

67. N. Romming, et al. Field-dependent size and shape of single magnetic skyrmions. *Phys. Rev. Lett.*, 114:177203, 2015.

68. U. K. Rößler, A. N. Bogdanov, and C. Pfleiderer. Spontaneous skyrmion ground states in magnetic metals. *Nature*, 442:797, 2006.

69. T. Sakakibara, H. Mollymoto, and M. Date. Magnetization and magnetoresistance of MnSi. *J. Phys. Soc. Japan*, 51(8):2439–2445, 1982.

70. J. Sampaio, et al. Nucleation, stability and current-induced motion of isolated magnetic skyrmions in nanostructures. *Nat. Nanotechnol.*, 8(11):839–844, 2013.

71. S. Schroeter and M. Garst. Scattering of high-energy magnons off a magnetic skyrmion. *Low Temp. Phys.* 41, 817 (2015).

72. N. L. Schryer and L. R. Walker. The motion of 180 domain walls in uniform dc magnetic fields. *J. Appl. Phys.*, 45:5406, 1974.

73. T. Schulz, et al. Emergent electrodynamics of skyrmions in a chiral magnet. *Nat. Phys.*, 8(4):301, 2012.

74. C. Schütte and M. Garst. Magnon-skyrmion scattering in chiral magnets. *Phys. Rev. B*, 90:094423, 2014.

Chapter 9

75. C. Schütte, J. Iwasaki, A. Rosch, and N. Nagaosa. Inertia, diffusion, and dynamics of a driven skyrmion. *Phys. Rev. B*, 90:174434, 2014.

76. T. Schwarze, et al. Universal helimagnon and skyrmion excitations in metallic, semiconducting and insulating chiral magnets. *Nat. Mater.*, 14:478, 2015.

77. S. Seki, S. Ishiwata, and Y. Tokura. Magnetoelectric nature of skyrmions in a chiral magnetic insulator Cu_2OSeO_3. *Phys. Rev. B*, 86:060403, 2012.

78. S. Seki, X. Z. Yu, S. Ishiwata, and Y. Tokura. Observation of skyrmions in a multiferroic material. *Science*, 336(6078):198, 2012.

79. K. Shibata, et al. Large anisotropic deformation of skyrmions in strained crystal. *Nat. Nanotechnol.*, 10(7):589–592, 2015.

80. T. H. R. Skyrme. A non-linear field theory. *Proc. R. Soc. A*, 260(1300):127–138, 1961.

81. T. H. R. Skyrme. A unified field theory of mesons and baryons. *Nucl. Phys.*, 31:556–569, 1962.

82. S. L. Sondhi, A. Karlhede, S. A. Kivelson, and E. H. Rezayi. Skyrmions and the crossover from the integer to fractional quantum hall effect at small Zeeman energies. *Phys. Rev. B*, 47:16419–16426, 1993.

83 G. Tatara and H. Kohno. Theory of current-driven domain wall motion: Spin transfer versus momentum transfer. *Phys. Rev. Lett.*, 92:086601, 2004.

84. G. Tatara, H. Kohno, and J. Shibata. Microscopic approach to current-driven domain wall dynamics. *Phys. Rep.*, 468:213, 2008.

85. A. A. Thiele. Steady-state motion of magnetic domains. *Phys. Rev. Lett.*, 30:230–233, 1973.

86. Y. Tokunaga, et al. A new class of chiral materials hosting magnetic skyrmions beyond room temperature. *Nat. Commun.*, 6, 2015.

87. V. Vlaminck and M. Bailleul. Current-induced spin-wave doppler shift. *Science*, 322(5900):410–413, 2008.

88. J. S. White, et al. Electric field control of the skyrmion lattice in Cu_2OSeO_3. *J. Phys.: Condens. Matter*, 24(43):432201, 2012.

89. J. S. White, et al. Electric-field-induced skyrmion distortion and giant lattice rotation in the magnetoelectric insulator Cu_2OSeO_3. *Phys. Rev. Lett.*, 113:107203, 2014.

90. D. C. Wright and N. D. Mermin. Crystalline liquids: The blue phases. *Rev. Mod. Phys.*, 61:385–432, 1989.

91. G. M. Wysin and A. R. Völkel. Normal modes of vortices in easy-plane ferromagnets. *Phys. Rev. B*, 52:7412–7427, 1995.

92. J. H. Yang, et al. Strong Dzyaloshinskii-Moriya interaction and origin of ferroelectricity in Cu_2OSeO_3. *Phys. Rev. Lett.*, 109:107203, 2012.

93. S. D. Yi, S. Onoda, N. Nagaosa, and J. H. Han. Skyrmions and anomalous Hall effect in a Dzyaloshinskii-Moriya spiral magnet. *Phys. Rev. B*, 80:054416, 2009.

94. X. Z. Yu, et al. Near room-temperature formation of a skyrmion crystal in thin-films of the helimagnet FeGe. *Nat. Mater.*, 10(2):106, 2011.

95. X. Z. Yu, et al. Skyrmion flow near room temperature in an ultralow current density. *Nat. Commun.*, 3:988, 2012.

96. X. Z. Yu, et al. Real-space observation of a two-dimensional skyrmion crystal. *Nature*, 465(7300):901, 2010.

97. X. Yu, M. et al. Magnetic stripes and skyrmions with helicity reversals. *PNAS*, 109(23):8856–8860, 2012.

98. J. Zang, M. Mostovoy, J. H. Han, and N. Nagaosa. Dynamics of skyrmion crystals in metallic thin films. *Phys. Rev. Lett.*, 107:136804, 2011.

99. B. Zhang, et al. Microwave-induced dynamic switching of magnetic skyrmion cores in nanodots. *Appl. Phys. Lett.*, 106(10):102401, 2015.

100. Y. Zhou and M. Ezawa. A reversible conversion between a skyrmion and a domain-wall pair in a junction geometry. *Nat. Commun.*, 5:4652, 2014.

10. Novel Topological Resonant Excitations of Coupled Skyrmions

Yingying Dai, Han Wang, and Zhidong Zhang

Chinese Academy of Sciences, Shenyang, People's Republic of China

Chapter 10

A skyrmion, a particle-like topological spin texture where the spins point in all directions covering a whole sphere surface, has nonlocal topological density distribution. When the skyrmion is driven by a magnetic field or a current, the nonlocal topological density distribution may obtain large deformation and lead to novel dynamical behaviors. In this chapter, we investigate the static and dynamical properties of coupled skyrmions obtained in Co/Ru/Co nanodisks, and discuss how the topological density distribution affects the dynamical behaviors.

10.1 Introduction

Skyrmions were introduced by Tony Skyrme to describe localized, particle-like configurations in the field of pion particles (Skyrme 1962). Since that paper, skyrmions have been developed in many fields, such as classic liquids (Cross and Hohenberg 1993), liquid crystal (Wright and Mermin 1989), Bose–Einstein condensates (Al Khawaja and Stoof 2001), quantum Hall magnets (Sondhi et al. 1993), and two-dimensional isotropic magnetic materials (Belavin and Polyakov 1975; Waldner 1986; Zaspel et al. 1995). Bogdanov et al. predict theoretically that skyrmion can exist in magnets without inversion symmetry, which is associated with the Dzyaloshinskii–Moriya interaction (DMI) (Bogdanov and Yablonskii 1989). In recent years, skyrmions have been discovered in different magnets, such as in helimagnets (Mühlbauer et al. 2009; Yu et al. 2010; Yu et al. 2011), a monolayer Fe film (Heinze et al. 2011), multiferroics (Seki et al. 2012), a doped antiferromagnet (Raičević et al. 2011), and Sc-doped barium ferrite with inversion symmetry (Yu et al. 2012). However, most of the skyrmions observed in experiment were induced by an external magnetic field and existed at low temperature, which limits the technological application of skyrmions (Kiselev et al. 2011). Therefore, searching skyrmion states at or above room temperature without an external magnetic field is an important issue.

Topologically, spin textures have attracted long-standing attention because of their peculiar spin configurations and dynamic behaviors for promising applications and, in particular, the rich physics involved (Martin et al. 2003). Extensive theoretical and experimental researches have revealed that their dynamical behaviors depend on their topological charges and spin textures (Kasai et al. 2006; Pribiag et al. 2007; Dussaux et al. 2010). Dynamics of skyrmions have shown many interesting phenomena, such as the magnetoelectric effect (Belesi et al. 2012; Okamura et al. 2013), the topological Hall effect and the skyrmion Hall effect (Neubauer et al. 2009; Kanazawa et al. 2011; Li et al. 2013; Nagaosa and Tokura 2013), and three spin-wave modes (Mochizuki 2012), which lead to an increasing attention on skyrmion dynamics. It is worthwhile to investigate other dynamical behaviors of skyrmions apart from the above. Unlike a vortex moving with local deformation at the core and a bubble with local deformation at the domain wall (Kammerer et al. 2011; Makhfudz et al. 2012; Petit-Watelot et al. 2012), a skyrmion may move with a large nonlocal deformation which leads to the mechanisms of skyrmion dynamics being unclear.

In this chapter, we report on an effort to find the criteria of the stability of a skyrmion ground state and the resonant excitations in a Co/Ru/Co nanodisk. The exposition is organized as follows: Section 10.2 discusses how to obtain coupled skyrmions in the nanodisk and shows the influence of nanodisk size and magnetic field on the formation of the coupled skyrmions. Section 10.3 investigates the gyrotropic motion of the coupled skyrmions. Section 10.4 presents resonant excitations of the coupled skyrmions in a dual-frequency or single-frequency microwave field and discusses the influence of frequency, frequency ratio, and the effective mass on dynamics of the coupled skyrmions. Section 10.5 investigates current-driven resonant excitations of the coupled skyrmions and shows the effect of the frequency and amplitude ratio and magnetostatic interaction on the resonant excitations of the skyrmions.

10.2   Skyrmion State in Co/Ru/Co Nanodisks

10.2.1   Formation of Coupled Skyrmions in Co/Ru/Co Nanodisks

To find the way to obtain skyrmion ground state at or above room temperature, we use the three-dimensional object-oriented micromagnetic framework (OOMMF) code (Donahue and Porter, 1999) to investigate a magnetic material with high Curie temperature and appropriate uniaxial magnetocrystalline anisotropy. Here, the sandwiched structural Co/Ru/Co nanodisk is chosen to be investigated. The energy considering in the Co/Ru/Co nanodisk is as follows (Albuquerque and Freitas 1997):

$$E = \int_V \left[A(\nabla \cdot \boldsymbol{m})^2 + K_u m_z^2 - \frac{1}{2}\mu_0 M_s (\boldsymbol{H}_d \cdot \boldsymbol{m}) \right] d^3 r$$

$$+ \int_S J_1 (1 - \boldsymbol{m} \cdot \boldsymbol{m}') \, dS, \tag{10.1}$$

where $\boldsymbol{m}(\boldsymbol{r}, t) = \boldsymbol{M}(\boldsymbol{r}, t)/M_s$ is the unit vector of local magnetization, $\boldsymbol{m}'$ the magnetization vector at the interface, A the exchange stiffness,

Chapter 10

K_u the uniaxial anisotropy constant, H_d the demagnetizing field, and J_1 the interfacial coupling strength. The first term in Equation 10.1 is the exchange energy, the uniaxial anisotropy energy, and the demagnetization energy, and the second term is the RKKY interaction between the two Co nanolayers.

The material parameters of cobalt are saturation magnetization $M_s = 1.4 \times 10^6$ A/m, exchange stiffness $A = 3 \times 10^{-11}$ J/m, and uniaxial anisotropy constant $K_u = 5.2 \times 10^5$ J/m^3 in the direction perpendicular to the nanodisk plane. Interfacial coupling coefficients of the adjacent surfaces for different thicknesses of Ru were consisted with Bloemen et al. (1994). The radius R of the Co/Ru/Co nanodisk was tuned from 80 to 350 nm. The thickness L of Co varied from 5 to 25 nm, and that of Ru varied from 1 to 20 nm. The cell size was $4 \times 4 \times 1$ nm^3 for simulating magnetic ground states, and reduced to $2 \times 2 \times 1$ nm^3 to test the stability of the obtained states. The dimensionless damping α was chosen to be 0.25 for rapid convergence.

Sketch of a Co (20 nm)/Ru (2 nm)/Co (20 nm) nanodisk is shown in Figure 10.1a. The micromagnetic simulation result of the nanodisk with $R = 100$ nm evolving from an out-of-plane-like initial state is demonstrated in Figure 10.1b. Both the top and bottom nanolayers are typical skyrmion-like configurations with opposite chiralities, that is, right-handed for the top one and left-handed for the bottom one. In the center of the nanodisk, the magnetization M is down (along the $-z$-axis), at the boundaries, it is up (along the $+z$-axis), and in the intermediate regions, it rotates gradually from the $-z$-axis to the $+z$-axis. The skyion number for one of the two nanolayers is calculated by (Rajaraman 1987):

$$S = \frac{1}{4\pi} \iint q d^2 r, q \equiv \boldsymbol{m} \cdot \left(\frac{\partial \boldsymbol{m}}{\partial x} \times \frac{\partial \boldsymbol{m}}{\partial y} \right), \tag{10.2}$$

where $\boldsymbol{m}(\boldsymbol{r})$ is the unit vector of local magnetization and q the topological density. For the top or the bottom nanolayer, S is calculated to be approximately -1, which shows the signature of skyrmion. Similar skyrmion configurations have been found in a patterned Co/Ru/Co nanodisk array, as shown in Figure 10.2. Two-dimensional periodic boundary condition is used. The distance between the centers of two nearest-neighboring nanodisks was 250 nm and R was 100 nm. In this case, the stray field of two nearest-neighboring nanodisks has little influence on the skyrmion configurations. Sun et al. (2013) have obtained similar artificial skyrmion lattice in ordinary magnets.

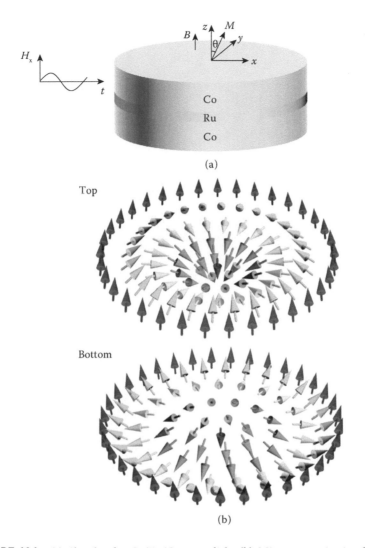

FIGURE 10.1 (a) Sketch of a Co/Ru/Co nanodisk. (b) Micromagnetic simulation result for a Co (20 nm)/Ru (2 nm)/Co (20 nm) nanodisk with the radius $R = 100$ nm. Arrows correspond to the directions of the local magnetization at every point. (From Dai YY, et al., *Phys. Rev. B.*, 88, 054403, 2013. With permission.)

The formation of a stable magnetic state follows the minimization of the Gibbs free energy of a magnetic system. Figure 10.3a illustrates the total energy, exchange energy, uniaxial anisotropy energy, demagnetization energy, and antiferromagnetic coupling energy as functions of time for the case shown in Figure 10.1. The exchange, uniaxial anisotropy, and demagnetization energies are two orders of magnitude higher than the interfacial antiferromagnetic coupling energy and, thus,

Chapter 10

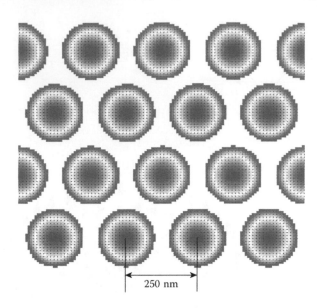

250 nm

FIGURE 10.2 Spin configurations of a Co (20 nm)/Ru (2 nm)/Co (20 nm) nanodisk array evolving from an out-of-plane-like initial state. Here, two-dimensional periodic boundary condition is used and only the result of the top nanolayers is shown.

contribute mainly to the formation of the skyrmions. Initially, the total energy is very high due to the significant demagnetization energy. To minimize the total energy, the demagnetization energy decreases significantly, while both the exchange and uniaxial anisotropy energy increase rapidly. At about 0.5 ns, a balance is reached and the total energy is almost unchanged. In the meantime, the skyrmion number S drops to about -1 for the top and bottom nanolayers, as shown in Figure 10.3b, indicating the formation of skyrmion configurations.

As illustrated in many articles, the DMI plays an important role in the formation of skyrmion states, as it favors canted spins. The DMI is defined as follows (Dzyaloshinskii 1958; Moriya 1960; Yu et al. 2010):

$$H_{DMI} = \iint DM \cdot (\nabla \times M) dx dy, \tag{10.3}$$

where D is the DMI constant. But in our Co/Ru/Co nanodisk, skyrmions are obtained without the DMI. As discussed above, the competition among different energy terms is crucial to the formation of skyrmions. To quantify the effect of the competition, we define a quantity Ψ to mimic the DMI:

$$\Psi = \iint M \cdot (\nabla \times M) dx dy, \tag{10.4}$$

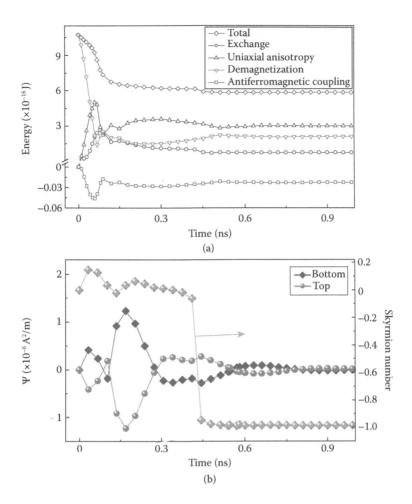

FIGURE 10.3 (a) Different energies as functions of time for a Co (20 nm)/Ru (2 nm)/ Co (20 nm) nanodisk with $R = 100$ nm. (b) Ψ and the skyrmion number as functions of time. (From Dai YY, et al., *Phys. Rev. B.*, 88, 054403, 2013. With permission.)

which is shown in Figure 10.3b as a function of time for the top and the bottom Co nanolayers. It shows that Ψ for both Co nanolayers significantly changes when the energies compete drastically with each other, and remains relatively unchanged when the system reaches equilibrium, which indicates that the competition among the energies creates an effect similar to that of the DMI to form the skyrmions.

10.2.2   Influence of Size on Formation of Skyrmions

To obtain the most stable magnetic states for different sizes, different initial magnetic states, like vortex (with the same or opposite chirality),

Chapter 10

in-plane and out-of-plane spin configurations are used. The dependence of the magnetic ground state on the thicknesses of ruthenium (t_{Ru}) and cobalt (t_{Co}) layers derived from micromagnetic simulations is shown in Figure 10.4 for t_{Ru} varies from 1 to 20 nm, t_{Co} from 5 to 25 nm, and R is fixed at 100 nm. There are four regions in the phase diagram: vortex, skyrmion, multidomain, and mixed states. In the mixed-state region more than one stable or metastable state was found. Stable skyrmions can be formed only in a small fraction of the phase diagram where t_{Co} is in the range of 13–23 nm and t_{Ru} smaller than 4 nm, which is determined by the balance between different energies, especially the interlayer magnetostatic interaction. A multidomain state appears when t_{Co} is set beyond 25 nm, regardless of t_{Ru}, to minimize the demagnetization energy. When t_{Ru} is chosen to be very thick, the interlayer magnetostatic interaction between the two Co nanolayers becomes weak and the intralayer demagnetization factor plays a more essential role, constraining the magnetization in the plane to form the vortex state, which minimizes the demagnetization energy. On the other hand, in setting t_{Co} smaller than 8 nm, the effect of the interfacial antiferromagnetic or ferromagnetic coupling becomes important.

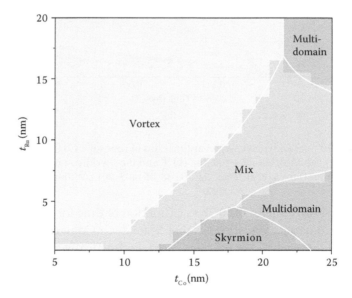

FIGURE 10.4 Magnetic phase diagram derived from micromagnetic simulations. The spin textures depending on the thickness of the Co (t_{Co}) layer and of the Ru (t_{Ru}) layer illustrate four regions: vortex state, skyrmion state, multidomain state, and mixed state. Phase boundaries are marked by lines. (From Dai YY, et al., *Phys. Rev. B.*, 88, 054403, 2013. With permission.)

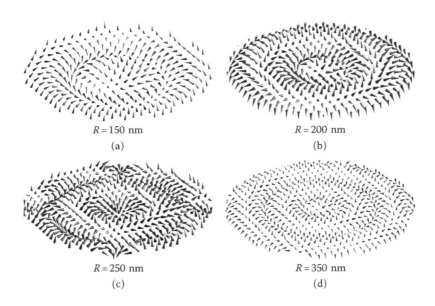

$R = 150$ nm

(a)

$R = 200$ nm

(b)

$R = 250$ nm

(c)

$R = 350$ nm

(d)

FIGURE 10.5   Influence of the diameter of the Co (20 nm)/Ru (2 nm)/Co (20 nm) nanodisk on stable magnetic states. Only the spin configurations of the top nanolayer are shown here.

When R is larger than 100 nm, we can obtain skyrmions with $|S|$ either 1 or 0, or even a multidomain state with a skyrmion core in the center, as shown in Figure 10.5. When R reaches 150 nm, θ twists 2π from the center to the boundary of the nanodisk, and thus $|S|$ becomes 0 (Choi et al. 2012). When R increases to 200 nm, θ twists 3π from the center to the boundary of the nanodisk and $|S|$ equals to 1. However, when R is equal or larger than 250 nm, the multidomain state appears with a skyrmion core in the center of the nanodisk, as demonstrated in Figure 10.5c and d.

10.2.3   Influence of an External Magnetic Field on the Skyrmions

Influence of an external magnetic field along the $+z$-axis on the skyrmions is shown in Figure 10.6. The magnetization increases almost linearly as the magnetic field increases until saturation at 0.64 T, at which the skyrmion is completely suppressed. Concurrently, the skyrmion numbers of both the top and bottom nanolayers change slightly, from approximately -1.0 to -0.9 initially, and then rise sharply to 0 when the field reaches 0.64 T. It indicates that the skyrmions can be

Chapter 10

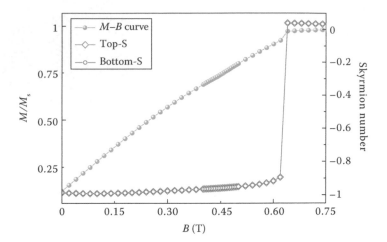

FIGURE 10.6 Normalized magnetization curve (*M–B* curve) of the Co (20 nm)/Ru (2 nm)/Co (20 nm) nanodisk with $R = 100$ nm and the corresponding skyrmion number as a function of the field for the top ("top-S") and bottom ("bottom-S") nanolayers. (From Dai YY, et al., *Phys. Rev. B.*, 88, 054403, 2013. With permission.)

stable in the external field below 0.64 T, and broken by the field higher than 0.64 T (a ferromagnetic state), which verifies that a skyrmion is not topologically equivalent to a ferromagnetic state.

10.3 Gyrotropic Motion of the Coupled Skyrmions

To investigate the dynamics of coupled skyrmions, R is chosen to be 100 nm, $L = 18$ nm, $\alpha = 0.02$, and the cell size is $2 \times 2 \times 2$ nm³. As discussed in Chapter 1, when a skyrmion moves, it may create a large nonlocal deformation of the topological density distribution. To consider the large nonlocal nature of the deformation, we use a guiding center (R_x, R_y) to depict its dynamics. The guiding center is defined by (Papanicolaou and Tomaras 1991):

$$R_x = \frac{\iint xq\,dx\,dy}{\iint q\,dx\,dy}, R_y = \frac{\iint yq\,dx\,dy}{\iint q\,dx\,dy}, \tag{10.5}$$

where q is the topological density defined in Equation 10.2. To investigate the gyrotropic motion of the coupled skyrmions, a pulsed magnetic field with 10-ns pulse width and 50-mT magnitude is applied along the +x-direction to the top nanolayer, as shown in Figure 10.7a.

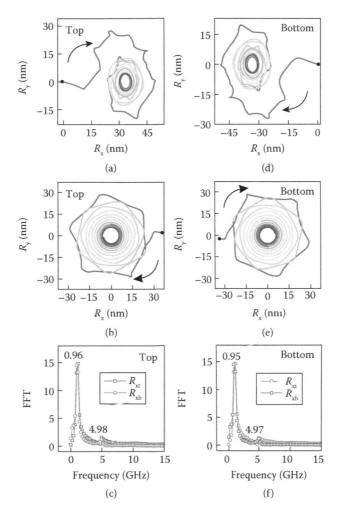

FIGURE 10.7 Influence of skyrmion chiralities on the gyrotropic motion of the guiding centers. (a, d) Motion of skyrmions with opposite chiralities in a pulsed magnetic field applied to the top or bottom nanolayer along the +x-direction, respectively. (b, e) Gyrotropic motion of skyrmions with opposite chiralities after the field is switched off. (c, f) The corresponding FFT spectra of the hexagonal trajectories in (b) and (e). (R_{xt}, R_{yt}) and (R_{xb}, R_{yb}) are the positions of the guiding center in the top and bottom nanolayers, respectively. (From Dai YY, et al., *Phys. Rev. B.*, 88, 054403, 2013. With permission.)

Under the action of the field, the guiding center of the top nanodisk moves and rotates toward its new equilibrium position along the field, which is different from the motion of the vortex core in a magnetic field (Schneider et al. 2000; Tanase et al. 2009). The trajectory of (R_x, R_y) is initially like a star and then damped to an elliptical orbit around the new equilibrium position (33.5 nm, 0 nm). Even though the bottom nanolayer is lack of a field, it is still stimulated by the motion of the top one through the strong interlayer magnetostatic interaction, and moves almost synchronously with the similar trajectory (Dai et al. 2013).

Once the field is switched off, (R_x, R_y) in both nanolayers begin to rotate back to the original position $(0, 0)$, as shown in Figure 10.7b. At the beginning, the trajectory of (R_x, R_y) is hexagons (before 15.24 ns), which are similar to hypocycloid involute. The corresponding fast Fourier transform (FFT) spectrum of the hexagonal trajectory has two eigenfrequencies, 0.96 and 4.98 GHz, with the ratio of about 1:5 as demonstrated in Figure 10.7c. Because of the opposite chiralities of the two skyrmions (the right-handed for the top nanolayer, and left-handed for the bottom one), the guiding centers move to opposite directions when a pulse magnetic field is applied to one of the two nanolayers, as shown in Figure 10.7a and d. After the applied field is switched off, the guiding centers gyrate around the nanodisk center and show hexagonal trajectories, as shown in Figure 10.7b and e. The hexagonal trajectories of (R_x, R_y) have never been observed in other topological spin configurations, such as a vortex or a bubble usually having circular or elliptical orbits. Note that a pentagram trajectory has been discovered in dynamics of a magnetic bubble, which results from the local deformation of the circular domain wall (Moutafis et al. 2009; Makhfudz et al. 2012). The trajectory describes the motion of the mean position (X, Y) defined by the moments of the magnetization, other than the moments of the topological density associating with the topological density distribution of the whole system. In addition, the motion of the skyrmions cannot be correctly described by the mean position (X, Y). Therefore, the hexagonal trajectories of (R_x, R_y) here are different from the pentagram trajectory of (X, Y) for the magnetic bubble, which is ascribed to the difference in spin textures. After 14.24 ns, the rotation orbits change to circular ones with the eigenfrequency of 1.15 GHz.

10.4   Microwave Field–Driven Resonant Excitations of Coupled Skyrmions

10.4.1   The Effective Mass of Coupled Skyrmions

Due to the smooth global spin configuration, the deformation of the topological density distribution of a skyrmion at resonance may be large and nonlocal. Considering the deformation's nature, the local magnetization $M(x,t)$ depends on both the guiding center's position $R(t)$ and its velocity $\dot{R}(t)$ (Wysin et al. 1993), we can get:

$$M(x,t) = M(x - R(t), \dot{R}(t)). \tag{10.6}$$

Then, the extended Thiele's equation for the guiding center can be obtained as follows:

$$-\partial U/\partial R + \mu\,H + G \times \dot{R} = \hat{e}_i M_{\text{eff}} \ddot{R}_j, \tag{10.7}$$

where G is the gyrovector, $\hat{e}_i$ the unit coordinate vector (i (j) = x, (y)), H an external magnetic field, μ a function of the structural and magnetic parameters (Guslienko 2006; Lee and Kim 2007), $\ddot{R}_j$ the acceleration of the guiding center, and M_{eff} the effective mass tensor with elements:

$$M_{\text{eff}} = -\gamma^{-1} M_s \int d^2 x m \left(\frac{\partial m}{\partial x_i} \times \frac{\partial m}{\partial \dot{R}_j} \right), \tag{10.8}$$

where γ is the gyromagnetic ratio and M_s the saturation magnetization. The potential energy of the guiding centers is:

$$U = \frac{1}{2} K R_t^2 + \frac{1}{2} K R_b^2 + U_{sky\text{-}sky}(d), \tag{10.9}$$

where K is the stiffness coefficient, $d = |R_t - R_b|$ the distance between two guiding centers, and $U_{sky-sky}(d)$ the magnetostatic coupling between two skyrmions. Considering dynamics of a skyrmion with the nonlocal deformation, the effective mass, associated with the time derivative of the topological density $\dot{q}$, has to be considered to comprehend the dynamical behavior of the skyrmion.

10.4.2   Microwave Field Applied to One of the Two Cobalt Nanolayers

10.4.2.1   Dual-Frequency Microwave-Driven Resonant Excitations

An in-plane microwave field is used to drive the resonant excitation of the coupled skyrmions obtained in Co/Ru/Co nanodisks. The waveforms of the microwave fields are $H_0 \sin(2\pi f t)$ with frequencies of

Chapter 10

1 or 5 GHz, and $H_1\sin(2\pi f_1 t) + H_2\sin(2\pi f_2 t)$ with the frequency ratio f_1/f_2 of 1/2, 1/3, 1/4, 1/5, 1/6 or the golden ratio. The frequency f and f_1 of the field are chosen to be equal or close to the eigenfrequencies of the system, which are obtained by the gyrotropic motion of the coupled skyrmions, as shown in Section 10.3. Otherwise, the dynamics of skyrmions does not show a strong resonance (Wang et al. 2014). The field is applied along the +x-direction to the top nanolayer.

The trajectories of (R_x, R_y) for the top nanolayer in the field $H_0\sin(2\pi f t)$ with the frequencies of 1 and 5 GHz are illustrated in Figure 10.8a and b, respectively. Both of the two steady orbits are approximately circular with the radius of 30.5 nm for 1 GHz (Figure 10.8a) and 3.3 nm for 5 GHz (Figure 10.8b). The rotation directions are opposite, clockwise (CW) for the lower-frequency mode and counterclockwise (CCW) for the higher-frequency mode. Similar to resonant excitation of a vortex (Park and Crowell 2005; Kammerer et al. 2011), the rotation direction for skyrmions depends on the frequency of the microwave field as well as on their polarities. The microwave absorption spectra of a two-dimensional model for a skyrmion crystal have been investigated, in which the skyrmion crystal is stabilized by the DMI (Mochizuki 2012). It was shown that for twofold spin-wave modes, the frequency difference between the two modes is small. Here, however, we find that due to the effect of the effective mass, Equation 10.8, the two resonant excitation modes are far from being degenerate and exhibit a large frequency difference.

When a dual-frequency microwave field $H_1\sin(2\pi f_1 t) + H_2\sin(2\pi f_2 t)$ with $f_1 = 1$ GHz and $f_2 = 5$ GHz is applied, the trajectory becomes a

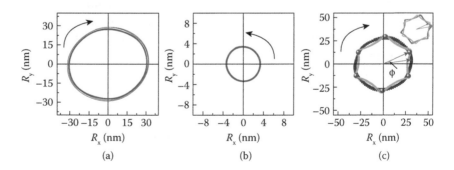

(a) (b) (c)

FIGURE 10.8 The trajectory of the skyrmion in the top nanolayer in a single-frequency field with the frequency $f = 1$ GHz (a), 5 GHz (b), and in a dual-frequency field with the frequency ratio f_1/f_2 of 1/5 and H_1/H_2 of 5/1 (c). The light and dark gray lines in (c) are the sum of the orbits at 1 and 5 GHz with amplitude ratios of 5/1 and 1/1, respectively. The curved arrows represent the directions of rotation. (From Wang H, et al., *RSC Adv.*, 4, 62179–62185, 2014. With permission.)

hypocycloid instead of a circle as shown in Figure 10.8c, which is unusual in the resonant excitation of topological spin configurations (Lee and Kim 2007). Interestingly, there are differences in amplitude and phase (about 12 degrees) when comparing the trajectory in the dual-frequency field to the sum trajectory of ones in the two single-frequency fields, which indicates that the two modes couple with each other. The inset in Figure 10.8c shows that the dynamical phase difference becomes much clearer and remains unchanged when the contribution of the higher-frequency mode is increased.

To comprehend the hexagonal resonant excitation, Figure 10.9 shows the topological density distribution in one period for the steady orbits

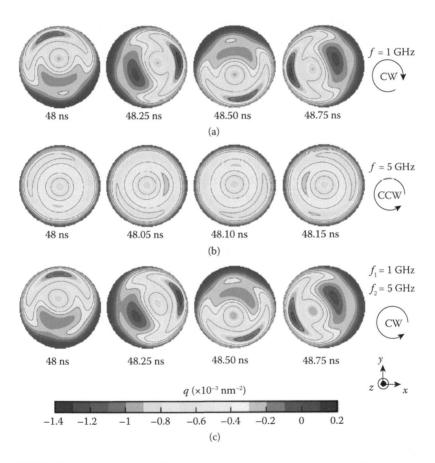

FIGURE 10.9 The topological density distribution for the top nanolayer in one period under the action of a single-frequency field with the frequency $f = 1$ GHz (a), 5 GHz (b), and in a dual-frequency field with the frequency ratio f_1/f_2 of 1/5 and H_1/H_2 of 5/1 (c). (From Wang H, et al., *RSC Adv.*, 4, 62179–62185, 2014. With permission.)

in Figures 10.8a through c. All the distributions show large nonlocal deformations, unlike small local deformations at the vortex cores or in the narrow domain walls of bubbles, suggesting that the effective mass related to the deformation of the topological density distribution shall be considered to describe the dynamics of the coupled skyrmions. The rotation direction of the lower-frequency mode is CW, while that of the higher-frequency one is CCW, which is consistent with the results in Figure 10.8a and b. Under the action of the dual-frequency microwave magnetic field, the rotation direction is also CW and the topological density distribution is similar to that in the low-frequency field with a slight difference, suggesting that the hexagon is the result of superposition of the CW and CCW modes and is dominated by the CW mode.

We have demonstrated that the trajectory of resonant excitation can be transformed from an approximate circle to a hexagon by replacing the single-frequency field by a dual-frequency one. Here, we further show in Figure 10.10 that varying the frequency ratio f_1/f_2 is also useful to control the skyrmion dynamics. The frequency f_1 is chosen to be 1.15 GHz, the same value as the eigenfrequency of the gyrotropic motion of the coupled skyrmions with the circular trajectory. Driven by a microwave field, the skyrmion shows a strong resonant excitation. On the basis of the strong resonant excitation, polygonal trajectories can be obtained by changing the frequency ratio f_1/f_2 of the dual-frequency field. Figure 10.10 shows that the trajectory of the resonance is indeed able to be changed from a hexagon to other polygons like a triangle, quadrangle, pentagon, and heptagon. The controllability of the resonant excitation by means of a dual-frequency field would be useful to manipulate skyrmions and design microwave devices.

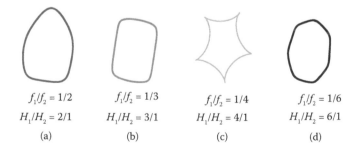

$f_1/f_2 = 1/2$ $f_1/f_2 = 1/3$ $f_1/f_2 = 1/4$ $f_1/f_2 = 1/6$

$H_1/H_2 = 2/1$ $H_1/H_2 = 3/1$ $H_1/H_2 = 4/1$ $H_1/H_2 = 6/1$

(a)   (b)   (c)   (d)

FIGURE 10.10   Polygonal trajectories of resonant excitations activated by a dual-frequency field with the waveform $H_1 \sin(2\pi f_1 t) + H_2 \sin(2\pi f_2 t)$. For (a), (b), (c), and (d), f_1 is chosen to be 1.15 GHz, H_1 is 50 Oe, f_1/f_2 is 1/2, 1/3 1/4, and 1/6, and the corresponding amplitude ratio H_1/H_2 is 2/1, 3/1, 4/1 and 6/1, respectively. Only the steady orbits are shown here. (From Wang H, et al., *RSC Adv.*, 4, 62179–62185, 2014. With permission.)

In a dual-frequency field with an incommensurate frequency ratio (i.e., the golden ratio $\left[\sqrt{5}-1\right]/2$), the resonant excitation of the coupled skyrmions shows quasiperiodic behavior as shown in Figure 10.11. As it is impossible to use a true irrational number in the simulations, we use a rational number with a precision up to 16 significant decimal digits to deal with the golden ratio, and find that the result is insensitive to the number of digits when the decimal digits go higher than 10. The trajectories in a microwave field with a rational-number frequency show periodic circles. But when introducing a field with an irrational-number frequency, the system begins to display quasiperiodicity. As H_1/H_2 increases, the trajectory becomes more uncertain because of an increasing influence of the irrational-number frequency.

To comprehend the polygonal resonant excitations of coupled skyrmions driven by a dual-frequency field, numerical solutions of Equation 10.7 were obtained by means of the Runge-Kutta method, some results are shown in Figure 10.12. As the moving of the coupled skyrmions is almost synchronous, we can use an effective potential energy to replace the total potential energy of one of the skyrmions. Equation 10.7 with a dissipation term for the top nanodisk can be rewritten as follows:

$$\mu H_x - KR_x - G\dot{R}_y - D\dot{R}_x = M_{\text{eff}}\ddot{R}_x, \tag{10.10}$$

$$-KR_y + G\dot{R}_x - D\dot{R}_y = M_{\text{eff}}\ddot{R}_y, \tag{10.11}$$

where K and D are the effective stiffness coefficient and the damping parameter, respectively (Guslienko 2006). Here, for numerical solutions to Equations 10.10 and 10.11, $D = 5.59 \times 10^{-14}$ J s m^{-2}, $G = 4\pi M_s L/\gamma = 1.8 \times 10^{-12}$ J s m^{-2}, and $\mu = \pi\mu_0 RLM_s\xi = 1.0 \times 10^{-14}$ kg m^2 (A^{-1} s^{-2}) with

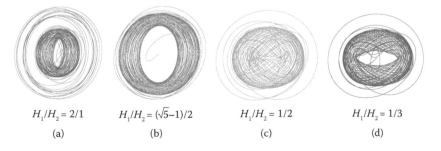

| $H_1/H_2 = 2/1$ | $H_1/H_2 = (\sqrt{5}-1)/2$ | $H_1/H_2 = 1/2$ | $H_1/H_2 = 1/3$ |
| (a) | (b) | (c) | (d) |

FIGURE 10.11 Quasiperiodic behavior of the coupled skyrmions driven by a dual-frequency field with $H_1 = 50$ Oe, $f_1 = 1.15$ GHz, and $f_1/f_2 = 1/$golden ratio. For (a), (b), (c), and (d) H_1/H_2 is 2/1, golden ratio, 1/2, and 1/3, respectively. (From Wang H, et al., *RSC Adv.*, 4, 62179–62185, 2014. With permission.)

Chapter 10

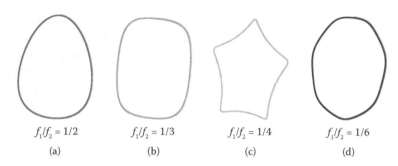

$f_1/f_2 = 1/2$ $f_1/f_2 = 1/3$ $f_1/f_2 = 1/4$ $f_1/f_2 = 1/6$

(a)   (b)   (c)   (d)

FIGURE 10.12   Numerical solutions to Equations 10.10 and 10.11 for the coupled skyrmions in a dual-frequency field $H_x = H_1\sin(2\pi f_1 t) + H_2\sin(2\pi f_2 t)$ with $f_1 = 1.15$ GHz. f_1/f_2 and H_1/H_2 are (a) 1/2 and 2/1, (b) 1/3 and 3/1, (c) 1/4 and 4/1, and (d) 1/6 and 6/1. (From Wang H, et al., *RSC Adv.*, 4, 62179–62185, 2014. With permission.)

$\xi \approx 0.93$ for the skyrmion here (Guslienko 2006). K (0.013 J m^{-2}) and M (7.08 × 10^{-23} kg) are calculated according to the eigenfrequencies of 0.96 and 4.98 GHz for a hexagonal trajectory and 1.15 GHz for a circular one (Makhfudz et al. 2012).

The polygonal resonant excitations can be activated by a dual-frequency field with $f_1 = 1.15$ GHz and $f_2 =$ an integral multiple of f_1, as shown in Figure 10.12a through d. The polygonal trajectories from the numerical solution are in best agreement with those from micromagnetic simulations (Figure 10.10), which indicates that the effective mass and the frequency ratio are vital to the polygonal resonant excitations.

Note that the twofold spin-wave mode excitations are theoretically anticipated but only the low-frequency mode has been experimentally observed, which may be attributed to the mixture of the high-frequency mode with the helical mode or a breathing mode because of their close resonance frequencies (Mochizuki 2012; Onose et al. 2012; Okamura et al. 2013). However, the two modes of the coupled skyrmions here have a larger frequency difference, which may lead to the possibility of observing experimentally both the two resonant modes.

10.4.2.2   Single-Frequency Microwave-Driven Resonant Excitations

A single-frequency microwave field with the waveform $H_0\sin(2\pi ft)$ is used to investigate the resonant excitations of the coupled skyrmions. The field is applied only to the top nanolayer. Two resonance excitation modes are observed: CW mode in the low-frequency region and CCW mode in the high-frequency region. The two modes are analytic solutions of the generalized Thiele equation. The forms of the solutions are (Makhfudz et al. 2012):

$$\omega_{\pm} = -G/2M_{eff} \pm \sqrt{\left(G/2M_{eff}\right)^2 + K/M_{eff}}, \tag{10.12}$$

where G, M_{eff}, and K are the gyrocoupling vector, effective mass, and stiffness coefficient, respectively. Equation 10.12 indicates that the CCW mode is more sensitive to mass change than the CW mode. Consequently, the frequency for mode flipping is close to 2 GHz, as shown in Figure 10.13. The field amplitude H_0 is chosen to be 50 Oe when $f < 1.5$ GHz, due to the strongest resonance, and 100 Oe as f is higher than 1.5 GHz. Figure 10.13 shows two pronounced peaks of R (the maximal radius at which the guiding center can reach): one at 1.13 GHz as the superposition of the two peaks with eigenfrequencies 1.00 and 1.15 GHz, and the other at 4.75 GHz, where a complicated dynamical behavior is found as shown in Figure 10.14. The resonance peak at 5.00 GHz may be indirectly concluded from the decreasing amplitude starting from 5.00 GHz, which is more or less hidden behind a pronounced peak around 4.60 GHz. As an overtone of 1.15 GHz, this

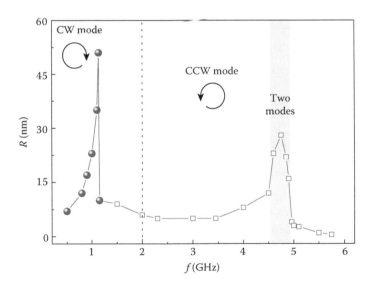

FIGURE 10.13 The maximal radius R (= sqrt($(R_x)^2 + (R_y)^2$)) which the guiding center can reach as a function of the frequency in a single-frequency field. The field is applied to the top nanolayer and the field amplitude is set to be 100 Oe expect for $f < 1.5$ GHz where 50 Oe is used because of a very strong resonance. The resonant mode is CW when the frequency is below 2.00 GHz, and CCW above 2.00 GHz. In the olive region, the two modes coexist. (From Dai YY, et al., *Sci. Rep.*, 4, 6153, 2014. With permission.)

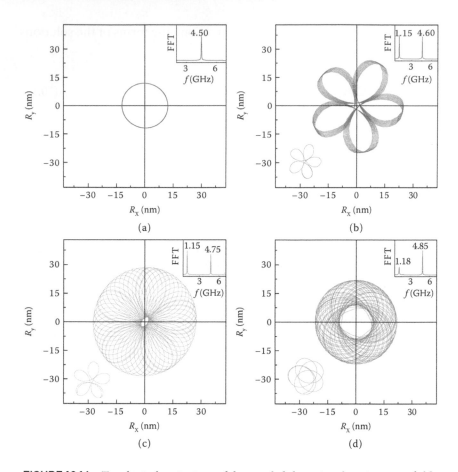

FIGURE 10.14 Topological excitations of the coupled skyrmions by microwave fields with different frequencies. The frequency of the field is (a) 4.50, (b) 4.60, (c) 4.75, and (d) 4.85 GHz, and H_0 is 100 Oe. Only the steady parts of trajectories in the range of 40–50 ns are shown here. The insets in the upper-right and lower-left corners show the corresponding FFT spectra and the trajectories in about one period, respectively. (From Dai YY, et al., *Sci. Rep.*, 4, 6153, 2014. With permission.)

frequency activates the base mode and results in the two modes coexisting in the frequency range of 4.60–4.85 GHz.

Now, we focus on the interesting frequency range 4.50–4.85 GHz. The trajectories of coupled skyrmions in the field with the frequency in the range are shown in Figure 10.14. It only shows the steady parts of trajectories in the range of 40–50 ns. When the frequency is at 4.50 GHz, the steady orbit is a circle with the radius of 12 nm, while in the 4.60–4.85 GHz range of frequencies, the trajectory transforms to a flower-like one with increased amplitude, which is quite unusual. The one-period trajectories shown in the insets in Figure 10.14b through d show five "petals." These petals evolve into ribbon- or flower-like trajectories because of the phase difference between different periods. As the frequency increases, the phase difference increases. The flower-like dynamics may be observed indirectly in an experiment: using Lorentz transmission electron microscopy (LTEM) (Yu et al. 2010) to obtain the magnetization distribution at different times and then calculating the position of the guiding center by means of Equation 10.5. The FFT spectra in the insets show that when the frequency is 4.50 GHz, only one resonant frequency is observed, whereas for other frequencies, an additional frequency around 1.15 GHz shows up.

To understand the flower-like trajectories better, a microwave field with f = 4.60 GHz and H_0 varying from 20 to 500 Oe is used. The results are presented in Figure 10.15. It shows only the steady parts of

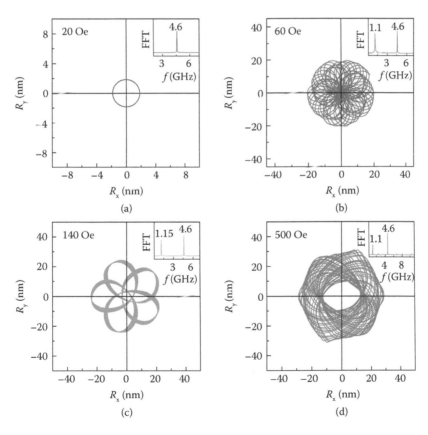

FIGURE 10.15 Influence of H_0 on the resonance of the coupled skyrmions. The frequency of the field is 4.60 GHz, and H_0 is (a) 20, (b) 60, (c) 140, and (d) 500 Oe. Only the steady parts of trajectories in the range of 40–50 ns are shown here. Insets show the corresponding FFT spectra. (From Dai YY, et al., *Sci. Rep.*, 4, 6153, 2014. With permission.)

trajectories in the range of 40–50 ns. Surprisingly, when the amplitude is 20 Oe, the trajectory is a circle with a small radius, less than 2 nm (Figure 10.15a), and the corresponding FFT spectrum presents only one peak, at 4.60 GHz. As the amplitude increases, the trajectory changes into flower-like style with an increasing maximum radius. With a further increase in the amplitude, from 60 Oe to 140 Oe, the flower-like trajectory changes from a sunflower to a bauhinia with a decreasing phase difference. When the amplitude reaches 500 Oe, the trajectory becomes distinct from the previous ones, having six "petals" instead of five in each period (Figure 10.15d). The corresponding FFT spectra in the insets in Figure 10.15b through d show that another peak, near 1.15 GHz, appears along the one at 4.60 GHz related to the external field.

Figures 10.14 and 10.15 illustrate that the flower-like trajectories are related to the excitation of an intrinsic mode near 1.15 GHz. To comprehend why the 1.15 GHz mode can be stimulated and exist stably in the field with the frequency in the range of 4.60–4.85 GHz, Figure 10.16 shows the topological density distribution of the skyrmions at the maximum deformation. For a static skyrmion, the topological distribution is radially symmetric, which can be broken by a microwave field. As the amplitude increases, the deformation of the topological density distribution increases and becomes global if the amplitude is larger than 20 Oe,

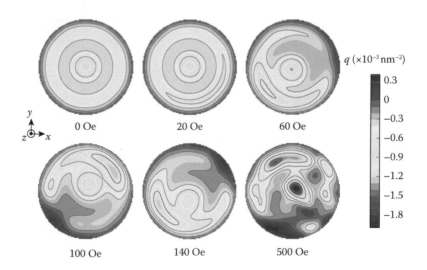

FIGURE 10.16 The topological density distribution of skyrmion in different fields. The frequency of the field is 4.60 GHz and the amplitude is from 0 to 500 Oe. The figure shows the distribution when the deformation of the topological density distribution is maximal in one period. (From Dai YY, et al., *Sci. Rep.*, 4, 6153, 2014. With permission.)

and turns very complex when the field is as large as 500 Oe. Figures 10.14 through 10.16 imply that the large nonlocal deformation of the topological density distribution gives rise to the stable existence of the 1.15 GHz mode, seen in the FFT spectra of Figures 10.14 and 10.15, and thus to the flower-like trajectories.

10.4.3 Microwave Field Applied to the Whole Co/Ru/Co Nanodisk

In Sections 10.4.1 and 10.4.2, the microwave field is applied to only one of the two nanolayers, which is difficult to realize in experiments. Figure 10.17 demonstrates the trajectories of the guiding centers when the microwave field is applied to the whole system. It shows only the trajectories in the range of 40–50 ns. The two guiding centers move to opposite directions and separate from each other in the action of the field due to the opposite chiralities, as shown in the insets in Figure 10.17. If there were no magnetostatic interactions between the two skyrmions, two independent microwave-field-induced flower-like trajectories in the top and bottom nanolayers would be expected. However, the separation changes the interlayer magnetostatic interaction, which has an opposite effect to the field and destroys the flower-like trajectories. This is different from the case when the microwave field is applied only to one nanolayer, and the magnetostatic interaction forces the other nanolayer to move in a synchronic flower-like trajectory. In addition, the

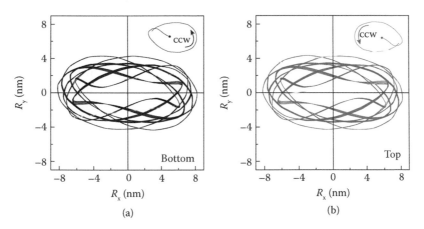

(a)   (b)

FIGURE 10.17 Trajectories of the guiding center for (a) the bottom and (b) the top nanolayers in a magnetic field with 200 Oe amplitude and 4.6 GHz frequency. The magnetic field is applied to the whole system. The insets in the upper-right corners are the initial trajectories. Dots and arrows denote the starting points and the rotation direction, respectively. (From Dai YY, et al., *Sci. Rep.*, 4, 6153, 2014. With permission.)

magnetostatic interaction is strong, giving rise to a smaller distance of motion than that in Figure 10.15.

10.5　Current-Driven Resonant Excitations of Coupled Skyrmions

In Section 10.4, we have shown that due to the opposite chiralities of the two skyrmions, to observe the polygonal or flower-like behavior of coupled skyrmions, the microwave field has to be applied to only one of the two nanolayers, which makes the experimental observation of the dynamics very difficult. Therefore, finding alternative effective method to manipulate the coupled skyrmions synchronously is important. Spin-transfer torque (STT), having been widely used to drive dynamics of vortices and the translational motion of skyrmions, is expected to manipulate the resonant excitations of the coupled skyrmions (Kasai et al. 2006; Kim et al. 2008; Fert et al. 2013; Zhang et al. 2015). To simulate the influence of STT generated by the in-plane current for this purpose, the Landau–Lifshitz–Gilbert (LLG) equation has to be modified as follows (Zhang and Li 2004):

$$\frac{\partial m}{\partial t} = -\gamma m \times H_{\mathrm{eff}} + \alpha m \times \frac{\partial m}{\partial t} + u m \times \left(m \times \frac{\partial m}{\partial x} \right) - \beta u m \times \frac{\partial m}{\partial x}, \quad (10.13)$$

$$u = \frac{JPg\mu_B}{2eM_s}, \quad (10.14)$$

where m is the unit vector of the local magnetization, H_{eff} the effective magnetic field, J the current density, e the electron charge, μ_B the Bohr magneton, P the spin-current polarization of the ferromagnet, g a constant known as the Landé g-factor, and β the nonadiabaticity factor.

Figure 10.18 shows the guiding-center component R_x as a function of time when the coupled skyrmions are driven by an AC current. The current is applied to the whole nanodisk instead of one of the two nanolayers. As expected, R_x in the top nanolayer is the same as that in the bottom nanolayer at any moment. Under the force of the current, the two guiding centers move in the same direction instead of in opposite directions, as occurs when the coupled skyrmions are driven by a magnetic field (Figure 10.17), which makes experimental observation of skyrmion dynamics much easier. Note that because the torque acted on a spin by the field is $-\gamma m \times H$, while that by the current is $u m \times \left(m \times \frac{\partial m}{\partial x} \right)$, for spins of the top and bottom nanolayers pointing in the same direction,

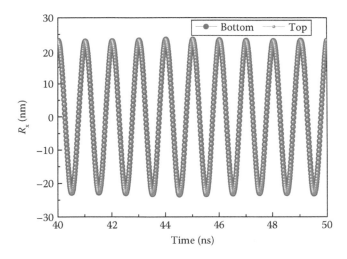

FIGURE 10.18 Time dependence of R_x for the top and bottom nanolayers when the coupled skyrmions are moved by an AC current $J\sin(2\pi f t)$ with $f = 1.0$ GHz and $J = 1.0 \times 10^8$ A/cm^2.

the torques forced by the field point to the same direction, while those forced by the current point to the opposite directions because of the opposite chiralities of the two skyrmions. Therefore, the actions driven by the field and the current are different.

Resonant excitations of the coupled skyrmions in a dual-frequency AC current are shown in Figure 10.19. It only shows the steady parts of trajectories in the range of 30–50 ns. As f_1 increases, not only the shapes of the trajectory change, for example from a hexagon to a circle, but also the maximum radius R increases, from 12 to 40 nm. In addition, with increasing f_1, the hexagonal trajectory transforms from a curtate hypocycloid into a hypocycloid, which indicates that contribution of these two modes (the CW and CCW modes) to the skyrmion dynamics varies with the frequency value. When f_1 increases to 1.1 GHz, the trajectory is a circle, suggesting that in the case of the strongest resonance at 1.1 GHz, the CW mode is prominent and the CCW mode can be ignored.

The influence of the amplitude ratio J_1/J_2 on the trajectories of coupled skyrmions is shown in Figure 10.20. It shows only the steady parts of trajectories in the range of 30–50 ns. When $J_2 > J_1$, the trajectory of the guiding center is complex. The insets in Figure 10.20a and b present the trajectory within 1 ns. For $J_1/J_2 = 1/3$, the trajectory consists of a hexagon, while for $J_1/J_2 = 1/2$, the trajectory consists of circles. In contrast, Figure 10.20c through f shows periodic behavior

Chapter 10

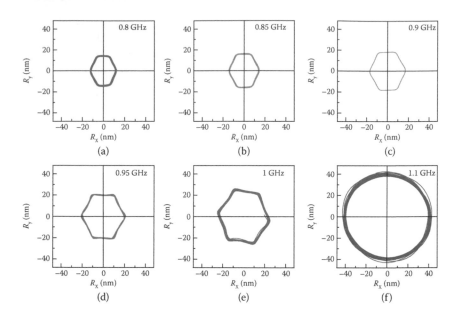

FIGURE 10.19 Dynamics of the coupled skyrmions driven by a dual-frequency AC current $J_1\sin(2\pi f_1 t) + J_2\sin(2\pi f_2 t)$. f_1/f_2 and J_1/J_2 are 1/5 and 5/1, respectively. In (a–f), f_1 varies from 0.8 to 1.1 GHz, and J_1 is 1.0×10^8 A/cm², except for the case of $f_1 = 1.1$ GHz where 5×10^7A/cm² is used. (From Dai YY, et al., *Int. J. Mod. Phys. B.*, 30, 1550254, 2016. With permission.)

after the guiding center reaches its steady orbit which is a hexagon. As J_1 increases, the maximum radius of the steady orbit increases from 23.0 to 26.7 nm.

Different trajectories in Figures 10.19 and 10.20 can be due to different contributions of the CW and CCW modes to the dynamics of the coupled skyrmions, as further shown in Figure 10.21. To quantify the contributions, we define the important parameters—I_{CW}, I_{CCW}, and I_{tot} as follows: First, we obtain the FFT spectra of all the trajectories in Figures 10.19 and 10.20. Second, integrating the FFT spectra to get the total integrated intensity I_{tot} in the whole frequency range, I_{CW} for $f < 2$ GHz and I_{CCW} for $f > 2$ GHz according to Figure 10.13. Finally, we can obtain the ratio I_{CCW}/I_{tot} as functions of the f_1 and the J_1/J_2.

Combining Figure 10.21a with Figure 10.19, we deduce that the change of hexagonal trajectory from a curtate hypocycloid to a hypocycloid is due to the increased contribution (above 10%) of the CCW mode. Similar results can be obtained by varying the amplitude ratio J_1/J_2, as shown in Figure 10.21b. When I_{CCW}/I_{tot} is larger than 0.1 but smaller than 0.3, the trajectory can stay hypocycloid. However, if $I_{CCW}/I_{tot} > 0.3$

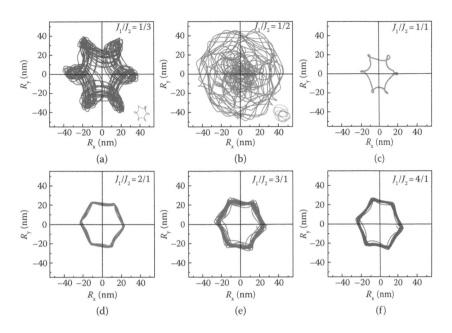

FIGURE 10.20 Influence of J_1/J_2 of a dual-frequency AC current on resonant excitation of the coupled skyrmions. The AC current has the waveform $J_1\sin(2\pi f_1 t)$ + $J_2\sin(2\pi f_2 t)$ with $f_1 = 1.0$ GHz and $f_1/f_2 = 1/5$. (a–f) J_1/J_2 is 1/3, 1/2, 1/1, 2/1, 3/1, and 4/1, respectively. In all cases, $J_1 = 1.0 \times 10^8$ A/cm². The insets in (a) and (b) show the trajectories of the guiding center in one period. (From Dai YY, et al., *Int. J. Mod. Phys. B.*, 30, 1550254, 2016. With permission.)

($J_1/J_2 < 1/1$), the contribution of the CCW mode is so large that the trajectory cannot become a steady one. The above results show that the fraction of CCW mode can be controlled by varying f_1 or J_1/J_2 to obtain the desired trajectory.

The influence of magnetostatic interaction on the dynamics of coupled skyrmions is shown in Figure 10.22. First, a magnetic field varying from 150 to 1500 Oe is applied to the whole system, which leads to the two guiding centers moving in opposite directions and decoupling with each other. As the field increases, the distance between the two new equilibrium positions (denoted by dots) increases from 6.5 to 65.0 nm. After 10 ns, a dual-frequency AC current is applied to the system. Driven by the AC current, the two guiding centers begin to rotate around their new equilibrium positions. When the field is 150 Oe (Figure 10.22a), the two guiding centers are separated by a small distance, and the trajectories remain hexagonal, similar to the one in Figure 10.19e. As the field increases, the hexagons elongate along the y-axis and scale down along the x-axis until the hexagons disappear. The maximum value of R_y decreases first from

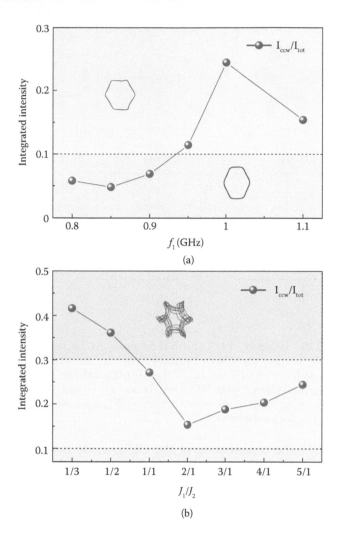

FIGURE 10.21   Integrated intensity ratio I_{CCW}/I_{tot} as a function of (a) f_1 and (b) J_1/J_2. The integrated intensity is extracted from the FFT spectra of the orbits in Figures 10.19 and 10.20. The integral is calculated with $f < 2$ GHz for the CW mode and with $f > 2$ GHz for the CCW mode according to Figure 10.13. (From Dai YY, et al., *Int. J. Mod. Phys. B.*, 30, 1550254, 2016. With permission.)

25.9 to 11.0 nm, and then increases to about 26.0 nm. There are three factors that determine the trajectories: the field which determines the distance of the new equilibrium positions, the dual-frequency AC current which leads to hexagons, and the magnetostatic interaction between the two skyrmions. Therefore, the change of the trajectories at different magnetic-field strength is attributed to the magnetostatic interaction.

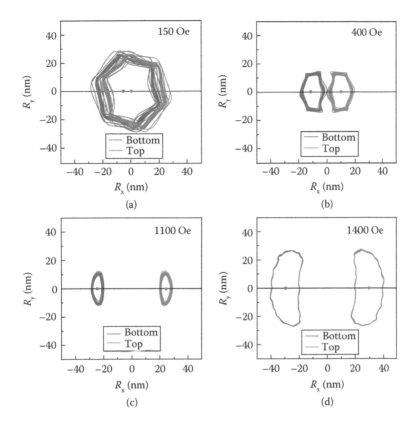

FIGURE 10.22 Influence of the magnetostatic interaction on the dynamics of the coupled skyrmions. First, a magnetic field is applied to the static coupled skyrmions, and then 10 ns later, a dual-frequency AC current with $f_1/f_2 = 1/5$ and $J_1/J_2 = 5/1$ is applied to the system. (a–d) The applied fields are 150, 400, 1100, and 1400 Oe, respectively. Dots denote the new equilibrium position of the guiding centers in the field. All figures only show trajectories after applying the AC current for 10 ns. (From Dai YY, et al., *Int. J. Mod. Phys. B.*, 30, 1550254, 2016. With permission.)

10.6 Summary

We have revealed that the coupled skyrmions with opposite chiralities can exist in Co/Ru/Co nanodisks without the chiral DMI. The competition of different energy terms acts similarly to the DMI and thus leads to the formation of the coupled skyrmions. The gyrotropic motion of the coupled skyrmions shows a star-like trajectory in the field and a hexagonal one after the field is switched off, which is distinguished from vortex or bubble.

Resonant excitations of the coupled skyrmions have two modes: the clockwise mode at a low frequency and the counterclockwise mode at

Chapter 10

a high frequency. Polygonal resonant excitation may be excited when the two modes are activated by a dual-frequency microwave field, and can be modulated from a triangle-like to a heptagon-like one by tuning the frequency ratio. Flower-like dynamics may be observed when the two modes are stimulated by a single-frequency microwave field, which is associated with the large nonlocal deformation of the topological density distribution of coupled skyrmions, and can be controlled systematically by the amplitude and frequency of the external magnetic field.

Resonant excitations of the coupled skyrmions can also be driven by STT. Better than the microwave field, STT can make the two skyrmions with opposite chiralities move in the same direction, making it easier to experimentally observe dynamics of the coupled skyrmions. The resonant excitations can be manipulated by varying the frequency or the amplitude ratio of the AC current. These findings contribute to a better understanding of skyrmion dynamics and may open a new avenue to the manipulation of skyrmions.

Acknowledgments

The work is supported by the National Natural Science Foundation of China (Grant No. 51331006 and 51590883), and the Key Research Program of Chinese Academy of Sciences (Grant No. KJZD-EW-M05-3).

References

Albuquerque GB and Freitas PP (1997). Micromagnetic modeling of spin-valve tape head. *Physica B: Condensed Matter.* 233:294–301.

Al Khawaja U and Stoof H (2001). Skyrmions in a ferromagnetic Bose-Einstein condensate. *Nature.* 411:918–920.

Belavin AA and Polyakov AM (1975). Metastable states of a two-dimensional isotropic ferromagnet. *Zhurnal Eksperimental'noi i Teoreticheskoi Fiziki, Pis'ma v Redaktsiyu.* 22:503–506.

Belesi M, et al. (2012). Magnetoelectric effects in single crystals of the cubic ferrimagnetic helimagnet Cu_2OSeO_3. *Physical Review B.* 85:224413.

Bloemen PJH, van Kesteren HW, Swagten HJM, and de Jonge WJM (1994). Oscillatory interlayer exchange coupling in Co/Ru multilayers and bilayers. *Physical Review B.* 50:13505–13514.

Bogdanov AN and Yablonskiĭ DA (1989). Thermodynamically stable 'vortices' in magnetically ordered crystals. The mixed state of magnets. *Soviet Physics—JETP.* 68:101–103.

Choi J-Y, et al. (2012). Imprinting skyrmion spin textures in spinor Bose-Einstein condensates. *New Journal of Physics.* 14:053013.

Cross MC and Hohenberg PC (1993). Pattern formation outside of equilibrium. *Reviews of Modern Physics*. 65:851–1112.

Dai YY, et al. (2013). Skyrmion ground state and gyration of skyrmions in magnetic nanodisks without the Dzyaloshinskii-Moriya interaction. *Physical Review B*. 88:054403.

Dai YY, Wang H, Yang T, and Zhang ZD (2016). Resonant excitation of coupled skyrmions by spin-transfer torque. *International Journal of Modern Physics B*. 30:1550254.

Dai YY, et al. (2014). Flower-like dynamics of coupled skyrmions with dual resonant modes by a single-frequency microwave magnetic field. *Scientific Reports*. 4:6153.

Donahue MJ and Porter DG (1999). *OOMMF User's Guide, Version 1.2a3*, Available at http://math.nist.gov/oommf.

Dussaux A, et al. (2010). Large microwave generation from current-driven magnetic vortex oscillators in magnetic tunnel junctions. *Nature Communications*. 1:8.

Dzyaloshinsky I (1958). A thermodynamic theory of "weak" ferromagnetism of antiferromagnetics. *Journal of Physics and Chemistry of Solids*. 4:241–255.

Fert A, Cros V, and Sampaio J (2013). Skyrmions on the track. *Nature Nanotechnology*. 8:152–156.

Guslienko KY (2006). Low-frequency vortex dynamic susceptibility and relaxation in mesoscopic ferromagnetic dots. *Applied Physics Letters*. 89:022510.

Heinze S, et al. (2011). Spontaneous atomic-scale magnetic skyrmion lattice in two dimensions. *Nature Physics*. 7:713–718.

Kammerer M, et al. (2011). Magnetic vortex core reversal by excitation of spin waves. *Nature Communications*. 2:279.

Kanazawa N, et al. (2011). Large topological Hall effect in a short-period helimagnet MnGe. *Physical Review Letters*. 106:156603.

Kasai S, et al. (2006). Current-driven resonant excitation of magnetic vortices. *Physical Review Letters*. 97:107204.

Kim S-K, Lee K-S, Yu Y-S, and Choi Y-S (2008). Reliable low-power control of ultrafast vortex-core switching with the selectivity in an array of vortex states by in-plane circular-rotational magnetic fields and spin-polarized currents. *Applied Physics Letters*. 92:022509.

Kiselev NS, Bogdanov AN, Schäfer R, and Rößler UK (2011). Chiral skyrmions in thin magnetic films: New objects for magnetic storage technologies? *Journal of Physics D: Applied Physics*. 44:392001.

Lee K-S and Kim S-K (2007). Gyrotropic linear and nonlinear motions of a magnetic vortex in soft magnetic nanodots. *Applied Physics Letters*. 91:132511.

Li Y, et al. (2013). Robust formation of skyrmions and topological Hall effect anomaly in epitaxial thin films of MnSi. *Physical Review Letters*. 110:117202.

Makhfudz I, Krüger B, and Tchernyshyov O (2012). Inertia and chiral edge modes of a skyrmion magnetic bubble. *Physical Review Letters*. 109:217201.

Martín JI, et al. (2003). Ordered magnetic nanostructures: Fabrication and properties. *Journal of Magnetism and Magnetic Materials*. 256:449–501.

Mochizuki M (2012). Spin-wave modes and their intense excitation effects in skyrmion crystals. *Physical Review Letters*. 108:017601.

Moriya T (1960). Anisotropic superexchange interaction and weak ferromagnetism. *Physical Review*. 120:91–98.

Moutafis C, Komineas S, and Bland JAC (2009). Dynamics and switching processes for magnetic bubbles in nanoelements. *Physical Review B*. 79:224429.

Mühlbauer S, et al. (2009). Skyrmion lattice in a chiral magnet. *Science*. 323:915–919.

Chapter 10

Nagaosa N and Tokura Y (2013). Topological properties and dynamics of magnetic skyrmions. *Nature Nanotechnology.* 8:899–911.

Neubauer A, et al. (2009). Topological Hall effect in the A phase of MnSi. *Physical Review Letters.* 102:186602.

Okamura Y, et al. (2013). Microwave magnetoelectric effect via skyrmion resonance modes in a helimagnetic multiferroic. *Nature Communications.* 4:2391.

Onose Y, et al. (2012). Observation of magnetic excitations of skyrmion crystal in a helimagnetic insulator Cu_2OSeO_3. *Physical Review Letters.* 109:037603.

Papanicolaou N and Tomaras TN (1991). Dynamics of magnetic vortices. *Nuclear Physics B.* 360:425–462.

Park JP and Crowell PA (2005). Interactions of spin waves with a magnetic vortex. *Physical Review Letters.* 95:167201.

Petit-Watelot S, et al. (2012). Commensurability and chaos in magnetic vortex oscillations. *Nature Physics.* 8:682–687.

Pribiag VS, et al. (2007). Magnetic vortex oscillator driven by d.c. spin-polarized current. *Nature Physics.* 3:498–503.

Raičević I, et al. (2011). Skyrmions in a doped antiferromagnet. *Physical Review Letters.* 106:227206.

Rajaraman R (1987). Solitons and Instantons. Amsterdam: North-Holland.

Schneider M, Hoffmann H, and Zweck J (2000). Lorentz microscopy of circular ferromagnetic permalloy nanodisks. *Applied Physics Letters.* 77:2909–2911.

Seki S, Yu XZ, Ishiwata S, and Tokura Y (2012). Observation of skyrmions in a multiferroic material. *Science.* 336:198–201.

Skyrme THR (1962). A unified field theory of mesons and baryons. *Nuclear Physics.* 31:556–569.

Sondhi SL, Karlhede A, Kivelson SA, and Rezayi EH (1993). Skyrmions and the crossover from the integer to fractional quantum Hall effect at small Zeeman energies. *Physical Review B.* 47:16419–16426.

Sun L, et al. (2013). Creating an artificial two-dimensional skyrmion crystal by nanopatterning. *Physical Review Letters.* 110:167201.

Tanase M, Petford-Long AK, Heinonen O, Buchanan KS, Sort J, and Nogués J (2009). Magnetization reversal in circularly exchange-biased ferromagnetic disks. *Physical Review B.* 79:014436.

Waldner F (1986). Two dimensional soliton energy and ESR in AFM. *Journal of Magnetism and Magnetic Materials.* 54–57:873–874.

Wang H, Dai Y, Yang T, Ren W, and Zhang Z (2014). Dual-frequency microwave-driven resonant excitations of skyrmions in nanoscale magnets. *RSC Advances.* 4:62179–62185.

Wright DC and Mermin ND (1989). Crystalline liquids: The blue phases. *Reviews of Modern Physics.* 61:385–432.

Wysin GM, Mertens FG, Völkel AR, and Bishop AR (1994). Mass and momentum for vortices in two-dimensional easy-plane magnets. In: *Nonlinear Coherent Structures in Physics and Biology* (Spatschek KH, Mertens FG, eds.). New York: Springer.

Yu X, et al. (2012). Magnetic stripes and skyrmions with helicity reversals. *Proceedings of the National Academy of Sciences.* 109:8856–8860.

Yu XZ, et al. (2010). Real-space observation of a two-dimensional skyrmion crystal. *Nature.* 465:901–904.

Yu XZ, et al. (2011). Near room-temperature formation of a skyrmion crystal in thin-films of the helimagnet FeGe. *Nature Materials.* 10:106–109.

Zaspel CE, Grigereit TE, and Drumheller JE (1995). Soliton contribution to the electron paramagnetic resonance linewidth in the two-dimensional antiferromagnetic. *Physical Review Letters.* 74:4539–4542.

Zhang S and Li Z (2004). Roles of nonequilibrium conduction electrons on the magnetization dynamics of ferromagnets. *Physical Review Letters.* 93:127204.

Zhang X, et al. (2015). Skyrmion-skyrmion and skyrmion-edge repulsions in skyrmion-based racetrack memory. *Scientific Reports.* 5:7643.

11. Skyrmion Stability in Magnetic Nanostructures

C. P. Chui
The University of Hong Kong, Pokfulam, Hong Kong

Long Yang, Wenqing Liu, and Yongbing Xu
Nanjing University, Nanjing, People's Republic of China

Yan Zhou
The Chinese University of Hong Kong, Shenzhen,
People's Republic of China

In this chapter, recent work on the stability of magnetic skyrmions have been reviewed for two types of low-dimensional magnetic nanostructures, that is, nanowires and nanocontact spin-transfer oscillators (NC-STOs) with perpendicular magnetic anisotropy (PMA) and

Chapter 11

Dzyaloshinskii–Moriya interaction (DMI) due to spin–orbit interaction at the interfaces. By means of micromagnetic simulations, various physical and geometrical conditions for stabilizing a skyrmion in low-dimensional nanostructures have been discussed. It is found that a combination of reasonably large PMA and DMI are essential for establishing magnetic skyrmions in a nanowire, especially at elevated temperatures. There are also some dimensional/geometrical constraints of the nanowires for skyrmion stability. A moderate damping parameter (larger than 0.1 for typical interface-induced DMI cases) is a prerequisite for maintaining a skyrmion without relaxing into a ferromagnetic or helical state. In addition, stability of skyrmions in nanodisks have also been briefly discussed. Finally, various perspectives on realistic applications based on skyrmions in low-dimensional systems including memory and microwave generators have been discussed.

11.1 Introduction

Magnetic skyrmion is a topologically protected magnetic texture that can be manipulated by external excitations such as electrical current and magnetic field. Due to lower Joule heating [1], it is an ideal candidate as information carrier because a current density as low as 10^2 A/cm^2 is sufficient to drive its motion [2]. Another reason for its suitability is the possibility of arranging skyrmions in the form of an array that represents binary digits [3–5]. Because of the potential of skyrmions in future applications, its stability in magnetic nanostructures such as nanowires and nanocontact spin-transfer oscillators (NC-STOs) has attracted increasing concern.

The magnetization in a nanowire and in an NC-STO depends largely on the material parameters. For example, DMI [6,7] helps to generate a tilted texture near the surface of a chiral magnet, where the inversion symmetry is broken [3,8–10]. PMA is another parameter that stabilizes a skyrmion without relying on an external magnetic field. The presence of PMA is highly beneficial to the stability of a skyrmion [11]. PMA is also found to inhibit the helical state that would hinder the stability of a skyrmion [12,13]. Theoretical analysis reveals that, in order to stabilize a skyrmion in a nanodot, the aspect ratio of the nanodot has to be adjusted in a way to accommodate to the PMA [14]. It has been shown recently that the formation of magnetic vortices due to DMI is inhibited by strong PMA [15,16], but the temperature effect on these parameters is yet to be studied.

The approach to sustaining a skyrmion inside magnetic nanostructures is under development. For example, a number of studies focus on storing a skyrmion in a nanowire [17–20]. The stability of a skyrmion inside circular and rectangular nanostructures has been studied numerically [21]. The finite thickness effect on skyrmions has been proposed theoretically on nanowires without constraints on planar dimensions [22]. By means of Monte Carlo simulations, the skyrmion core is allowed to modulate by adjusting the nanowire thickness [23]. It has been found that skyrmions can be stabilized even in the absence of DMI and dipolar interaction, with dynamical processing property between the vortex-like and hedgehog spin textures [24]. Recently, skyrmions with a topological number of 2 (known as high-Q skyrmions) can be generated by inputting spin-polarized currents in the presence of DMI [25]. The nucleation process of high-Q skyrmions is achievable from magnetic bubbles (Q = 0) by generating two Bloch points successively.

Attempts to realize skyrmion-based nanotracks have been made, the focus of which lies in issues such as nucleating, annihilating, and identifying skyrmions as information carriers. To eliminate the skyrmion Hall effect that bends the skyrmion trajectory toward the edges, two antiferromagnetically exchange-coupled bilayer nanotracks have been employed to accommodate two skyrmions of opposite skyrmion numbers (known as bilayer-skyrmions) [26]. Spin waves can govern the movement of a skyrmion along a nanotrack and different types of corners [27]. This study finds that damping hinders the skyrmion mobility. It also shows that spin waves can drive a skyrmion around an L-corner, T-junction, and Y-junction, with the turning direction depending on the sign of the skyrmion number. The length, separation, and period of skyrmion chains along a nanotrack can be modulated by varying the microwave frequency or the current density [28]. Skyrmion-based dynamic magnonic crystals can be formed on nanotracks, with controllable creation and destruction of a skyrmion achieved by changing the current density and applied field strength through nanocontacts [29]. By this configuration, fast switching between full transmission and full rejection of the spin waves is possible by varying the driving current density.

Another direction of skyrmion application is its use in spintronic devices. Skyrmions can act as information carriers in a transistor, in which a voltage gate applies an electric field to control the PMA of the gate region [30]. Then the presence or absence of a skyrmion at the drain can be sensed by the change in the tunnel magnetoresistance of a magnetic tunnel junction reader. Another study has simulated the skyrmionic AND and OR gates [31], which involve the conversion between domain walls and skyrmions along nanowires of varying widths [20].

Chapter 11

The damping parameter is essential for determining the relaxation time to result in a stabilized skyrmion. A large exchange interaction is also helpful to the stabilization of a skyrmion in a magnetic film [32,33]. The interplay between these two parameters is of interest to the design of materials that can sustain a skyrmion.

Skyrmions in NC-STOs is, on the other hand, a more recent topic than in nanowires. Except a micromagnetic study which demonstrates the possibility of orbiting skyrmions around the nanocontact to generate microwave frequency [34], we may have to rely on our understanding of conventional NC-STOs in order to realize the underlying principles of skyrmion-based STOs.

The NC-STOs have gained extensive attention because of their ability to generate microwave frequencies [35]. Studies of conventional NC-STOs are abundant. Here, we review a number of these findings. For the fixed layer, it is found that an oblique polarization angle can achieve precession of magnetization under zero-field conditions [36]. The polarization angle can also govern the static or dynamic mode of magnetization [37]. Investigation of the role of the spacer includes the effect of the spacer materials on the in-plane torque and field-like torque. The in-plane torque is found to oppose the damping torque, but the effect of the field-like torque is still unclear [38]. Besides, the size of the nanocontact is crucial for determining the mode of spin waves according to theoretical analysis [39], followed by experimental [40] and computational [41] findings.

In conventional STOs, one can change the precession frequency by adjusting the applied magnetic field and current density. It has been shown experimentally that the frequency of an STO can be increased by a stronger magnetic field and a weaker current density [42]. A magnetic field not totally perpendicular to the free layer can increase the operating frequency of an STO [43]. The magnetic field at certain angles can transform the spin wave of an STO between the propagating mode and the bullet mode [44]. Indeed, the applicability of the above findings of conventional STOs to skyrmion-based STOs still needs to be determined.

As a step toward realizing the favorable conditions for establishing a skyrmion, we present our recent studies of the stability of skyrmions in a nanowire [45] and in an NC-STO [46]. A number of magnetic-phase diagrams and snapshots are displayed to investigate the effect of tuning the nanowire and STO parameters on the successful formation of a skyrmion. We expect the readers to gain some insights into the appropriate choice of materials that can maintain a skyrmion for industrial production of skyrmion-carrying devices.

The following is the organization of this chapter. Section 11.2 is the description of the micromagnetic simulation employed in our study. Section 11.3 shows our results related to a skyrmion in a nanowire. Section 11.4 displays our results when a skyrmion stays inside an NC-STO. Section 11.5 discusses our findings, addressing the most suitable conditions for stabilizing a skyrmion in the nanostructures. Section 11.6 is the summary of our study.

11.2   Methods

11.2.1   Model

Micromagnetic simulations have been implemented extensively for investigating a skyrmion in a nanowire and in an NC-STO. We employ MuMax3 [47]—an open-source simulation tool that runs on graphics processing units in all our computations. When highly separated simulation cells that are numerically uncorrelated to each other are processed concurrently, the speed of computation can raise by about 100 times. We employ the well-known Landau–Lifshitz–Gilbert–Slonczewski equation to calculate the magnetization dynamics:

$$\frac{\partial \mathbf{M}(\mathbf{r},t)}{\partial t} = \frac{\gamma}{1+\alpha^2}\left\{\mathbf{M}(\mathbf{r},t)\times\mathbf{B}_{\text{eff}} +\alpha\left[\mathbf{M}(\mathbf{r},t)\times\left(\mathbf{M}(\mathbf{r},t)\times\mathbf{B}_{\text{eff}}\right)\right]\right\}$$
$$+ C\frac{\varepsilon-\alpha\varepsilon'}{1+\alpha^2}\left[\mathbf{M}(\mathbf{r},t)\times\left(\mathbf{M}_P\times\mathbf{M}(\mathbf{r},t)\right)\right] - C\frac{\varepsilon'-\alpha\varepsilon}{1+\alpha^2}\left(\mathbf{M}(\mathbf{r},t)\times\mathbf{M}_P\right), \tag{11.1}$$

$$C = \frac{J_z\hbar}{M_{sat}el}, \tag{11.2}$$

$$\varepsilon = \frac{P\Lambda^2}{\left(\Lambda^2+1\right)+\left(\Lambda^2-1\right)\left(\mathbf{M}(\mathbf{r},t).\,\mathbf{M}_P\right)}. \tag{11.3}$$

In Equation 11.1, $\mathbf{M}(\mathbf{r},t)$ is the magnetization at position $\mathbf{r}$ and time t, γ is the gyromagnetic ratio, and α is the Gilbert damping parameter. $\mathbf{M}_p$ is the polarizer magnetization. The effective magnetic field experienced by the free layer is expressed in Equation 11.4 as:

$$\mathbf{B}_{\text{eff}} = \mathbf{B}_{\text{ext}} + \mathbf{B}_{\text{anis}} + \mathbf{B}_{\text{demag}} + \mathbf{B}_{\text{exch}} + \mathbf{B}_{\text{th}}, \tag{11.4}$$

Chapter 11

where $\mathbf{B}_{ext}$ is the external magnetic field, $\mathbf{B}_{anis}$ the field contributed by the PMA, $\mathbf{B}_{demag}$ the demagnetizing field, and $\mathbf{B}_{exch}$ the exchange field. For finite-temperature simulations, the random thermal field of the Brown form $\mathbf{B}_{th}$ was included in the effective field.

In Equation 11.1, the term $\mathbf{M}(\mathbf{r}, t) \times \mathbf{B}_{eff}$ controls the precession of magnetization, whereas the term $\mathbf{M}(\mathbf{r}, t) \times (\mathbf{M}(\mathbf{r}, t) \times \mathbf{B}_{eff})$ controls the damping of magnetization toward the effective magnetic field $\mathbf{B}_{eff}$. These two terms are sufficient to describe the equilibrium magnetization of a nanowire without current injection.

In NC-STOs, the last two terms on the right-hand side of Equation 11.1 are required. From these two terms, the vector product $\mathbf{M}(\mathbf{r}, t) \times (\mathbf{M}_p \times \mathbf{M}(\mathbf{r}, t))$ is the spin-transfer torque (STT), and the vector product $\mathbf{M}(\mathbf{r}, t) \times \mathbf{M}_p$ is the field-like torque. ε is the parameter that controls the amount of the in-plane torque, whereas $\acute{\varepsilon}$ is the secondary STT parameter that represents the magnitude of the field-like torque.

In Equation 11.2, $\hbar$ is the reduced Planck constant, M_{sat} is the saturation magnetization, e is the electron charge, and l is the thickness of the free layer. J_z is the current density along the z-direction, which is defined as positive when electric current passes from the fixed layer to the free layer.

In Equation 11.3, Λ is the Slonczewski asymmetry parameter, which is set to unity. P is the spin polarization of the fixed layer.

11.2.2   Simulation Procedure

The procedure of simulation is as follows. The nanowire and the free layer of an STO are based on the parameters of Pt/Co [19]. By default, the material parameters are set as follows. The PMA K_u is 0.8 MJ/m^3, whereas the saturation magnetization M_{sat} is 580 kA/m. The exchange stiffness is fixed as 15 pJ/m, and the interfacial DMI D_{ind} is fixed at 3 mJ/m^2. The Gilbert damping parameter α is 0.3. Unless otherwise specified in later sections, the material parameters remain unchanged. The dimensions of the nanowire are varied in different situations, which are indicated in the upcoming sections.

There are in general two different types of DMI responsible for the stability of skyrmions. Bulk DMI in B20 materials supports chiral skyrmion structure (or Bloch-type skyrmion configuration), whereas interfacial DMI supports hedgehog skyrmion structure (or Néel-type skyrmion) [17,48]. For example, a multilayer structure of Pt/Co thin film (or similarly W/Fe, Ir/Fe, etc.) is considered, where the interface provides a sizeable DMI because of the strong spin–orbit coupling of the heavy metal. It has been shown by both first-principle calculations and experimental measurements that such an interfacial DMI due to broken mirror symmetry can be very strong in thin films ranging from 1 to 8 mJ/m^2 as described in the literature [49–52]. It is worth noting that such an interfacial DMI has been demonstrated in a wide range of materials such as Pt/Co, W/Fe, Ir/Fe, Pt/CoFe/MgO, Pt/Co/Ni/Co/TaN, and Pt/Co/AlOx [49,51].

The Pt/Co parameters are employed in the STOs to model the free layer, with additional parameters given below. The thickness of the cylindrical free layer l is 1 nm, with the diameter being 200 nm. The Slonczewski asymmetry parameter Λ is 1, such that the STO is symmetric. The secondary STT parameter $\acute{\varepsilon}$ is 0. The fixed layer is characterized by its polarization $\mathbf{M}_p$, which makes an angle $\beta = 90°$ with the fixed-layer plane by default. In this case, the spin polarization is perpendicular to the fixed layer. The magnitude of spin polarization P is set at 1. Two inputs are included. The current density along the z-axis, J_z, is -3×10^8 A/cm^2, and the external magnetic field $\mathbf{B}_{ext}$ of 0.1 T strength is directed at a polar angle θ of 90°. The negative sign of J_z refers to the current injection from the free layer to the fixed layer. Two NC leads with a 50-nm diameter are attached to the fixed layer and the free layer. Figure 11.1a shows the default setup of an NC-STO, with these parameters listed on it. Starting from Section 11.4, parameters other than the two to be varied in each magnetic phase diagram remain the values shown in Figure 11.1a.

Unless otherwise stated, a Bloch skyrmion with counterclockwise chirality and downward polarity has been put at the center of the nanowires and NC-STOs as the initial state. Regardless of the physical dimensions of the nanowires and the STO, the samples are partitioned into a number of simulation cells such that each simulation mesh size is 1–2 nm. No periodic boundary conditions have been implemented in all samples. Both studies of nanowires and STOs at zero temperature employ the Dormand-Prince method with variable-time step around 1 fs. In cases of investigating the temperature effect of skyrmions, the Heun solver with a time step of 4 fs is adopted. When no temperature is involved, the simulation time for each nanowire is varied, such that equilibrium magnetization is achieved. For the temperature-induced behavior of skyrmions, 20 ns of simulation time is attempted. On the other hand, the simulation time for each STO is set at 10 ns.

We have noticed that M_z can vary with time while keeping the average M_x and M_y constant. Therefore, we decide to obtain the breathing frequency from the frequency spectra of M_z. Because variable time steps

Chapter 11

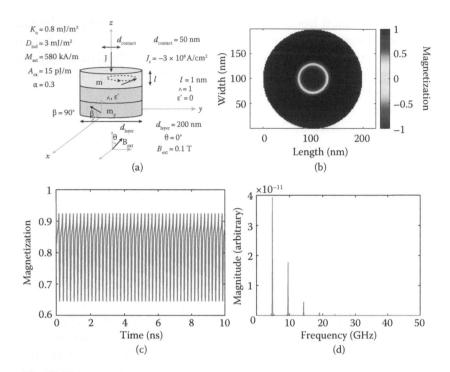

FIGURE 11.1 (a) The schematic setup of an NC-STO, from which we tune the parameters for our study. (b) The M_z plot of the resulting magnetization from (a) for the simulation time of 10 ns. The color bar represents the scale of M_z. The free layer magnetization results from the reflection of the electrons by the fixed layer. (c) The time evolution of M_z. (d) The frequency spectrum of M_z for the 10-ns simulation time. (Reprint of figure 1 in Chui, C.P., and Zhou, Y., *AIP Adv.*, 5, 097126, 2015. With permission.)

are involved, the Lomb-Scargle periodogram is employed to obtain the frequency spectra.

During the 10-ns simulation, the skyrmion undergoes breathing in the NC-STO. Figure 11.1b is the resulting out-of-plane magnetization along the z-axis after 10 ns by using the default Pt/Co parameter. The skyrmion can be stabilized in the given conditions, with its core polarity remaining downward as the initial condition. Figure 11.1c is the time-dependent magnetization trace along z-axis, M_z for the first 10 ns, from which we realize that the waveform is not sinusoidal. Figure 11.1d is the frequency spectrum of this NC-STO obtained by the Lomb-Scargle method on M_z. The main frequency of breathing is about 5 GHz with other harmonic frequencies. This demonstrates that the breathing is nonlinear.

11.3   Skyrmion in a Nanowire

11.3.1   Finite Size Effect

First, we investigate the stability of a skyrmion when the length L, width W, and thickness T of a nanowire are varied. We change the length and width between 10 and 100 nm, and adjust the thickness between 1 and 4 nm. By using the material parameters of Pt/Co, we obtain the phase diagrams for $T = 1 - 4$ nm in Figure 11.2a through d. One can identify a number of magnetic phases, namely the spin-up (SU) state, spin-down (SD) state, helical (HE) state, and skyrmion (SK) state. Their M_z plots are given in Figure 11.2e through h, and the gray bar of M_z is shown in Figure 11.2i. The DMI and PMA in a nanowire are of interfacial type, so these two quantities have been scaled by the inverse of the nanowire thickness ($1/T$) before simulation takes place. By this way, the inverse relation between thickness and DMI or PMA can be modeled. As indicated by the arrows in the M_z plots, an initial Bloch skyrmion has transformed to a Néel skyrmion whose in-plane magnetization radiates from the core.

One can summarize some characteristics of skyrmion stability as the nanowire dimensions change. The threshold planar dimensions required for stabilizing a skyrmion increases with nanowire thickness. This trend is more apparent when $T = 4$ nm, where only the spin-down state can remain in the nanowire. As a second observation, the helical phase region is suppressed by increasing the nanowire thickness. By combining these findings, we realize that a thick nanowire is detrimental to the stability of both SK and HE states. Our results is a realization of the dilution effect that inhibits the stabilization of a skyrmion as the nanowire thickness increases.

The in-plane magnetization of the magnetic phases is shown in Figure 11.2j through l. The edges of a nanowire constrains the homogeneity of the in-plane magnetization and limits the expansion of the skyrmion. The tilting of magnetization along the nanowire edges is attributed to the DMI [53], and our findings can verify this result.

The radius of the skyrmions formed as the dimensions vary is plotted in Figure 11.3. In these figures, the skyrmion radius is measured from the core to the magnetization region where $M_z = 0$. We find that a larger skyrmion can be obtained from a thicker nanowire. In addition, the semi-major axis and semi-minor axis have approximately equal length, meaning that the skyrmions established are generally circular.

When different initial conditions are applied, the equilibrium magnetization would result from the energy minima of the corresponding

Chapter 11

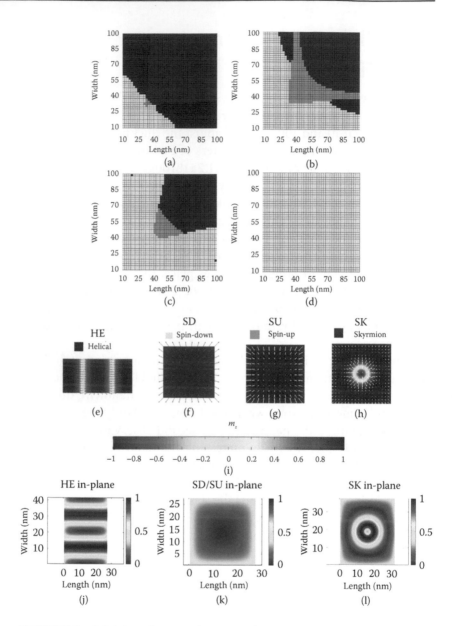

FIGURE 11.2 *L–W* phase diagrams for various thickness: (a) 1 nm, (b) 2 nm, (c) 3 nm, and (d) 4 nm. Four states exist: HE (red), SD (pale green), SU (pale blue), and SK (dark blue). (e–h) The various out-of-plane magnetization (M_z) patterns from the *L–W* phase diagram. The arrows indicate the in-plane component of magnetization. (i) The color bar of M_z. (j–l) In-plane component of magnetization with color scale for the phases obtained from (a). (Reprint of figure 1 in Chui, C.P., et al., *AIP Adv.*, 5, 047141, 2015. With permission.)

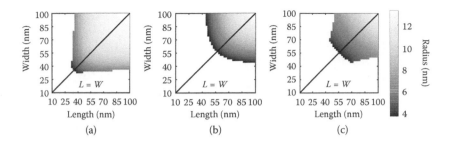

FIGURE 11.3 Radius of a skyrmion along the length dimension for thickness t of (a) 1 nm, (b) 2 nm, and (c) 3 nm. The gray bar shows the radius in nanometer, which is the distance between the skyrmionic core and the phase region where $M_z = 0$. The line $L = W$ is the axis of symmetry, along which the nanowire length equals the nanowire width. This line reflects the isotropy of the simulations. (Reprint of figure 2 in Chui, C.P., et al., *AIP Adv.*, 5, 047141, 2015. With permission.)

magnetization phase space. In this regard, we also tried Bloch skyrmions with a radius of 2 units and Néel skyrmions of default size with charge −1 and downward polarity in nanowires of 1-nm thickness to obtain different energy minima and hence different magnetic phase diagrams. The corresponding results are displayed in Figure 11.4a–b, respectively. One can realize from Figure 11.4a that a larger Bloch skyrmion needs to be stabilized in a nanowire with larger dimensions. However, a Néel skyrmion does not result in drastic changes of the phase diagram compared to the Bloch skyrmion case in Figure 11.4b. Figure 11.4c and d shows that the equilibrium radius of a skyrmion varies with the initial magnetization conditions as well.

11.3.2   PMA and DMI

It is known that the DMI encourages magnetic vortex formation, while the PMA enforces a uniform magnetization along some direction. One can expect that a suitable choice of these two parameters is necessary to stabilize a skyrmion in a nanowire.

To study the competition between PMA and DMI, we use nanowires of $L = 160$ nm, $W = 40$ nm, and $T = 1$. The PMA, K_u, is tuned between 0 and 2 MJ/m³ in the step size of 0.1 MJ/m³. Besides, the DMI, D_{ind}, is tuned between 0 and 9 mJ/m² in the step size of 0.5 mJ/m². Other parameters of Pt/Co remain the same as specified in Section 11.2.

Two initial conditions have been attempted. The first one is a Bloch skyrmion with a radius of 2 units and the other is a Néel skyrmion of default size. Figure 11.5a and b displays the resulting magnetic phase diagrams due to initial Bloch skyrmions with a radius of 2 units and

Chapter 11

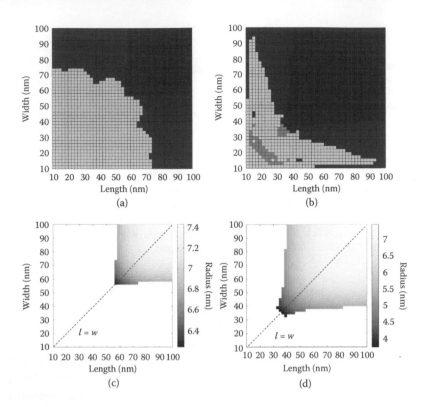

FIGURE 11.4 (a) The magnetic phase diagram obtained from an initial Bloch skyrmion with twice the radius as the ones used in Figure 11.2. The nanowire thickness in this case is 1 nm. (b) The magnetic phase diagram whose initial condition is a Néel skyrmion with charge –1 and downward polarization. The nanowire thickness in Figure 11.4b is also 1 nm. The gray scheme of the phase diagrams is the same as those in Figure 11.2e through h. (c) Stabilized skyrmion radius obtained from (a). (d) Stabilized skyrmion radius obtained from Figure 11.4b. Circular skyrmions are obtained from our two initial conditions. (Reprint of figure 3 in Chui, C.P., et al., *AIP Adv.*, 5, 047141, 2015. With permission.)

a Néel skyrmion of default size, respectively. We find from these two figures that the in-plane (IP) magnetization occurs at low D_{ind} and low K_u. Other magnetic phases include the SU (pale blue), SD (pale green), SK (dark blue), and HE (red) phases. A Bloch skyrmion with a radius of 2 units is unable to stabilize at varying K_u and D_{ind}. However, a Néel skyrmion can be stabilized for the ranges of K_u and D_{ind} being investigated. The failure of the stabilization of a larger Bloch skyrmion can be attributed partly to the limited planar dimensions of the nanowire that limits the full winding of skyrmion magnetization.

The dipolar field (also known as the demagnetizing field) has insignificant effect on the stability of skyrmions. We obtain the phase

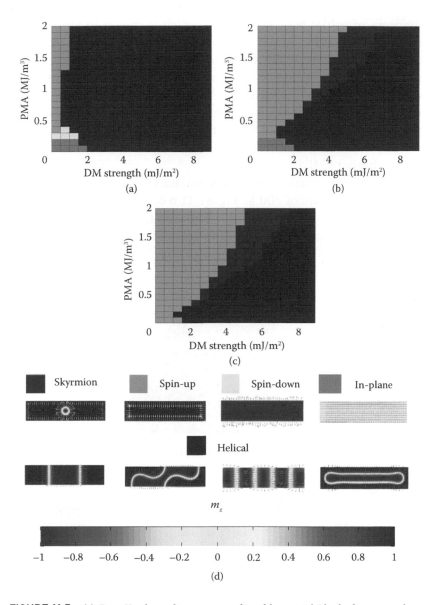

FIGURE 11.5 (a) D_{ind}–K_u phase diagrams produced by initial Bloch skyrmions having radius of 2 units. In the given D_{ind} and K_u intervals, no skyrmions are stabilized. (b) D_{ind}–K_u phase diagrams obtained from initial Néel skyrmion of charge −1 and downward polarization. (c) D_{ind}–K_u phase diagram by repeating (b) without the demagnetizing field. (d) The legend of the magnetic phase points with typical M_z plots and the color bar. (Reprint of figure 4 in Chui, C.P., et al., *AIP Adv.*, 5, 047141, 2015. With permission.)

diagram of initial Néel skyrmions, which is shown in Figure 11.5c by deactivating the dipolar field. The legend of the magnetic phases are listed in Figure 11.5d. One can compare it with Figure 11.5b to realize the contribution of the dipolar field, and see that the region of the SK phase has increased slightly when no dipolar field is present. In addition, the in-plane phase can no longer be established without the dipolar field. In fact, the IP phase is induced by the dipolar field along the thickness dimension, where the shape anisotropy is higher [54].

As a further investigation, we simulate the skyrmion stability at a finite temperature of 300 K. Figure 11.6 displays the D_{ind}–K_u phase diagram when a Bloch skyrmion is placed in a nanowire at 300 K. These phase diagrams use the same gray scheme as Figure 11.5. As time progresses from 4 ns (Figure 11.6a) to 10 ns (Figure 11.6b) and

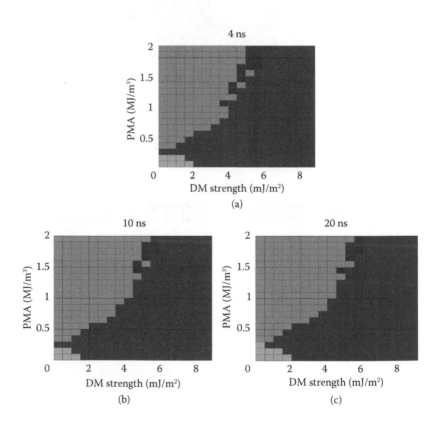

FIGURE 11.6 D_{ind}–K_u phase diagram at 300 K after a simulation time of (a) 4 ns, (b) 10 ns, and (c) 20 ns, indicating the shrinkage of the SK phase region with time. The gray scheme of Figure 11.5 applies to this figure. (Reprint of figure 5 in Chui, C.P., et al., *AIP Adv.*, 5, 047141, 2015. With permission.)

then to 20 ns (Figure 11.6c), the phase region of skyrmions diminishes under room temperature. The SK phase region that exists at zero temperature (Figure 11.5b) is replaced by either the SU state or the HE state. In general, we can realize that a skyrmion at higher DMI and higher PMA can withstand finite temperatures for a longer simulation time. Three representative (D_{ind}, K_u) points are chosen to show the variation of M_z with time. As shown in Figure 11.7a through c, the phase point $(D_{ind}, K_u) = (4.5, 1.4)$ can sustain a skyrmion at 300 K for 20 ns. In Figure 11.7d through f, however, the initial skyrmion at the phase point $(D_{ind}, K_u) = (4.5, 1.4)$ cannot survive for 20 ns. Instead, a spin-up state remains. At $(D_{ind}, K_u) = (6.0, 1.9)$ shown in Figure 11.7g through i, the initial skyrmion becomes a helical state after 20 ns.

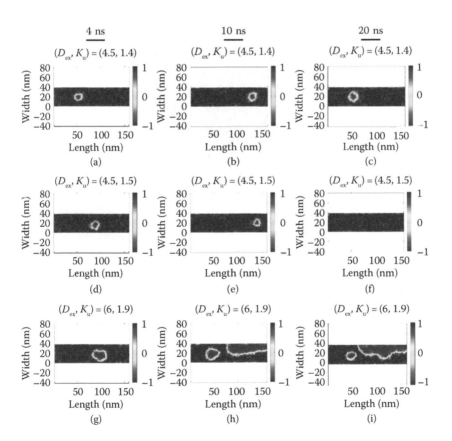

FIGURE 11.7 (a–c) Snapshots of M_z for $(D_{ind}, K_u) = (4.5, 1.4)$, showing the SK state at 20 ns. (d–f) Snapshots of M_z for $(D_{ind}, K_u) = (4.5, 1.5)$, showing the SU state at 20 ns. (g–i) Snapshots of M_z for $(D_{ind}, K_u) = (6.0, 1.9)$, showing the HE state at 20 ns. (Reprint of figure 6 in Chui, C.P., et al., *AIP Adv.*, 5, 047141, 2015. With permission.)

Chapter 11

11.3.3   Exchange Stiffness and Gilbert Damping Parameter

A large exchange stiffness prefers alignment of the magnetization, while a small Gilbert damping parameter retards the rate of such alignment. Accordingly, the exchange stiffness can compete with the Gilbert damping parameter in the stability of skyrmions. We study their effect on a skyrmion, which is placed at the center of a nanowire of length $L = 160$ nm, width $W = 40$ nm, and thickness $T = 1$ nm. The exchange stiffness A_{ex} varies from 0 to 20 pJ/m in 0.5 pJ/m step size, whereas the Gilbert damping parameter α varies from 0.01 to 0.4 in 0.01 step size. Other material parameters of Pt/Co still apply to this subsection. The α–A_{ex} phase diagram is given in Figure 11.8a, which displays three types of magnetic phases. The phase region in red represents the helical

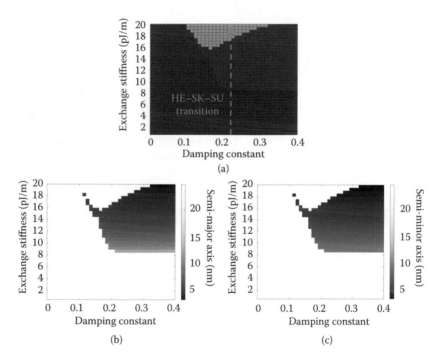

FIGURE 11.8   (a) The α–A_{ex} phase diagram where $0.01 \le \alpha \le 0.4$ and $0 \le A_{ex} \le 20$ pJ/m, with HE state (red), SK state (dark blue), and SU state (pale blue). The M_z plots obtained in the α–A_{ex} phase diagram have the same pattern as those obtained in the D_{ind}–K_u phase diagram in Figure 11.5d. As the DMI is fixed at 3 mJ/m^2, the phase transition from HE to SK and finally to SU state occurs as the exchange stiffness increases. (b–c) The semi-major and semi-minor axes of the elliptical skyrmion in equilibrium. The skyrmion dimensions decrease with increasing α and A_{ex}. (Reprint of figure 7 in Chui, C.P., et al., *AIP Adv.*, 5, 047141, 2015. With permission.)

(HE) state. The phase region in dark blue is the skyrmion (SK) state. The final phase region in pale blue is the spin-up (SU) state. An initial skyrmion can be stabilized if α is no less than 0.12 and A_{ex} is at least 8.2 pJ/m. According to our choice of α range, A_{ex} seems to have an upper limit that allows for the stability of a skyrmion. If A_{ex} further increases, the skyrmion would transform to the SU state. In spite of the upper bound of A_{ex} that would hamper the stability of a skyrmion, both large PMA and large α encourage the stability of a skyrmion.

Figure 11.8b and c provides two diagrams that indicate the semi-major and semi-minor axes of the skyrmions. The definition of the axes follow that of the skyrmion radius. We evaluate, along the length and width dimensions, the distance between the core where $M_z = -1$ and the magnetization where $M_z = 0$ in the course of magnetization winding. One can realize that the semi-major and semi-minor axes obtained from a (α, A_{ex}) phase point are considerably different; therefore, the skyrmions obtained from our phase diagram are generally elliptical. Skyrmions near the SU–SK phase boundary tend to be more circular, while those near the HK–SK phase boundary are elliptical. We can realize that the skyrmion growth in dimensions is limited by the edge effect of our rectangular plane, such that the skyrmions are more likely to reach equilibrium at an elliptical shape.

By referring to the α–A_{ex} phase diagram, the skyrmion state is the intermediate state between the helical state and the spin-up state when one fixes α and increases A_{ex}. This phenomenon is understandable because A_{ex} helps to maintain the homogeneous magnetization along the thickness dimension, and hinders the skyrmion growth which requires full winding of magnetization. A subtle balance between α and A_{ex} is thus crucial for the maintenance of a skyrmion. Our findings suggest that a minimum value of α is necessary for stabilizing a skyrmion, though a theoretical understanding of the role of α is still lacking in the literature. Nevertheless, our simulation result of the α–A_{ex} phase diagram is consistent with other studies. First, numerical simulations suggest that the generation of isolated skyrmions rely on strong damping under circular electric currents [55]. Also, damping is found to be necessary for dynamical skyrmion transformation [24] and skyrmion nucleation processes [19]. As an observation of our findings, the multidomain state prevails if α is below a minimum value. We expect that damping is a critical parameter that is responsible for the nucleation phase diagram and that more in-depth understanding of this mechanism is to be unraveled.

Chapter 11

11.4   Skyrmion in an NC-STO

11.4.1   Varying the Fixed–Layer Parameters

In the first place, we study the effect of varying the polarization angle of the fixed layer on the magnetization. The $J_z - \beta$ phase diagram composed of STO snapshots at 10 ns is given in Figure 11.9a. This diagram is obtained by using the Pt/Co parameters except J_z is adjusted from -5×10^8 to $+5 \times 10^8$ A/cm² in steps of 0.5×10^8 A/cm² and β is adjusted from $0°$ to $90°$ in steps of $5°$. From this phase diagram, one can realize the change in the circular nature of the initial skyrmion as the polarization angle slightly deviates from the out-of-plane configuration of $90°$ (e.g., $\beta = 85°$).

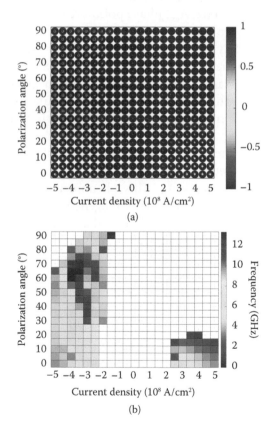

(a)

(b)

FIGURE 11.9   (a) The magnetic phase diagram of the J_z–β phases of NC-STOs, composed of snapshots at 10 ns. The color bar represents the M_z value of the STOs. (b) The breathing frequency as the parameters used in part (a) changes.

(Continued)

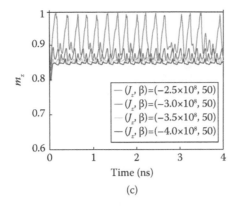

(c)

FIGURE 11.9 (Continued)   (c) The time evolution of M_z for four representative STOs after 4 ns, showing that the waveforms of M_z is highly nonsinusoidal. (Reprint of figure 2 in Chui, C.P., and Zhou, Y., *AIP Adv.*, 5, 097126, 2015. With permission.)

Although it has been shown that the breathing mode can be achieved by an alternating magnetic field that is perpendicular to the free layer [56], we show that it is already achievable at a constant magnetic field. The main frequency having the highest power as J_z and β vary is plotted in Figure 11.9b. The frequency generated by STOs of varying tilt angles is of gigahertz order. However, the variation of the frequency does not follow a specific trend because the M_z evolution is far from sinusoidal (see Figure 11.9c). This means that a majority of the STOs in the phase diagram have strong higher harmonics. In this case, we find difficult to specify the most representative frequency of an STO.

11.4.2   Varying the Spacer Parameters

In this subsection, we obtain the snapshots of magnetization for the STOs where $0.1 \leq \Lambda \leq 2$ and $0 \leq \dot{\varepsilon} \leq 0.2$. The Slonczewski asymmetry parameter Λ has a step size of 0.1, whereas the secondary STT term $\dot{\varepsilon}$ has a step size of 0.01. The Λ–$\dot{\varepsilon}$ phase diagram at 10 ns is shown in Figure 11.10a. For our Pt/Co parameters, both Λ and $\dot{\varepsilon}$ can have a wider choice in a way that a skyrmion can be maintained in an NC-STO. Since the effect of the field-like torque is believed to be insignificant [57], the value of $\dot{\varepsilon}$ brings about little effect on the stability of a skyrmion. Our result in Figure 11.10a coincides with the literature because $\dot{\varepsilon}$ does not need to be extremely small (less than about 0.08) in order to stabilize a skyrmion. The asymmetry parameter Λ needs to be above 0.5 to stabilize a skyrmion in an NC-STO, meaning that an asymmetric system can also allow for skyrmion stability.

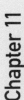

Chapter 11

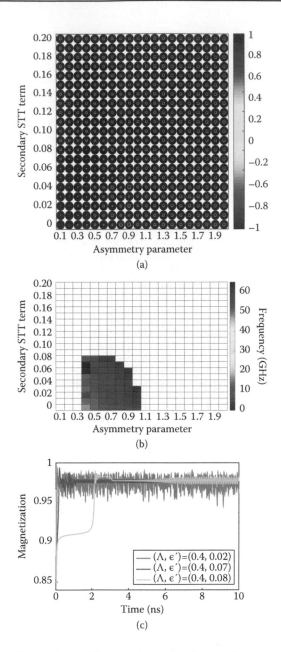

FIGURE 11.10 (a) Snapshots of the Λ–ɛ̇ phase after the simulation time of 10 ns. The color bar shows the M_z scale of the STOs. (b) The frequency of breathing by using the range of parameters in Figure 11.10a. (c) The temporal variation of M_z for three STOs, demonstrating the instability of the breathing mode at Λ = 0.4. (Reprint of figure 3 in Chui, C.P., and Zhou, Y., *AIP Adv.*, 5, 097126, 2015. With permission.)

Though skyrmions can be stabilized in a large portion of the Λ–$\dot{\varepsilon}$ phase diagram, only a minority of them can perform breathing. As can be seen from Figure 11.10b, STOs with $0.2 \leq \Lambda \leq 1$ and $0 \leq \dot{\varepsilon} \leq 0.09$ can return the breathing frequency. Also, the range of frequency obtained is quite narrow in the order of 1 GHz, except some particular points of comparatively higher frequency. At $\Lambda = 0.4$, the main frequency is higher because the M_z evolution is highly nonsinusoidal. Some of these M_z waveforms at $\Lambda = 0.4$ are provided in Figure 11.10c, from which we suspect that $\Lambda = 0.4$ is the intermediate state between the static mode and the breathing mode.

11.4.3   Varying the Free-Layer Parameters

A number of free layer parameters are compared, in order to realize the proper adjustment of them for the stability of a skyrmion in an NC-STO.

First, we study the exchange field by tuning the saturation magnetization M_{sat} from 50 to 1,000 kA/m and the exchange stiffness A_{ex} from 1 to 20 pJ/m. The step size of M_{sat} is 50 kA/m, and the step size of A_{ex} is 1 pJ/m. The snapshots of magnetization after 10 ns are plotted in Figure 11.11a, from which we can specify the lower limit of A_{ex} that allows for skyrmion stability. If the lower limit of A_{ex} is not met, an initial skyrmion would become a chiral liquid after 10 ns.

The main frequency with the largest power as M_{sat} and A_{ex} vary is plotted in Figure 11.11b. Most of these frequencies are the fundamental frequency of the M_z evolution. We notice from Figure 11.11b that a large portion of the phase diagram allows for skyrmion stability. However, only those STOs whose free layer has a smaller M_{sat} and a larger A_{ex} can perform breathing. We can see that the operating frequency increases with A_{ex} and decreases with M_{sat}. If the free layer has a large M_{sat}, it requires a large A_{ex} to keep a skyrmion in the breathing mode. The frequency of the breathing skyrmion evaluated in our ranges of M_{sat} and A_{ex} is usually below 20 GHz, except those STOs operating at about 80 GHz which have an extremely low M_{sat}. The frequency spectra of three STOs are shown in Figure 11.11c, which indicates the suppression of higher harmonics with the increase in fundamental frequency. In other words, increasing the exchange stiffness while maintaining a fixed saturation magnetization can result in suppression of higher harmonics.

Second, we investigate the effect of the DMI and PMA on skyrmion stability. This time, we vary the induced DMI, D_{ind}, between 0 and 9 mJ/m^2 and the PMA, K_u, between 0 and 2 MJ/m^3. The step size of D_{ind} is 0.5 mJ/m^2, and the step size of K_u is 0.1 MJ/m^3. The snapshots of the STOs at 10 ns are shown in Figure 11.12a, whose pattern is similar

Chapter 11

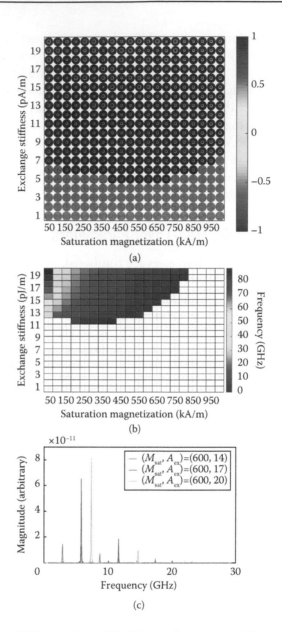

FIGURE 11.11 (a) The snapshots of the M_{sat}–A_{ex} phases at 10 ns. The color bar refers to the M_z value of the STOs. (b) The frequency plot of the STOs, from which we find that the highest frequencies occur at extremely low M_{sat}. (c) The frequency spectra of three STOs obtained from the time evolution of M_z, showing that the power of higher harmonics decreases with the fundamental frequency as M_{sat} is kept constant. (Reprint of figure 4 in Chui, C.P., and Zhou, Y., *AIP Adv.*, 5, 097126, 2015. With permission.)

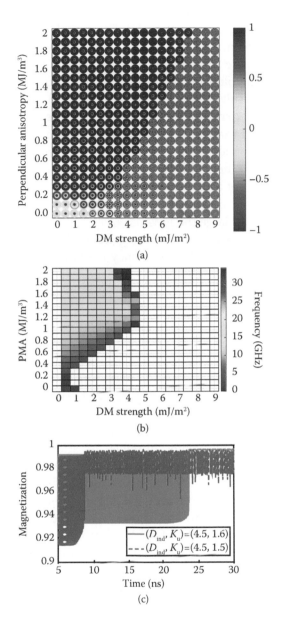

FIGURE 11.12   (a) The snapshots of D_{ind}–K_u phases diagram at 10 ns. The color scale of M_z is shown on the right. (b) The main frequency of the STOs. (c) The time evolution of M_z of two STOs with $(D_{ind}, K_u) = (4.5, 1.5)$ and $(D_{ind}, K_u) = (4.5, 1.6)$ near the static/breathing boundary, indicating an abrupt change of M_z amplitude and frequency after certain times. (Reprint of figure 5 in Chui, C.P., and Zhou, Y., *AIP Adv.*, 5, 097126, 2015. With permission.)

to the case of nanowires in Ref. [16]. Same as the nanowire case, the NC-STOs require large DMI and large PMA to stabilize a skyrmion. If both parameters are small, the skyrmion core can be maintained after 10 ns, with its surrounding magnetization becoming the in-plane configuration. The result shows that the spin-polarized current controls the magnetization underneath the NC region. Another implication of the phase diagram is that as the PMA is small, the induced DMI largely determines the winding of magnetization. The presence of both an external magnetic field and spin-polarized current does not guarantee the formation of a skyrmion. Conversely, an excessively large DMI returns a distorted magnetization which cannot be turned back to a skyrmion, however large the PMA is.

Figure 11.12b shows the breathing frequency of the STOs in our D_{ind}–K_u phase diagram. The frequency is obtained in the vicinity of 10 ns, the end of our simulation time. We see that the range of operating frequency is between 1 and 35 GHz. The operating frequency generally increases with PMA for a given DMI. In these STOs, the fundamental frequency has the largest power.

It is important to specify the time of recording the operating frequency because the amplitude of M_z evolution might possibly fluctuate after our default simulation time of 10 ns. As an example, we further increase the simulation time of some STOs. Figure 11.12c shows the M_z evolution of two STOs with $(D_{ind}, K_u) = (4.5, 1.5)$ and $(D_{ind}, K_u) = (4.5, 1.6)$. From these two STOs, we can realize that at large DMI and large PMA, the frequency output can become unstable. As another observation, the amplitude of M_z is smaller when D_{ind} increases. This result suggests that the breathing mode would decline and switch to the static mode as D_{ind} further increases. When the DMI and PMA combination results in the static/breathing phase boundary shown in red, the operating frequency is appreciably higher because of the M_z instability.

Third, we investigate the effect of the nanocontact diameter $d_{contact}$ and the free layer diameter d_{layer} on the stability of a skyrmion. We adjust $d_{contact}$ between 20 and 200 nm in steps of 10 nm, and vary d_{layer} between 50 and 500 nm in steps of 25 nm. The $d_{contact}$–d_{layer} phase diagram at 10 ns is shown in Figure 11.13a, in which only those STOs satisfying $d_{contact} \leq d_{layer}$ are given. It indicates that a large number of STOs can keep a skyrmion after the simulation time of 10 ns. The skyrmionic core area also increases with $d_{contact}$, so we suggest that the NC area is able to control the skyrmion size. It is important to ensure that the area between the NC and the edge of the free layer permits a full winding of magnetization before a skyrmion can be stabilized in an NC-STO.

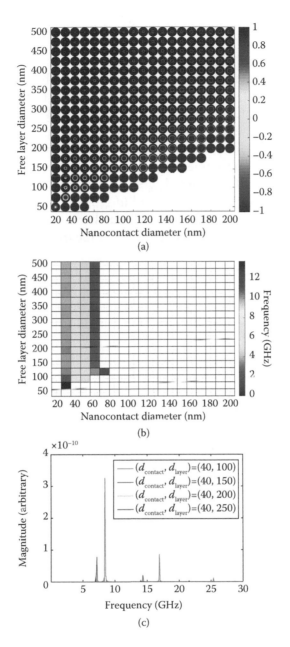

FIGURE 11.13 (a) The snapshots of the $d_{contact} - d_{layer}$ phases at 10 ns. The M_z scale is shown on the color bar. (b) The main frequency of the STOs. (c) The frequency spectra of M_z for four phase points (40, 100), (40, 150), (40, 200), and (40, 250), indicating the insignificant effect on the fundamental frequency as the nanocontact diameter increases, at least numerically. (Reprint of figure 6 in Chui, C.P., and Zhou, Y., *AIP Adv.*, 5, 097126, 2015. With permission.)

Chapter 11

The breathing frequency of the STOs is plotted in Figure 11.13b. By comparing the frequency chart with Figure 11.13a, we notice that a majority of the STOs in our settings remain at the static mode after 10 ns. We realize from the frequency chart that breathing occurs at STOs, roughly satisfying 30 nm $\leq d_{contact} \leq$ 60 nm for free layers of all diameters attempted. In addition, the frequency of breathing due to a given $d_{contact}$ is generally independent of d_{layer} if $d_{contact}$ is not sufficiently close to d_{layer}, at least numerically. The frequency spectra in Figure 11.13c verify this phenomenon. Figure 11.13c displays four spectra at $d_{contact}$ = 40 nm and 100 nm $\leq d_{layer} \leq$ 250 nm. We can see that the spectra for $(d_{contact}, d_{layer})$ = (40, 200) and $(d_{contact}, d_{layer})$ = (40, 200) overlap appreciably. Therefore, we suggest that further increase in the free-layer diameter results in insignificant change in the frequency as well as the power of the harmonics.

If the nanocontact diameter equals the free layer diameter, the STO becomes a nanopillar. For example, the NC-STOs where $(d_{contact}, d_{layer})$ = (100, 100) and $(d_{contact}, d_{layer})$ = (200, 200) are nanopillars, and the resulting spin-down state after 10 ns is consistent with the previous observation that the NC area controls the skyrmion core area.

In the literature, we have been told that the spin wave mode in a conventional NC-STO is controlled by the threshold current I_{th} experienced by the free layer [58,59]. The threshold current is dependent on the NC diameter as well as the material parameters of the free layer, and is independent of the dimensions of the free layer. Accordingly, if the nanocontact diameter is unchanged, the operating frequency is unaffected by increasing the size of the free layer.

For STOs with fixed-current density and larger NC diameter (70 nm in our study), more current is injected into the free layer. At that time, the static mode replaces the breathing mode at a smaller NC diameter.

The effect of varying the damping constant is also investigated. In Figure 11.14, we lower the damping constant α to become smaller than 0.3, the value used in our schematic setup. All other simulation parameters remain the same as in Figure 11.1a. After 2 ns, a smaller α of 0.1 would turn the initial skyrmion into a magnetic droplet, as shown in Figure 11.14a. In Figure 11.14b if a larger α of 0.2 is used, breathing of a skyrmion begins to be possible. We can realize frequency generation from the time evolution of M_z in Figure 11.14c. If α = 0.1, the oscillation of magnetization is no longer sustained after 2 ns. This shows that such a small damping constant disallows frequency generation. However, a larger α = 0.2 permits oscillation of M_z during the simulation. It means that frequency generation by breathing is possible. We can see that a

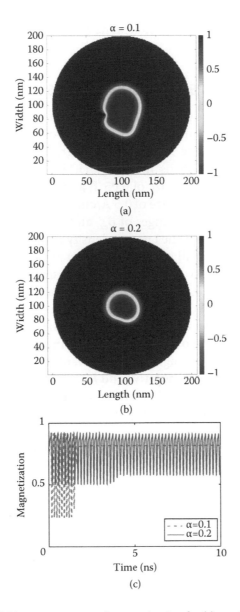

(a)

(b)

(c)

FIGURE 11.14   (a) The z-component of magnetization for (a) $\alpha = 0.1$ and (b) $\alpha = 0.2$. A magnetic droplet is formed as $\alpha = 0.1$. (c) The time variation of M_z demonstrating the generation of frequency at a larger damping constant instead of a smaller one. (Reprint of figure 7 in Chui, C.P., and Zhou, Y., *AIP Adv.*, 5, 097126, 2015. With permission.)

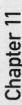

larger damping constant facilitates the stability of a skyrmion and the breathing mode.

11.4.4 Varying the External Parameters

In this subsection, we consider the effect of the external magnetic field B_{ext} and the current density J_z on the stability of a skyrmion in an STO.

First, we provide the B_{ext}–J_z phase diagram composed of magnetization snapshots at 10 ns in Figure 11.15a. Here, B_{ext} ranges from -0.3 to $+0.3$ T in the step size of 0.05 T, and J_z ranges from -5×10^8 to $+5 \times 10^8$ A/cm² in the step size of 0.5 A/cm². Large and small skyrmions can be established in the ranges of B_{ext} and J_z. If the external magnetic field is nonnegative and the current density is small in magnitude, the initial skyrmion of small size can remain static and unchanged in size after 10 ns. This observation indicates that a small J_z fails to adjust the damping torque and facilitate magnetization relaxation. If B_{ext} becomes positive and J_z rises above $+2.5 \times 10^8$ A/cm², the skyrmion would degenerate to a spin-up state entirely. At that time, the current density results in an STT which acts parallel to the damping torque. The STT reinforces the alignment of magnetization toward the direction of the effective field $\mathbf{B}_{eff}$. If B_{ext} is nonnegative and $J_z \leq -1.5 \times 10^8$ A/cm², skyrmions of larger size than the initial condition can be stabilized. It means that the current density is sufficiently negative to excite the change in the spin wave mode.

If the applied magnetic field is negative, the magnetization would be entirely different. The spin-down state is the dominant configuration in this case, especially for a negative current density. The initial skyrmion can persist under a positive current density of moderate magnitude, until the magnitude of the current density further increases. When the current density increases, the spin-up state replaces the static skyrmion mode. In simple words, the skyrmion can remain in an STO if the current density does not oppose the applied magnetic field too severely.

The breathing frequency of skyrmions under the B_{ext}–J_z phase diagram is plotted in Figure 11.15b. The range of current density that results in breathing skyrmions widens with the external magnetic field. At a stronger magnetic field, a given current density can produce a larger breathing frequency. This observation of skyrmion-based STOs coincides with that of the conventional STOs [42]. The increase in breathing frequency with the external magnetic field strength and its decrease with the magnitude of current density is consistent with an analytical study on conventional STOs [60]. The breathing

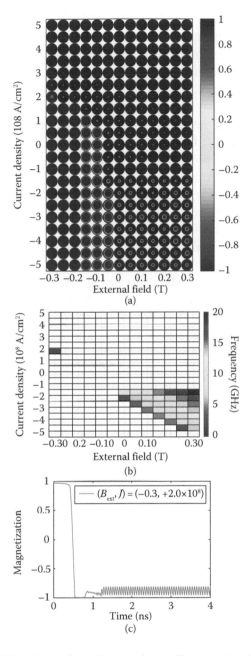

FIGURE 11.15 (a) The B_{ext}–J_z phase diagram obtained by grouping the magnetization snapshots at 10 ns. The value of M_z is given in the color bar. (b) The main frequency of the STOs. (c) The time variation of M_z of phase point $(B_{ext}, J_z) = (-0.3, +2.0 \times 10^8)$ for the first 4 ns, generating a breathing skyrmion with opposite polarity to the initial skyrmion. (Reprint of figure 8 in Chui, C.P., and Zhou, Y., *AIP Adv.*, 5, 097126, 2015. With permission.)

Chapter 11

frequency of a skyrmion decreases with the magnitude of current density because the spin wave mode covers a larger free layer area, and thus requires more time to complete a full rotation of the skyrmionic topological density [61]. The range of the frequency shown in Figure 11.15b is spread more evenly as B_{ext} increases and J_z increases its magnitude.

Even though a negative external magnetic field would return the spin-down state for most of our simulated STOs, a breathing skyrmion having a positive polarity can be established at $B_{ext} = 0.3$ T and $J_z = +2.0 \times 10^8$ A/cm². Figure 11.15c is the M_z evolution of the STO with time at this external condition for the first 4 ns. For the first 0.5 ns, the average magnetization still lies along the external field direction. Soon afterwards, the average magnetization reverses its sign, followed by breathing characterized by the oscillation of the M_z amplitude. This special case demonstrates that a reversed magnetic field does not always destroy the skyrmion mode. A breathing skyrmion with reversed core polarity can be an equilibrium mode instead.

Next, we investigate the effect of the strength of the external magnetic field B_{ext} and the polar angle θ of the applied field on skyrmion stability. Here, we adjust B_{ext} between -0.3 and $+0.3$ T and tune θ between 0° and 180°. The step size of B_{ext} is 0.05 T, whereas the step size of θ is 15°. The resulting B_{ext}–θ phase diagram at 10 ns is shown in Figure 11.16a. In general, skyrmions can be stabilized in an NC-STO if the external magnetic-field vector $\mathbf{B}_{ext}$ and the skyrmionic core magnetization $\mathbf{M}_{core}$ follow the relation $\mathbf{B}_{ext} \cdot \mathbf{M}_{core} \leq 0$, such that the angle between these two vectors is obtuse. Figure 11.16b plots the most powerful frequency of the breathing skyrmions, and all of them are the fundamental frequency. When the polar angle approaches the out-of-plane direction ($\theta \to 0°$) or ($\theta \to 180°$), the fundamental frequency shifts to a higher value. If the magnetic field strength is fixed, decreasing the polar angle results in the blue shift of the main frequency and the declining power of the higher harmonics. On the other hand, a magnetic field perpendicular to the free layer ($\theta = 0°$, 180°) cannot suppress the higher harmonics completely. When the magnetic field is in the in-plane direction ($\theta = 90°$), the skyrmion becomes the static mode. The result shows that an oblique polar angle is constructive of skyrmion breathing.

The phase diagram in Figure 11.16c is symmetric about the phase point (0, 90) at the center. This is understandable because the sign reversal of B_{ext} is equivalent to reflecting the perpendicular component of the vector $\mathbf{B}_{ext}$ about the free layer.

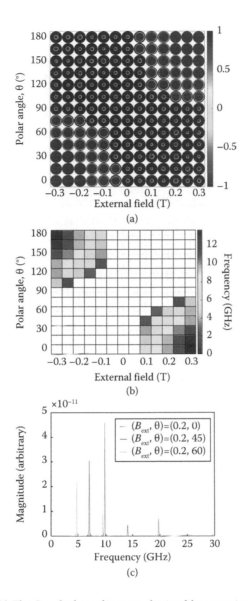

FIGURE 11.16 (a) The B_{ext}–θ phase diagram obtained by grouping the snapshots at 10 ns. The color bar represents the scale of M_z. (b) The main frequency of the STOs as the time gets close to 10 ns. (c) The frequency spectra of M_z for three phase points (0.2, 0), (0.2, 45), and (0.2, 60), from which we realize the increase in the fundamental frequency as the polar angle θ decreases. (Reprint of figure 9 in Chui, C.P., and Zhou, Y., *AIP Adv.*, 5, 097126, 2015. With permission.)

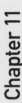

11.5   Discussion

First, we provide further comments on the stability of skyrmions in a nanowire. Then, we further illustrate our findings of the skyrmion-based NC-STOs.

For nanowires, our $L–W$ phase diagram in Figure 11.2a indicates the variation of magnetization when we increase the length and width of a nanowire. If the planar dimensions are small and the thickness is fixed, the equilibrium magnetization of the whole nanowire would follow the initial skyrmion core magnetization, resulting in the SD state in our case. As the planar dimensions increase slightly, magnetization reversal toward the SU state occurs, such that the nanowire magnetization is parallel to our initial skyrmion rim. A complete magnetization winding required for establishing a skyrmion can be achieved by further increasing the planar dimensions. Another important result is that thicker nanowires necessitate larger planar areas to sustain a skyrmion.

Our $L–W$ phase diagram in Figure 11.2a suggests an external method to manipulate the magnetic state of a nanowire without relying on the intrinsic parameters of the nanowire materials. As a skyrmion enters a progressively narrower nanowire, it would encounter changes in the boundary conditions. As a result, the equilibrium magnetization changes as well. It is, however, noted that our result is unable to deduce that a helical state that is stabilized in a narrower nanowire can return to the skyrmion state if it is driven toward the wider portion of a nanowire.

The $D_{ind}–K_u$ phase diagram for nanowires at absolute zero temperature has already been recorded in Ref. [16], which also demonstrates the requirement of large DMI and large PMA in stabilizing a skyrmion. According to our $D_{ind}–K_u$ phase diagram with temperature effect included (Figure 11.6), we can realize that even larger DMI and PMA are essential for sustaining a skyrmion. Otherwise, a skyrmion at finite temperature would degenerate to the helical state or the ferromagnetic state.

From Figure 11.8, we observe that both α and A_{ex} are important for maintaining a skyrmion in a nanowire. We can also notice that the skyrmion dimensions decrease with increasing α and A_{ex}, suggesting the role of both parameters in promoting parallel alignment of magnetization.

The competition between A_{ex} and D_{ind} can be inferred from the $\alpha–A_{ex}$ phase diagram. We know that the value of D_{ind} is 3 mJ/m². When we fix $\alpha > 0.12$ and increase A_{ex} from 0 to 20 pJ/m, the transition from the HE to SK state and then to the SU state is noticeable. With this trend, the

SK state can be regarded as the intermediate state between the helical and ferromagnetic state in the course of increasing A_{ex} [62].

For the fixed layer of an NC-STO, if the polarization angle does not follow the current-perpendicular-to-plane (CPP) configuration, higher harmonics of high power would result. This makes the frequency plot lose an apparent trend. In fact, the conventional STOs do not exhibit this highly unpredictable property. In conventional ones, a change in the polarization angle can drive a gradual transition of the operating frequency [63]. In short, one cannot obtain a stable skyrmion by tilting the fixed layer polarization.

According to Figure 11.9a, the skyrmion cores are distorted in a direction that follows the fixed layer polarization. It is known from Equation 11.1 that the polarization M_p can change the in-plane torque as well as the field-like torque. With an oblique polarization, the skyrmions cannot maintain the full magnetization winding in the presence of a spin-polarized current. The NC-STOs with polarization angle $\beta \leq 40°$ are the most prominent examples of such a condition. In short, an NC-STO relies on the out-of-plane polarization to preserve the topological protection possessed by a skyrmion.

The spacer of an NC-STO seems to have a less important role of skyrmion stability, whose reason is understandable. According to our Λ–$\acute{\varepsilon}$ phase diagram in Figure 11.10a, the most important criterion of a spacer is the passage of a spin-polarized current. If this condition is satisfied, the in-plane torque can be produced. The magnetization of the free layer can then be generated.

In addition, the variation of spacer parameters seems to provide a limited range of the breathing frequency, if any. Therefore, one might better tune the free layer parameters and the external parameters for the desired operating frequency. The spacer can be no more than a component that ensures electronic passage between the fixed layer and the free layer.

From the M_{sat}–A_{ex} phase diagram in Figure 11.11a, we can observe some phase points at small M_{sat} and large A_{ex} that can generate an operating frequency of decades of gigahertz. These ranges of M_{sat} and A_{ex} imply a large exchange field, expressed as:

$$\mathbf{B}_{exch} = \frac{2A_{ex}}{M_{sat}} \Delta \mathbf{M}, \tag{11.5}$$

where $\Delta \mathbf{M}$ is the aggregate change of the neighboring magnetization of a simulation cell. Equation 11.5 implies that one needs a strong exchange field to generate a high breathing frequency. The high frequency generated by STOs with large exchange field is applicable to satellite communications. It is possible that skyrmion-based STOs can find applications in this area.

Our result indicates that materials with low saturation magnetization are also a candidate of skyrmion-based NC-STOs, further suggesting free-layer materials that are not highly ferromagnetic ones. Indeed, some commercial ferrofluids can provide such a low saturation magnetization as 32 kA/m [64].

We have shown in Figure 11.12 that the breathing mode would degenerate to the static mode as the induced DMI, D_{ind}, increases further. This result suggests that, in order to generate microwave frequency by breathing, the DMI should be lower. We also realize that the magnetization dynamics could change after a longer simulation time if both the DMI and the PMA, K_u, are large. It means that unstable frequency could be generated when both of these parameters are large. Therefore, for stable frequency output, the use of the phase boundary points between the breathing mode and the static mode should be discouraged.

Our D_{ind}–K_u phase diagram verifies the analytical results of Ref. [14], which suggests that a large PMA without any DMI can result in the stability of a skyrmion. A skyrmion can exist in the presence of an external magnetic field together with a spin-polarized current. Even though a static skyrmion can remain in a free layer without DMI, no breathing skyrmions can exist without DMI. This observation follows from the far left column of the frequency plot in Figure 11.12b, which indicates that no frequency can be generated without DMI. In short, the presence of DMI is still important for skyrmion-based STOs which serve as frequency generators.

In addition to nanowires without external magnetic field and current, the skyrmion mode remains at the static mode [16]. The temporal instability of the operating frequency of STOs at the static/breathing boundary points deserves further experimental evidence.

In Figure 11.13a, the $d_{contact}$–d_{layer} phase diagram demonstrates the range of the NC diameters that can result in skyrmion breathing. It is found that the range derived from our Pt/Co settings is approximately between 30 and 60 nm. We also notice that the area of the free layer does not affect the frequency too significantly because it is the free layer region covered by the nanocontact that controls the magnitude of electronic current and hence controls the spin wave mode. Therefore, the free layer can be adjusted to the smallest possible size that permits the desired spin wave excitation.

We also see that the nanopillar configuration, once electrically conducted, destroys the magnetization winding of an initial skyrmion by changing the skyrmion into a highly ferromagnetic state. Accordingly, we should not employ this configuration to skyrmion-based STOs that produce microwave frequency.

The breathing frequency plot of the B_{ext}–J_z phase diagram (Figure 11.15b) can be analyzed using Equation 11.1. For a more negative J_z, the in-plane torque acts harder to drive the magnetization away from the free layer polarization along +z-axis. The in-plane torque would lead to a lower operating frequency because of the conservation of angular momentum of classical spins. The field-like torque cannot affect the total STT significantly because of its small magnitude [57].

We have shown that the frequency plot of the B_{ext} – J_z phase diagram shown in Figure 11.15a is consistent with the case of conventional STOs. In other words, the control of the spin wave mode in conventional STOs can apply to skyrmion-based STOs.

Figure 11.16 shows us that, as the polar angle approaches the out-of-plane direction, the higher harmonics of the M_z evolution can have a decreased power. However, the blue-shift of operating frequency would occur. In order to produce a low operating frequency as well as small power of higher harmonics, the choice of materials for the free layer is necessary.

Our findings about the increase in operating frequency as the polar angle θ approaches 0° or 180° is opposite to our understanding of conventional STOs [42,65]. In addition, our result shows that an in-plane magnetic field (θ = 90°) can only result in a static skyrmion without breathing. If one requires switching between the static mode and the breathing mode, an in-plane magnetic field can achieve it.

When all our phase diagrams pertaining to the NC-STOs are considered, we see that the static mode is easier to achieve than the breathing mode. One can change the frequency of breathing most easily by varying the external magnetic field vector as well as the magnitude of the current density. If a higher operating frequency (tens of gigahertz) is required, one can choose materials that have low saturation magnetization, high PMA, and strong DMI.

11.6 Summary

By means of micromagnetic simulations, we have discussed the favorable physical conditions for stabilizing a skyrmion in a nanowire [45]

Chapter 11

and an NC-STO [46]. It is important that our findings are based on various initial magnetization states pertaining to particular simulation settings.

First, we summarize the material parameters that are common to both nanowires and NC-STOs. The combination of a large PMA and a moderate DMI is essential for establishing a stable skyrmion in a nanowire or an NC-STO. If a breathing skyrmion is desired, the large DMI is still crucial. The role of DMI and PMA is more evident at a finite temperature, at which both parameters have to be even larger for skyrmion stability. Though the exchange stiffness of a nanowire should have an upper limit in order to maintain a stable skyrmion, a larger exchange stiffness is beneficial to the formation of skyrmions in STOs.

Then, we summarize the conditions of skyrmion stability specifically to nanowires. There are some dimensional constraints of the nanowires that are suitable for skyrmion stability. In addition, the shrinkage of nanowires can alter the magnetic state. A large damping parameter is a prerequisite for maintaining a skyrmion without relaxing into a ferromagnetic or helical state.

Finally, we focus on the parameters exclusive for NC-STOs. The oblique polarization angle of current injection distorts the skyrmion topology severely, so that it is not recommended for practical applications. A small saturation magnetization increases the breathing frequency. The spacer can allow for more flexible choices, such that a skyrmion stays at the static or breathing mode under a bias current. The nanocontact diameter of an STO is responsible for controlling the breathing rate of a skyrmion, given that the current density is fixed. Skyrmions can persist in the breathing mode in the presence of current density and applied magnetic field. The operating frequency increases with the field strength and decreases with the current density. In general, the method of controlling the operating frequency of conventional STOs applies to skyrmion-based STOs as well.

References

1. F. Jonietz, et al. Spin transfer torques in MnSi at ultralow current densities. *Science*, 330:1648–1651, 2010.
2. X. Z. Yu, et al. Skyrmion flow near room temperature in an ultralow current density. *Nature Communications*, 3:988, 2012.
3. S. Mühlbauer, et al. Skyrmion lattice in a chiral magnet. *Science*, 323(5916):915–919, 2009.
4. S. Heinze, et al. Spontaneous atomic-scale magnetic skyrmion lattice in two dimensions. *Nature Physics*, 7:713–718, 2011.
5. A. Tonomura, et al. Real-space observation of skyrmion lattice in helimagnet MnSi thin samples. *Nano Letters*, 12(3):1673–1677, 2012.

6. I. Dzyaloshinskii. A thermodynamic theory of "weak" ferromagnetism of anti-ferromagnetics. *Journal of Physics and Chemistry of Solids*, 4(4):241–255, 1958.
7. T. Moriya. Anisotropic superexchange interaction and weak ferromagnetism. *Physical Review*, 120:91–98, 1960.
8. X. Z. Yu, et al. Real-space observation of a two-dimensional skyrmion crystal. *Nature*, 465:901–904, 2010.
9. C. Pfleiderer and A. Rosch. Condensed-matter physics: Single skyrmions spotted. *Nature*, 465:880–881, 2010.
10. C. Pfleiderer. Magnetic order: Surfaces get hairy. *Nature Physics*, 7:673–674, 2011.
11. H. Du, et al. Highly stable skyrmion state in helimagnetic MnSi nanowires. *Nano Letters*, 14(4):2026–2032, 2014.
12. A. B. Butenko, A. A. Leonov, U. K. Rößler, and A. N. Bogdanov. Stabilization of skyrmion textures by uniaxial distortions in noncentrosymmetric cubic helimagnets. *Physical Review B*, 82:052403, 2010.
13. X. Z. Yu, et al. Near room-temperature formation of a skyrmion crystal in thin-films of the helimagnet FeGe. *Nature Materials*, 10:106–109, 2011.
14. K. Y. Guslienko. Skyrmion state stability in magnetic nanodots with perpendicular anisotropy. *IEEE Magnetic Letters*, 6:4000104, 2015.
15. M. N. Wilson, A. B. Butenko, A. N. Bogdanov, and T. L. Monchesky. Chiral skyrmions in cubic helimagnet films: The role of uniaxial anisotropy. *Physical Review B*, 89:094411, 2014.
16. J.-W. Yoo, S.-J. Lee, J.-H. Moon, and K.-J. Lee. Phase diagram of a single skyrmion in magnetic nanowires. *IEEE Transactions on Magnetics*, 50(11):1500504, 2014.
17. A. Fert, V. Cros, and J. Sampaio. Skyrmions on the track. *Nature Nanotechnology*, 8:152–156, 2013.
18. J. Iwasaki, M. Mochizuki, and N. Nagaosa. Current-induced skyrmion dynamics in constricted geometries. *Nature Nanotechnology*, 8:742–747, 2013.
19. J. Sampaio, V. Cros, S. Rohart, A. Thiaville, and A. Fert. Nucleation, stability and current-induced motion of isolated magnetic skyrmions in nanostructures. *Nature Nanotechnology*, 8:839–844, 2013.
20. Y. Zhou and M. Ezawa. A reversible conversion between a skyrmion and a domain-wall pair in a junction geometry. *Nature Communications*, 5:4652, 2014.
21. M. Beg, et al. Finite size effects, stability, hysteretic behaviour, and reversal mechanism of skyrmionic textures in nanostructures. Preprint, 9 pages, available online *arXiv*:1312.7665v2, 2014.
22. A. N. Bogdanov and U. K. Rößler. Chiral symmetry breaking in magnetic thin films and multilayers. *Physical Review B*, 87:037203, 2001.
23. H. Du, W. Ning, M. Tian, and Y. Zhang. Magnetic vortex with skyrmionic core in a thin nanodisk of chiral magnets. *Europhysics Letters*, 101:37001, 2013.
24. Y. Zhou, et al. Dynamically stabilized magnetic skyrmions. *Nature Communications*, 6:8193, 2015.
25. X. Zhang, Y. Zhou, and M. Ezawa. High-topological-number magnetic skyrmions and topologically protected dissipative structure. *Physical Review B*, 93:024415, 2016.
26. X. Zhang, Y. Zhou, and M. Ezawa. Magnetic bilayer-skyrmions without skyrmion Hall effect. *Nature Communications*, 7:10293, 2016.
27. X. Zhang, et al. All-magnetic control of skyrmions in nanowires by a spin wave. *Nanotechnology*, 26:225701, 2015.
28. F. Ma, M. Ezawa, and Y. Zhou. Microwave field frequency and current density modulated skyrmion-chain in nanotrack. *Scientific Reports*, 5:15154, 2015.

Chapter 11

29. F. Ma, Y. Zhou, H. B. Braun, and W. S. Lew. Skyrmion-based dynamic magnonic crystal. *Nano Letters*, 15:4029–4036, 2015.
30. X. Zhang, Y. Zhou, M. Ezawa, G. P. Zhao, and W. Zhao. Magnetic skyrmion transistor: Skyrmion motion in a voltage-gated nanotrack. *Scientific Reports*, 5:11369, 2015.
31. X. Zhang, M. Ezawa, and Y. Zhou. Magnetic skyrmion logic gates: Conversion, duplication and merging of skyrmions. *Scientific Reports*, 5:9400, 2015.
32. Ar. Abanov and V. L. Pokrovsky. Skyrmion in a real magnetic film. *Physical Review B*, 58:R8889(R), 1998.
33. A. S. Kirakosyan and V. L. Pokrovsky. From bubble to skyrmion: Dynamic transformation mediated by a strong magnetic tip. *Journal of Magnetism and Magnetic Materials*, 305(2):413–422, 2006.
34. S. Zhang, et al. Current-induced magnetic skyrmions oscillator. *New Journal of Physics*, 17:023061, 2015.
35. S. I. Kiselev, et al. Microwave oscillations of a nanomagnet driven by a spin-polarized current. *Nature*, 425:380–383, 2003.
36. Y. Zhou, C. L. Zha, S. Bonetti, J. Persson, and J. Åkerman. Spin-torque oscillator with tilted fixed layer magnetization. *Applied Physics Letters*, 92:262508, 2008.
37. Y. Zhou, H. Zhang, Y. Liu, and J. Åkerman. Macrospin and micromagnetic studies of tilted polarizer spin-torque nano-oscillators. *Journal of Applied Physics*, 112:063903, 2012.
38. Y. Zhou. Effect of the field-like spin torque on the switching current and switching speed of magnetic tunnel junction with perpendicularly magnetized free layers. *Journal of Applied Physics*, 109:023916, 2011.
39. J. C. Slonczewski. Excitation of spin waves by an electric current. *Journal of Magnetism and Magnetic Materials*, 195(2):L261–L268, 1999.
40. S. Bonetti, et al. Experimental evidence of self-localized and propagating spin wave modes in obliquely magnetized current-driven nanocontacts. *Physical Review Letters*, 105:217204, 2010.
41. Y. Liu, H. Li, Y. Hu, and A. Du. Oscillation frequency of magnetic vortex induced by spin-polarized current in a confined nanocontact structure. *Journal of Applied Physics*, 112:093905, 2012.
42. W. H. Rippard, M. R. Pufall, S. Kaka, S. E. Russek, and T. J. Silva. Direct-current induced dynamics in Co90Fe10/Ni80Fe20 point contacts. *Physical Review Letters*, 92:027201, 2004.
43. W. H. Rippard, M. R. Pufall, and S. E. Russek. Comparison of frequency, line-width, and output power in measurements of spin-transfer nanocontact oscillators. *Physical Review B*, 74:224409, 2006.
44. G. Consolo, et al. Micromagnetic study of the above-threshold generation regime in a spin-torque oscillator based on a magnetic nanocontact magnetized at an arbitrary angle. *Physical Review B*, 78:014420, 2008.
45. C. P. Chui, F. Ma, and Y. Zhou. Geometrical and physical conditions for skyrmion stability in a nanowire. *AIP Advances*, 5:047141, 2015.
46. C. P. Chui and Y. Zhou. Skyrmion stability in nanocontact spin-transfer oscillators. *AIP Advances*, 5:097126, 2015.
47. A. Vansteenkiste, et al. The design and verification of MuMax3. *AIP Advances*, 4:107133, 2014.
48. N. Nagaosa and Y. Tokura. Topological properties and dynamics of magnetic skyrmions. *Nature Nanotechnology*, 8:899–911, 2013.

49. K. Di, et al. Asymmetric spin-wave dispersion due to Dzyaloshinskii-Moriya interaction in an ultrathin Pt/CoFeB film. *Applied Physics Letters*, 106(5):052403, 2015.

50. C. Moreau-Luchaire, et al. Skyrmions at room temperature: From magnetic thin films to magnetic multilayers. *arXiv*:1502.07853v1, 2015.

51. S. Woo, et al. Observation of room temperature magnetic skyrmions and their current-driven dynamics in ultrathin Co films. *Nature Materials*, 15:501–506, 2016.

52. A. Hrabec, et al. Measuring and tailoring the Dzyaloshinskii-Moriya interaction in perpendicularly magnetized thin films. *Physical Review B*, 90:020402(R), 2014.

53. S. Rohart and A. Thiaville. Skyrmion confinement in ultrathin film nanostructures in the presence of Dzyaloshinskii-Moriya interaction. *Physical Review B*, 88:184422, 2013.

54. H. Y. Kwon, K. M. Bu, Y. Z. Wu, and C. Won. Effect of anisotropy and dipole interaction on long-range order magnetic structures generated by Dzyaloshinskii-Moriya interaction. *Journal of Magnetism and Magnetic Materials*, 324(13):2171–2176, 2012.

55. Y. Tchoe and J. H. Han. Skyrmion generation by current. *Physical Review B*, 85:174416, 2012.

56. M. Mochizuki. Spin-wave modes and their intense excitation effects in skyrmion crystals. *Physical Review Letters*, 108:017601, 2012.

57. A. Brataas, A. D. Kent, and H. Ohno. Current-induced torques in magnetic materials. *Nature Materials*, 11:372–381, 2012.

58. A. N. Slavin and P. Kabos. Approximate theory of microwave generation in a current-driven magnetic nanocontact magnetized in an arbitrary direction. *IEEE Transactions on Magnetics*, 41(4):1264–1273, 2005.

59. V. Puliafito, et al. Micromagnetic modeling of nanocontact spin-torque oscillators with perpendicular anisotropy at zero bias field. *IEEE Transactions on Magnetics*, 44(11):2512–2515, 2008.

60. M. A. Hoefer, T. J. Silva, and M. W. Keller. Theory for a dissipative droplet soliton excited by a spin torque nanocontact. *Physical Review B*, 82:054432, 2010.

61. G. Finocchio, et al. Nanoscale spintronic oscillators based on the excitation of confined soliton modes. *Journal of Applied Physics*, 114:163908, 2013.

62. S. Scki, X. Z. Yu, S. Ishiwata, and Y. Tokura. Observation of skyrmions in a multiferroic material. *Science*, 336:198–201, 2012.

63. Y. Zhou, C. L. Zha, S. Bonetti, J. Persson, and J. Åkerman. Microwave generation of tilted-polarizer spin torque oscillator. *Journal of Applied Physics*, 105:07D116, 2009.

64. R. Ravaud and G. Lemarquand. Mechanical properties of a ferrofluid seal: Three-dimensional analytical study based on the Coulombian model. *Progress in Electromagnetics Research B*, 13:385–407, 2009.

65. S. Bonetti, P. Muduli, F. Mancoff, and J. Åkerman. Spin torque oscillator frequency versus magnetic field angle: The prospect of operation beyond 65 GHz. *Applied Physics Letters*, 94:102507, 2009.

Chapter 11

12. Racetrack–Type Applications of Magnetic Skyrmions

Xichao Zhang[1], Jing Xia[1], Guoping Zhao[2,3], Hans Fangohr[4], Motohiko Ezawa[5], and Yan Zhou[1]

[1]*The Chinese University of Hong Kong, Shenzhen, People's Republic of China*
[2]*Sichuan Normal University, Chengdu, People's Republic of China*
[3]*Chinese Academy of Sciences, Ningbo, People's Republic of China*
[4]*University of Southampton, Southampton, United Kingdom*
[5]*University of Tokyo, Hongo, Japan*

Chapter 12

This chapter is devoted to a review on recent numerical studies dealing with racetrack-type applications of magnetic chiral skyrmions, where topologically stable skyrmions are used to store binary information. Since the theoretical prediction and experimental observation of skyrmions in magnetic materials, inspired by the domain wall (DW)-based racetrack memories, a number of racetrack-type applications of skyrmions have been proposed and investigated, such as the skyrmion-based racetrack memories, logic gates, and transistor-like devices. An overview is provided on the skyrmion-based racetrack memories where skyrmions are demonstrated to be driven and controlled by spin-polarized currents and spin waves (SWs). A review is then given on the logic computing and transistor-like functions of skyrmions in racetracks manipulated by spin-polarized currents and electric fields. Finally, a section of this chapter is devoted to a recent novel approach that is proposed to improve the transport reliability of skyrmions by eliminating the detrimental skyrmion Hall effect (SkHE) in racetrack-type applications of skyrmions.

12.1 Introduction

The concept of magnetic racetrack memory (RM) was proposed 10 years ago (Parkin 2004; Parkin et al. 2008), in which the DWs act as the carriers of the digital data, as shown in Figure 12.1 (Parkin and Yang 2015). Compared to the existing storage and magnetic random access memory (Parkin et al. 1999; Parkin et al. 2003), the transmission with low-energy consumption and the stability against thermal fluctuation are the merits of DW-based RM (Parkin and Yang 2015). In the early stage,

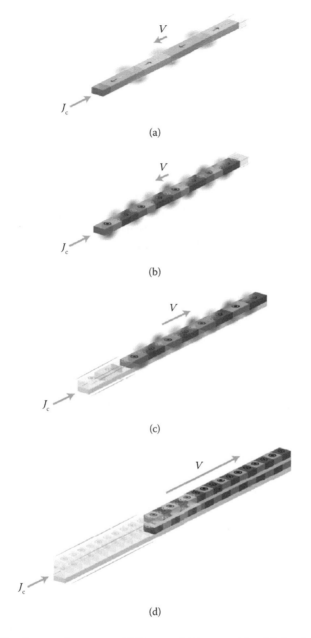

FIGURE 12.1   Racetrack memory. (a) Racetrack Memory 1.0, with in-plane magnetized racetracks. (b) Racetrack Memory 2.0, with perpendicularly magnetized racetracks. (c) Racetrack Memory 3.0: chiral spin torque drives DWs at high velocities. (d) Racetrack Memory 4.0, a giant exchange coupling torque drives DWs in synthetic antiferromagnets racetracks at extremely high velocities. (Reprinted from Parkin, S., and Yang, S.H., *Nat. Nanotech.*, 10, 195–198, 2015. With permission.)

Chapter 12

the DWs move along the in-plane magnetized racetracks, as shown in Figure 12.1a, which are magnetically soft, with a velocity of ~100 m/s for driving current density on the order of 10^8 A/cm^2 (Hayashi et al. 2008; Vlaminck and Bailleul 2008). Due to the shape anisotropy of the soft magnetic material, the width of the DWs is rather large. Moreover, the DWs are flexible and can expand to many times their equilibrium size as the torque provided by the current is applied. The materials with perpendicular magnetic anisotropy (PMA), such as Co/Ni superlattices (Daichi et al. 2010), are used to construct the racetracks (Figure 12.1b), thereby inhibiting the expansion of the DWs. As a result, the narrower DWs can be obtained and moved at roughly the same speed as those in permalloy. In 2011, much faster DW velocity is reported in the ultra-thin layers of cobalt deposited on a platinum layer (Miron et al. 2011), as shown in Figure 12.1c. The PMA derived from strong spin–orbital coupling at the interfaces between these layers and underlying heavy-metal layers and the Dzyaloshinskii–Moriya interaction (DMI) derived from the interfaces between magnetic layers and the heavy-metal layers contribute to the high velocity of DWs (Parkin and Yang 2015). The velocity of DWs driven by a current density of 10^8 A/cm^2 can reach around 350 m/s (Miron et al. 2011; Ryu et al. 2013, 2014). In 2015, the DW velocity has been demonstrated as high as 750 m/s at the driving current density of the order of 10^8 A/cm^2 in the synthetic antiferro-magnetic (SAF) racetrack (Yang et al. 2015). As shown in Figure 12.1d, the SAF racetrack is formed with two perpendicularly magnetized sub-racetracks, which are antiferromagnetically coupled. The net magnetic moment can be tuned to zero, and the DW velocity is around five times larger than that of an identical structure but with ferromagnetic (FM) coupling in the two sub-racetracks.

In the RM mentioned above, the information is encoded with the DW. Recently, it has been demonstrated that the digital information can be encoded in the magnetic skyrmions (Fert et al. 2013; Iwasaki et al. 2013a; Nagaosa and Tokura 2013; Romming et al. 2013; Tomasello et al. 2014; Zhang et al. 2015c), which are nanoscale magnetic excitations induced by the DMI arisen from spin–orbit coupling in lattices or at the interface of magnetic thin films. The magnetization configuration of magnetic skyrmion is topologically protected, which has been experi-mentally discovered in certain magnetic bulks, films, and nanowires since 2009 (Mühlbauer et al. 2009; Yi et al. 2009; Butenko et al. 2010; Münzer et al. 2010; Pfleiderer et al. 2010, 2011, 2013, 2014; Yu et al. 2010; Heinze et al. 2011; Seki et al. 2012; Romming et al. 2013; Du et al. 2014). As for skyrmion lattices, they have been observed in MnSi, FeCoSi, and other B20 transition metal compounds (Mühlbauer et al. 2009; Yi et al.

2009; Butenko et al. 2010; Münzer et al. 2010; Pfleiderer et al. 2010; Yu et al. 2010, 2011, 2014; Heinze et al. 2011; Seki et al. 2012; Yu et al. 2014) as well as in helimagnetic MnSi nanowires (Yu et al. 2013; Du et al. 2014). Individual skyrmions have been created using a spin-polarized current at 2013 (Romming et al. 2013). Most recently, the current-driven conversion from stripe domains to isolated magnetic skyrmions has been realized in a constricted geometry (Jiang et al. 2015).

Magnetic skyrmions are promising candidates as bit carriers for future data storage and logic devices consuming extremely low energy, with their motion driven by small electrical current (Jonietz et al. 2010; Yu et al. 2012, 2014; Tomasello et al. 2014; Zhou and Ezawa 2014; Zhang et al. 2015b). In this chapter, the applications of the magnetic skyrmions in the racetrack-type devices, in which the magnetic skyrmions are driven by spin current or SW, is reviewed.

12.2   RM Based on Current-Driven Magnetic Skyrmions

In the skyrmion-based RM (Fert et al. 2013), information is encoded in the form of magnetic skyrmions in magnetic nanotracks. For instance, as illustrated in Figure 12.2, the skyrmionic bits are driven by a vertical spin current and move at a velocity of 57 m/s. Figure 12.3 (Tomasello et al. 2014) shows more comprehensive scenarios for the design of a skyrmion RM, where the Bloch-type and Néel-type skyrmions driven by the spin-transfer torque or by the spin Hall effect in magnetic nano-tracks are discussed.

In this section, we first present and review the impacts of skyrmion–skyrmion repulsion and the skyrmion–edge repulsion on the feasibility of the skyrmion-based racetrack. In addition, we describe the numerical results on the elimination of skyrmionic bits by utilizing a notch structure at the terminal of the racetrack that has been proposed

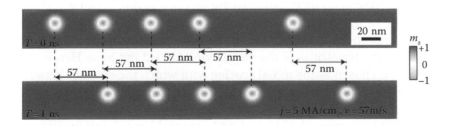

FIGURE 12.2 Chain of five skyrmions driven by a vertical current on the racetrack. (Reprinted from Fert, A., et al., *Nat. Nanotechnol.*, 8, 152–156, 2013. With permission.)

Chapter 12

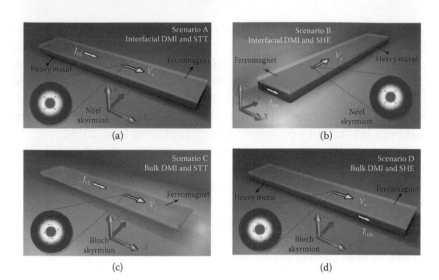

FIGURE 12.3 Four different scenarios for the design of a skyrmion racetrack memory. (Reprinted from Tomasello, R., et al., *Sci. Rep.*, 4, 6784, 2014. With permission.)

recently (Zhang et al. 2015c), which is an important strategy for avoiding the clogging of the skyrmionic bits as well as assisting the data erasing process.

12.2.1   Spacing between Skyrmionic Bits

As reported by Lin et al. (2013), the interaction between two skyrmions, which is induced by the overlap between both spin textures, decays exponentially with the spacing between them. Due to the interaction between skyrmions, the spacing between them changes as they move when the distance between them is small. The simulated result with the typical material parameters (Fert et al. 2013; Sampaio et al. 2013; Zhang et al. 2015c) shows that the distance between skyrmions increases significantly in the first 1 ns as the initial distance $d_i = 30$ nm. When $d_i = 57$ nm, the rate of increase in the distance is much lower. When d_i increases to 62 nm, the rate of increase in the distance is basically not detectable for the timescales of the simulations carried out here. Hence, $d_i = 62$ nm is chosen as the initial spacing with which the distance between skyrmions stays practically constant in motion. Therefore, under the assumption of the typical material parameters (Fert et al. 2013; Sampaio et al. 2013; Zhang et al. 2015c), $d_i \geq 62$ nm would be an ideal initial spacing for writing consecutive skyrmionic bits as well as the identification spacing for reading consecutive skyrmionic bits, that is, the bit length. And $d_i \geq 57$ nm can also be regarded as reliable and practicable initial spacing for consecutive skyrmionic bits, which has been verified by Fert et al. (2013) and Sampaio et al. (2013), where $d_i = 60$ nm is employed. More generally, we can relate these distances to the DMI helix length L_D (Beg et al. 2015), that is, $L_D = 4\pi A/|D|$, which equals to 62.8 nm with the typical material parameters of the material, and the practical spacing is in the range of $\sim L_D$ and $\sim 0.9 L_D$.

12.2.2   Elimination of Skyrmionic Bits

In DW-based RM, DW bits are annihilated simply by pushing them out of the racetrack with the driving current (Hayashi et al. 2008; Parkin et al. 2008; Parkin 2009). Similarly, once the skyrmionic bits in skyrmion-based RM are pushed beyond the reading element they can no longer be accessed and so should be eliminated at the end of the racetrack.

Figure 12.4 shows the motion of a skyrmionic bit chain with $d_i = 60$ nm at the end of the racetrack. As shown in Figure 12.4a and b, the skyrmionic bits experience congestion at the end of the racetrack because of the skyrmion–skyrmion (Zhang et al. 2015c) and skyrmion–edge repulsions (Iwasaki et al. 2013a). The spacing between the skyrmionic bits as well as the skyrmion size then reduces significantly. Besides, the skyrmions cannot easily exit from the track at this current density ($\sim 10^{10}$ A/m^2). As the current density increases to be larger than a threshold [$\sim 10^{11}$ A/m^2 in Zhang et al. (2015c)], the skyrmionic bits can exit the racetrack because the driving force from the current can overcome the repulsive force from the edge at the end of the racetrack, as demonstrated in Figure 12.5 by Iwasaki et al. (2013a). Figure 12.5a through d shows the case with a small current density, in which the skyrmion cannot reach the boundary because it cannot overcome the repulsive force from the edge at the end of the racetrack. In Figure 12.5e through h, the larger current density enables a skyrmion to exit the racetrack. Since the extremely lower power consumption is preferred, a triangular notch at the end of the racetrack has been proposed by Zhang et al. (2015c) to avoid the observed clogging of skyrmions in racetracks that does not require increasing the current density, as shown in Figure 12.4c through l, which can be realized by high-resolution nanolithography (Harriott 2001; Pavel et al. 2014). Figure 12.4d through l shows the process of the elimination of a skyrmion within 0.2 ns. The moving skyrmionic bit evolves

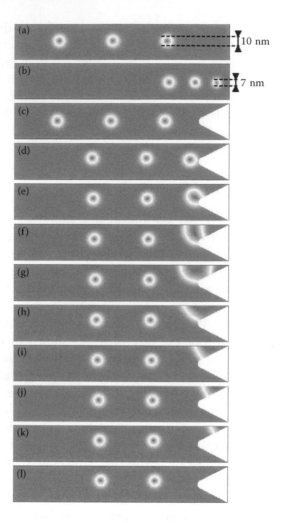

FIGURE 12.4 Vertical current–driven motion of a skyrmionic bit chain at the end of the 40-nm-wide racetrack without any notch at (a) $t = 0$ ns, (b) $t = 10$ ns, and at the end of the racetrack with a notch at (c) $t = 0$ ns, (d) $t = 0.95$ ns, (e) $t = 0.975$ ns, (f) $t = 1$ ns, (g) $t = 1.025$ ns, (h) $t = 1.05$ ns, (i) $t = 1.075$ ns, (j) $t = 1.1$ ns, (k) $t = 1.125$ ns, and (l) $t = 1.15$ ns. (Reprinted from Zhang, X., et al., *Sci. Rep.*, 5, 7643, 2015c. With permission.)

a DW pair (Zhou and Ezawa 2014) when touching the notch edge and is eventually cleared away by the current. As we can see from this process, the skyrmionic bit chain is not compressed at the end of the racetrack, in which all skyrmions move together coherently, which implies the technical reliability of the skyrmion-based RM in transporting and erasing data.

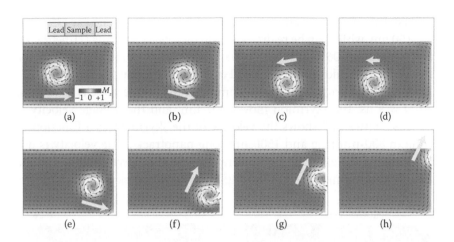

FIGURE 12.5 Two types of skyrmion motion near the edge of the magnetic region. (a)–(h), Snapshots of dynamical spin configurations at selected times near the end of the magnetic region for two different current densities j. In-plane components of the magnetic moments at sites (i_x, i_y) are represented by black arrows when $\mathrm{mod}(i_x, 3) = \mathrm{mod}(i_y, 3) = 1$. The color scale represents the z-components of the magnetic moments. Current density j is 1.0×10^{11} A/m^2 for (a)–(d) and 3.0×10^{11} A/m^2 for (c)–(h). Times corresponding to respective figures are $t = 1.95 \times 10^{-9}$ s (a), $t = 2.28 \times 10^{-9}$ s (b), $t = 2.60 \times 10^{-9}$ s (c), $t = 2.93 \times 10^{-9}$ s (d), $t = 8.13 \times 10^{-10}$ s (e), $t = 9.10 \times 10^{-10}$ s (f), $t = 9.75 \times 10^{-9}$ s (g) and $t = 10.08 \times 10^{-9}$ s (h). Inset in (a): schematic of the system with junctions of the magnetic region and non-magnetic leads. (Reprinted from Iwasaki, J., et al., *Nat. Nanotechnol.*, 8, 742–747, 2013a. With permission.)

12.3   RM Based on SW-Driven Magnetic Skyrmions

Magnetic skyrmions in the racetrack can be driven not only by the spin current but also by the SWs (Zhang et al. 2015a). This section is devoted to a review on the skyrmion dynamics driven by SWs in constricted magnetic nanowires, such as magnetic nanotracks, L-corners, and Y- and T-junctions.

12.3.1   A Magnetic Skyrmion Driven by SWs in a Nanotrack

The motion of a magnetic skyrmion driven by SWs in a narrow nanotrack has been recently reported by Zhang et al. (2015a). As shown in Figure 12.6 (Zhang et al. 2015a), the width (along y) of the nanotrack is 40 nm and the length (along x) is 800~1500 nm. The SW is injected via the applied magnetic field pulse applied along the lateral axis of the nanotrack. The width of the pulse element equals the width of the nanotrack, and the length is fixed to be 15 nm. Initially, one skyrmion is created and relaxed to a stable/

Chapter 12

metastable state at $x = 200$ nm. One magnetic pulse is applied at 135 nm < $x < 150$ nm, which can be realized with a microwave antenna (Hillebrands and Thiaville 2006; Bance et al. 2008; Iwasaki et al. 2014). The excited SW drives the skyrmion toward the end of the nanotrack at the same time. At $t = 5$ ns, the skyrmion moves 266 nm with an average velocity of 53 m/s. At $t = 9$ ns, the skyrmion moves 457 nm, and the average velocity is 51 m/s. Figure 12.7 shows the velocity of the skyrmion as a function of time in 800-nm-long and 1500-nm-long nanotracks. The velocity is not uniform, which experiences acceleration and then deceleration. For the case of motion in the 800-nm-long nanotrack, the skyrmion performs

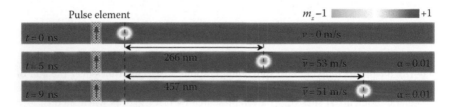

FIGURE 12.6   The propagation of a skyrmion driven by SW in nanotrack. The pattern green boxes denote the pulse elements. The color scale presents the out-of-plane component of the magnetization m_z. (The source of the material Zhang, X., et al., 225701, 2015a, and IOP Publishing is acknowledged.)

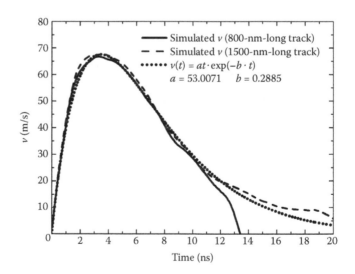

FIGURE 12.7   The velocity of a skyrmion as a function of time in 800-nm-long and 1500-nm-long nanotracks with the same spin wave injection. The dotted curve denotes the fitting function of the velocity versus time. (The source of the material Zhang, X., et al., All-magnetic control of skyrmions in nanowires by a spin wave, *Nanotechnology*, 26, 225701, 2015a, and IOP Publishing is acknowledged.)

accelerated motion in the first $\sim$ 3 ns, and then slows down. Finally, the skyrmion stops at a location balanced by the repulsion force and the SW driving force. In the 1500-nm-long nanotrack, the skyrmion behaves similarly. The skyrmion stops when t = 20 ns at the location which is far away from both the source of the SW and the end of the nanotrack. A skyrmion is at rest initially. Once the SW arrives at the skyrmion, it starts the accelerated motion $\mathbf{v}^{(s)} = at$, where a corresponds to the acceleration of the skyrmion. After long enough time, the velocity of the skyrmion becomes the same as that of SW: $\mathbf{v}^{(s)}(t) = \mathbf{v}^{(d)}(t)$. The velocity of a skyrmion also decays exponentially, $\mathbf{v}^{(s)}(t) \propto e^{-bt}$, since SW decays exponentially, $\mathbf{v}^{(d)}(x) = ce^{-dx}$. Accordingly, the fitting function $\mathbf{v}^{(s)}(t) = ate^{-bt}$ can be obtained, as shown in Figure 12.7. In addition, the effect of the DMI strength D, PMA constant K, exchange stiffness A, saturation magnetization M_S, amplitude of the magnetic pulse, and frequency of the pulse on the velocity has been investigated by Zhang et al. (2015a). The fitting parameters are shown in Table 12.1. It is found that $a \propto c$ and $b \propto d$. The former relation implies that the initial acceleration is proportional to the amplitude of SW, while the latter relation implies the exponential decay of the velocity of a skyrmion due to the SW decay. The acceleration a is proportional to the magnitude of the SW and the radius of the skyrmion, $R_{Sk} = \dfrac{D\pi^2}{\dfrac{s}{\pi}\mu_0 H + 2K\pi}$, in which

$\mu_0 H$ is the applied magnetic field in the $+z$-direction (Zhang et al. 2015a). The radius of the skyrmion increases for the cases of the increasing D and decreasing K (Sampaio et al. 2013; Zhang et al. 2015c), resulting in larger a. Since the larger amplitude and frequency of the SW result in higher energy, a is larger, correspondingly.

12.3.2 A Magnetic Skyrmion Driven by SWs in L-Corners and T- and Y-Junctions

To realize a skyrmionic logic circuit, the motion of a skyrmion in constricted geometries is necessary, such as L-corners, and T- and Y-junctions. For example, as shown in Figure 12.8, the skyrmion is initially placed at the left side of the L-corner, where the width of the branches for L-corners is 40 nm. Moreover, to avoid the destruction of the skyrmion, the 90-degree corners have been cut to 135-degree corners. When the magnetic pulse is applied, the skyrmion is turned to the other branch of the L-corner driven by the excited SW.

Figure 12.9 shows the motion of a skyrmion in the T- and Y-junctions. The width of the nanotracks of T- and Y-junctions is increased to 60 nm, and the DMI constant is decreased to 3.5 mJ/m², which

Chapter 12

Table 12.1 Constants of the Fitting Functions of the Velocity Curves with Different Parameters

$D\,(mJ/m^2)$	3.25	3.50	3.75	4.00
a	21.7807	36.4434	46.7089	53.0071
b	0.1973	0.2578	0.2827	0.2885
$K\,(MJ/m^3)$	0.800	0.825	0.850	0.900
a	53.0071	51.3847	37.3514	31.0713
b	0.2885	0.3215	0.3231	0.3267
$A\,(pJ/m)$	14	15	16	17
a	58.4507	53.0071	48.7456	40.7391
b	0.2951	0.2885	0.2908	0.2948
$M_S\,(kA/m)$	560	580	600	620
a	41.3331	53.0071	42.9825	35.3391
b	0.3161	0.2885	0.2659	0.2603
Amplitude (mT)	450	500	550	600
a	31.0865	37.6614	46.6487	53.0071
b	0.2939	0.2931	0.2974	0.2885
Frequency (GHz)	3.125	6.25	12.5	25
a	6.9819	16.1533	27.8073	53.0071
b	0.2385	0.2573	0.2889	0.2885

Source: The source of the material Zhang, X., et al., All-magnetic control of skyrmions in nanowires by a spin wave. *Nanotechnology,* 26, 225701, 2015a, and IOP Publishing is acknowledged.

broadens the channel and reduces the size of the skyrmion, leading to a better effect of the skyrmion turning at the junction. For the case of motion in T-junction, a skyrmion with a positive topological number, which is calculated with $Q = -\int d\mathbf{x}\, \frac{1}{4\pi}\mathbf{m}(\mathbf{x})\cdot\left(\partial_x \mathbf{m}(\mathbf{x}) \times \partial_y \mathbf{m}(\mathbf{x})\right)$ (Zhou and Ezawa 2014; Zhang et al. 2015b, 2015d), has been generated. The skyrmion turns from the central C-branch to the left L-branch, as shown in Figure 12.9a. In Figure 12.9b, the skyrmion moves from the left L-branch to the right R-branch. In Figure 12.9c, the skyrmion turns from the R-branch to the central C-branch. The simulated results show that the skyrmion always turns left. For the Y-junction, this behavior has also been found, as shown in Figure 12.9d, where the skyrmion turns from the C-branch to the L-branch. For the cases of motions in

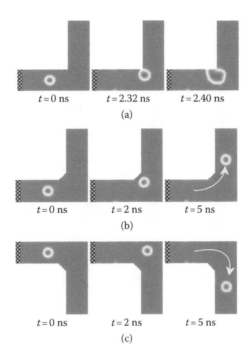

FIGURE 12.8 Snapshots of the SW-driven motion of a skyrmion ($Q = 1$) in L-corners. The magnetic pulse is applied along the lateral axis in the patterned light gray region. (a) The skyrmion is destroyed by the corner due to the tilts of magnetization at the corner edge. Hence, we cut the 90-degree corner into two 135-degree corners, and the skyrmion smoothly turns into two 135-degree corners, and the skyrmion smoothly turns left at the L-corners in (b) and turns right in (c). (The source of the material Zhang, X., et al., All-magnetic control of skyrmions in nanowires by a spin wave, *Nanotechnology*, 26, 225701, 2015a and IOP Publishing is acknowledged.)

T- and Y-junctions, it can be seen that the skyrmion always turns left when the skyrmion number is positive ($Q = 1$). By contrast, the skyrmion with $Q = -1$ always turns right.

12.4   Magnetic Skyrmion Logic Gates: Conversion, Duplication, and Merging of Skyrmions in Racetracks

The conversion between a DW pair and a skyrmion by connecting narrow and wide nanowires, which has been proposed and demonstrated by Zhou and Ezawa (2014), enables the information transmission between a DW-based device and a skyrmion-based device. This section is devoted to an overview of the conversion, duplication, and merging

Chapter 12

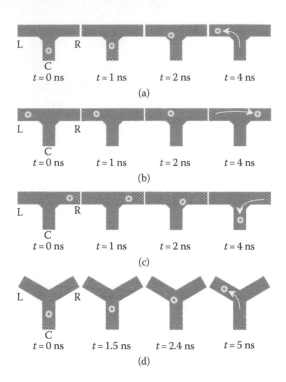

FIGURE 12.9 Snapshots of the SW-driven motion of a skyrmion ($Q = 1$) in T- and Y-junctions. (a) The skyrmion turns left from the C-branch into the L-branch. (b) The skyrmion goes straight from the L-branch to the R-branch. (c) The skyrmion turns left from R-branch into the C-branch. (d) The skyrmion turns left from the C-branch into the L-branch. (The source of the material Zhang, X., et al., All-magnetic control of skyrmions in nanowires by a spin wave, *Nanotechnology*, 26, 225701, 2015a, and IOP Publishing is acknowledged.)

of skyrmions. Furthermore, the novel magnetic logic gates based on the manipulation of magnetic skyrmions are shown.

A skyrmion is characterized by three numbers: the skyrmion number Q_s, the vorticity Q_v, and the helicity Q_h (Ezawa 2010; Zhang et al. 2015b). The profile of a skyrmion can be described with $\mathbf{n}(r) = [\sin(\theta(r)) \cos(Q_v\varphi + Q_h), \sin(\theta(r)) \sin(Q_v\varphi + Q_h), \cos(\theta(r))]$. $\theta(r)$ is the radius function, which determines the z-component of $\mathbf{n}$. The vorticity of a skyrmion Q_v is defined by the winding number of the spin configurations projected onto the s_x–s_y plane. The skyrmion number Q_s can be calculated with

$$Q_s = -\frac{1}{4\pi} \int dx\,dy \left(\mathbf{n}(r) \cdot \left(\partial_x \mathbf{n}(r) \times \partial_y \mathbf{n}(r) \right) \right)$$

$$= \frac{Q_v}{2} \left[\lim_{r \to \infty} \cos(\theta(r)) - \cos(\theta(0)) \right],$$

which is determined by the product of the vorticity and the difference between the spin direction of the core and the tail of the skyrmion. It is called a skyrmion (antiskyrmion) when this skyrmion number Q_s is positive (negative). It shows that it is possible to convert a skyrmion into an antiskyrmion or vice versa by changing the spin direction at the core and the tail (Figure 12.10c). On the other hand, the helicity does not contribute to the topological number, which is uniquely determined by the type of the DMI. A skyrmion with the helicity of 0 or π corresponds to the Néel-type skyrmion, while a skyrmion with the helicity of $\pi/2$ or $3\pi/2$ corresponds to the Bloch-type skyrmion.

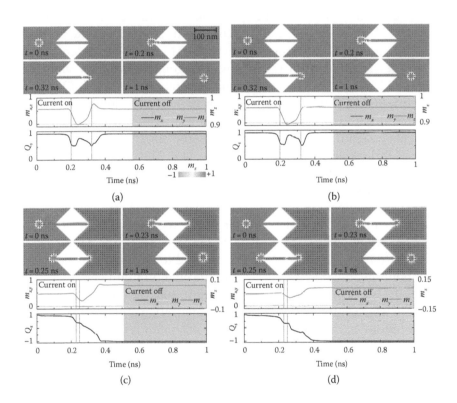

FIGURE 12.10 Conversions between skyrmions and antiskyrmions. The top panels show the snapshots of the magnetization configuration at four selected times corresponding to the vertical lines in the middle and bottom panels; the middle panels show the time evolution of the average spin components m_x, m_y, and m_z; the bottom panels show the time evolution of the skyrmion number Q_s. (a) Conversion between a skyrmion and a skyrmion with identical out-going helicity. (b) Conversion between a skyrmion and a skyrmion with opposite in-going helicity. (c) Conversion between a skyrmion and an antiskyrmion with opposite in-going helicity. (d) Conversion between a skyrmion and an antiskyrmion with identical out-going helicity. (From Zhang, X., et al., *Sci. Rep.*, 5, 9400, 2015b. With permission.)

The Néel-type skyrmion with the helicity of 0 can be transformed into a skyrmion with the helicity of π (Figure 12.10b).

12.4.1   Conversion of Magnetic Skyrmions

In the setup in Zhang et al. (2015b), the left input and right output wide nanowires are connected by a narrow nanowire, as shown in Figure 12.10. In Figure 12.10a, a skyrmion driven by the injected current moves toward the right nanowire. In the central narrow nanowire region, the skyrmion is converted into a DW pair. As the DW pair continues its motion toward to the right side, it is converted into a skyrmion. At the same time, the quantum numbers (Q_s, Q_v, Q_h) change as $(1, 1, 0) \rightarrow (0, 0, 0) \rightarrow (1, 1, 0)$. As the sign of DMI is opposite between the left (positive) and right (negative), the skyrmion can be converted into the skyrmion with the opposite in-going helicity, as shown in Figure 12.10b. In this process, the quantum numbers (Q_s, Q_v, Q_h) change as $(1, 1, 0) \rightarrow (0, 0, 0) \rightarrow (1, 1, \pi)$. Moreover, the conversion between skyrmion and anti-skyrmion can be realized in the system where the magnetization in the left is opposite to the one in the right with the same DMI, as shown in Figure 12.10c and d. The opposite orientation of the magnetization can be realized via applying external magnetic field pulses with opposite directions in the left and right region at first. For the skyrmion, the spin tail must have the same direction with the background magnetization. Accordingly, the skyrmion is stable in the left region, while the anti-skyrmion is stable in the right region. In Figure 12.10c, the skyrmion is converted to the anti-skyrmion with the opposite in-going helicity, where the quantum number changes as $(1, 1, 0) \rightarrow (0, 0, 0) \rightarrow (-1, 1, \pi)$. In Figure 12.10d, the conversions in the system where both the direction of magnetization and the sign of DMI are opposite have been illustrated. The skyrmion is converted into the antiskyrmion with the identical out-going helicity, and the quantum number changes as $(1, 1, 0) \rightarrow (0, 0, 0) \rightarrow (-1, 1, 0)$.

12.4.2   Duplication and Merging of Magnetic Skyrmions in Branched Racetracks

For the skyrmion-based devices, the duplication of the information is a crucial process. Based on the reversible conversion between a skyrmion and a DW pair, the duplication can be realized in the designed system as shown in Figure 12.11. In Figure 12.11a, a skyrmion is first converted into a DW pair. As it approaches the Y-junction, which is implemented as a fan-out element, one DW pair is split into two

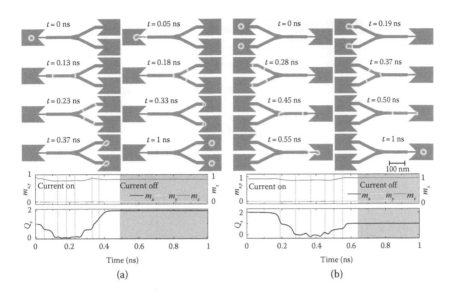

FIGURE 12.11 Duplication and merging of skyrmions. The top panels show the snapshots of the magnetization configuration at eight selected times corresponding to the vertical lines in the middle and bottom panels; the middle panels show the time evolution of the average spin components m_x, m_y, and m_z; the bottom panels show the time evolution of the skyrmion number Q_s. (a) Duplication of a skyrmion: the initial background magnetization of the sample points $+z$. (b) Merging of two skyrmions: the initial background magnetization of the sample points $+z$. (From Zhang, X., et al., *Sci. Rep.*, 5, 9400, 2015b. With permission.)

DW pairs. Then, they are converted into two skyrmions at the right side of the nanowire. Accordingly, the skyrmion number of the system changes from 1 to 2. Using this geometry, the information carried by skyrmions can be duplicated. The merging of skyrmion can also be realized using this geometry, as shown in Figure 12.11b. Initially, there are two skyrmions on the left side of the nanowire. They are converted into two DW pairs. As they approach the junction, the DW pairs are merged into one DW pair. Consequently, only one skyrmion is formed on the right side. The skyrmion number of the system changes from 2 to 1.

12.4.3   Skyrmion Logical Operations

Upon the realization of the duplication and merging of the magnetic skyrmions, the skyrmionic logic can be realized, as proposed and demonstrated by Zhang et al. (2015b).

The OR gate (Figure 12.12) is an operation of $0 + 0 = 0$, $1 + 0 = 1$, $0 + 1 = 1$, and $1 + 1 = 1$. In the skyrmionic logic, 0 represents the absence

Chapter 12

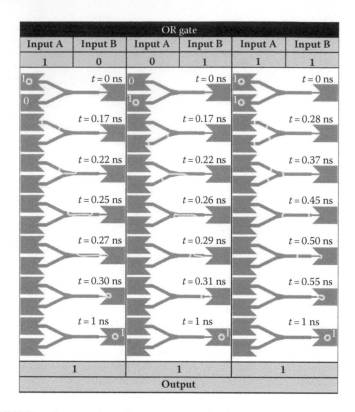

OR gate					
Input A	Input B	Input A	Input B	Input A	Input B
1	0	0	1	1	1
$t = 0$ ns		$t = 0$ ns		$t = 0$ ns	
$t = 0.17$ ns		$t = 0.17$ ns		$t = 0.28$ ns	
$t = 0.22$ ns		$t = 0.22$ ns		$t = 0.37$ ns	
$t = 0.25$ ns		$t = 0.26$ ns		$t = 0.45$ ns	
$t = 0.27$ ns		$t = 0.29$ ns		$t = 0.50$ ns	
$t = 0.30$ ns		$t = 0.31$ ns		$t = 0.55$ ns	
$t = 1$ ns		$t = 1$ ns		$t = 1$ ns	
1		1		1	
Output					

FIGURE 12.12 Skyrmion logical OR operation. The skyrmion represents logical 1, and the ferromagnetic ground state represents logical 0. Left panel: the basic operation of OR gate $1 + 0 = 1$: there is a skyrmion in the input A and no skyrmion in the input B at initial time, which represents input $= 1 + 0$. Middle panel: the basic operation of the OR gate $0 + 1 = 1$: there is a skyrmion in the input B side and no skyrmion in the input A side at initial time, which represents input $= 0 + 1$. Right panel: the basic operation of the OR gate $1 + 1 = 1$: there is a skyrmion in both the input A side and the input B side, which represents input $= 1 + 1$. (From Zhang, X., et al., *Sci. Rep.*, 5, 9400, 2015b. With permission.)

of a skyrmion and 1 denotes the presence of a skyrmion. The process of $0 + 0 = 0$ means that there is no input as well as there is no output. In the left panel of Figure 12.12, there is a skyrmion in input A branch and no skyrmion in input B branch initially. The skyrmion is converted into a DW pair in the narrow region, and one skyrmion is obtained at the right output nanowire. This is the operation of $1 + 0 = 1$. Similarly, in the middle panel of Figure 12.12, with no skyrmion in input A branch and a skyrmion in input B branch, one skyrmion is obtained at the right output nanowire. The operation of $0 + 1 = 1$ has been realized. In the right panel of Figure 12.12, the operation of $1 + 1 = 1$ has been realized. There is a skyrmion in both input A and B branches, and one skyrmion is obtained in the output nanowire.

With the similar designed device, the AND gate (Figure 12.13), $0 + 0 = 0$, $0 + 1 = 0$, $1 + 0 = 0$, and $1 + 1 = 1$, can be realized. The processes $0 + 0 = 0$ and $1 + 1 = 1$ are the same as the case of OR gate. For the process of $0 + 1 = 0$, there is a skyrmion in input A branch and no skyrmion in input B branch, as shown in the left panel of Figure 12.13. A skyrmion is converted into a meron in the central region and annihilated from the corner. The mechanism is the same for the case of $1 + 0 = 0$ as shown in the middle panel of Figure 12.13. In the right panel of Figure 12.13, there are skyrmions in input A and B branches,

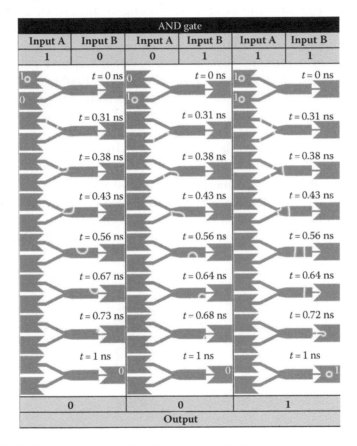

AND gate					
Input A	**Input B**	**Input A**	**Input B**	**Input A**	**Input B**
1	0	0	1	1	1
$t = 0$ ns		$t = 0$ ns		$t = 0$ ns	
$t = 0.31$ ns		$t = 0.31$ ns		$t = 0.31$ ns	
$t = 0.38$ ns		$t = 0.38$ ns		$t = 0.38$ ns	
$t = 0.43$ ns		$t = 0.43$ ns		$t = 0.43$ ns	
$t = 0.56$ ns		$t = 0.56$ ns		$t = 0.56$ ns	
$t = 0.67$ ns		$t = 0.64$ ns		$t = 0.64$ ns	
$t = 0.73$ ns		$t = 0.68$ ns		$t = 0.72$ ns	
$t = 1$ ns		$t = 1$ ns		$t = 1$ ns	
0		0		1	
Output					

FIGURE 12.13   Skyrmion logical AND operation. The skyrmion represents logical 1, and the ferromagnetic ground state represents logical 0. Left panel: the basic operation of AND gate $1 + 0 = 0$: there is a skyrmion in the input A branch and no skyrmion in the input B branch initially, which represents input $= 1 + 0$. Middle panel: the basic operation of the AND gate $0 + 1 = 0$: there is a skyrmion in the input B branch and no skyrmion in the input A branch initially, which represents input $= 0 + 1$. Right panel: the basic operation of the AND gate $1 + 1 = 1$: there is a skyrmion in both the input A and the input B branches, which represents input $= 1 + 1$. (From Zhang, X., et al., *Sci. Rep.*, 5, 9400, 2015b. With permission.)

Chapter 12

which are converted into a DW pair, leading to the generation of only one skyrmion in the output branch. This is the operation of $1 + 1 = 1$.

12.5   Skyrmion Transistor: Skyrmion Motion in Voltage-Gated Racetracks

To realize the commercialization of the *Skyrmionics* for future electronics, various challenges need to be solved such as creation and annihilation of skyrmions (Tchoe and Han 2012; Nagaosa and Tokura 2013; Sampaio et al. 2013; Zhang et al. 2015c), conversion of skyrmions with different helicity and vorticity (Zhou and Ezawa 2014; Zhang et al. 2015b), efficient transmission, and guiding and read-out of skyrmions (Iwasaki et al. 2013b; Nagaosa and Tokura 2013; Sampaio et al. 2013; Tomasello et al. 2014; Koshibae et al. 2015; Upadhyaya et al. 2015). This section is devoted to a review on recent numerical studies dealing with the design of the skyrmion transistor (Zhang et al. 2015d), in which the PMA in the gate region is locally controlled by an applied electric field due to the charge accumulations (Kim et al. 2014; Verba et al. 2014).

The design of the skyrmion transistor ($600 \times 100 \times 1$ nm^3) is shown in Figure 12.14a, where a gate voltage is applied in the center of a magnetic nanotrack. The skyrmion is created in the left side of the nanotrack ($x = 90$ nm) with MTJ and moves toward the right side driven

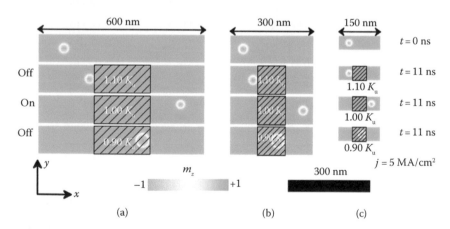

FIGURE 12.14   The top-view of the nanotracks with different sizes under the same working conditions at selected times. The sizes of the nanotracks are: (a) $600 \times 100 \times 1$ nm^3; (b) $300 \times 100 \times 1$ nm^3; and (c) $150 \times 50 \times 1$ nm^3. The color scale denotes the out-of-plane component of the magnetization. The black-line shadows represent the voltage-controlled PMA region. (From Zhang, X., et al., *Sci. Rep.*, 5, 11369, 2015d. With permission.)

by the spin current. The PMA value of the gate region (200 nm $< x <$ 400 nm) can be controlled by locally applied electric field E based on a linearly changing relationship, $K_{uv} = K_u + \Delta K_{uv}E$ (Schellekens et al. 2012; Kim et al. 2014), where ΔK_{uv} is a constant determining the strength of the voltage-induced anisotropy change. This linear relationship is demonstrated by Shiota et al. (2011) experimentally.

12.5.1   The Dependence of Working Status on DMI

Figure 12.14a shows the working conditions for $j = 5$ MA/cm^2 with different K_{uv}. Initially, the spin current and the local electric field are turned off, and the skyrmion is located at the left side of the nanotrack. When the electric field is turned off, $K_{uv} = K_u$, the skyrmion driven by the spin current moves from the left to the right side of the nanotrack within 11 ns, which means that the working condition of the voltage gate is in the *On* state. When the spin current and the local electric field are turned on, the skyrmion moves along the nanotrack driven by the spin current and the PMA of the gate region changes due to the applied local electric field. For $K_{uv} = 1.10\ K_u$, the skyrmion is driven by the spin current, and pinned when it approaches the first border of the voltage-gated region. The skyrmion cannot pass the gate region, and this is the *Off* working condition of the voltage gate. For $K_{uv} = 0.90\ K_u$, the skyrmion passes the first border of the gate region while pinned at the second border of the gate region, which means that the working condition of the voltage gate is in the *Off* state. However, when j is larger than a certain threshold value, the skyrmion can pass the voltage-gated region even when $K_{uv} \neq K_u$. Table 12.2 shows the working status of the voltage-gated skyrmion transistor with different spin current densities. When the local electric field is turned

Table 12.2   Working Status of Voltage-Gated Skyrmion Transistor at Different Voltage-Controlled PMA

	j (MA/cm^2)					
K_{uv}/K_u	5	6	7	8	9	10
1.10	*Off*	*Off*	*Off*	*Off*	*Off*	*Off*
1.05	*Off*	*On*	*On*	*On*	*On*	*On*
1.00	*On*	*On*	*On*	*On*	*On*	*On*
0.95	*On*	*On*	*On*	*On*	*On*	*On*
0.90	*Off*	*Off*	*Off*	*Off*	*Off*	*On*

Source: Zhang, X., et al., *Sci. Rep.*, 5, 11369, 2015d. With permission.

Chapter 12

off, the voltage-gated transistor is always in the *On* status, where the skyrmion driven by the spin current can pass the gate region and reach the right side of the nanotrack. When the local electric field is turned on and the PMA of the gate region is increased to $1.05K_u$, the transistor is *Off* when the spin current density $j = 5$ MA/cm². When the spin current density increases, $j = 6\sim10$ MA/cm², the transistor switches to the *Off* status. When the PMA of the gate region increases to $1.1K_u$, the transistor keeps *Off* status even as the spin current density increases to 10 MA/cm². Moreover, when the local electric field is turned on and the PMA of the gate region decreases to 0.95 K_u, the skyrmion can pass the gate region and the transistor keeps *On* status for $j = 5$–10 MA/cm². As the PMA in the gate region further reduces to 0.90-K_u, the transistor switches from *Off* to *On* when the spin current density increases to 10 MA/cm².

The dependence of the working condition on the DMI is shown in Tables 12.3 through 12.5. It can be seen from Table 12.3 that when $K_{uv} = 0.95$ K_u, the transistor is in the *Off* state for $D = 3.3 \sim 3.4$ mJ/m² and $j = 5$ MA/cm². When $D > 3.4$ mJ/m² or $j > 5$ MA/cm², the transistor is working in the *On* state. Similarly, as shown in Table 12.4, if $K_{uv} = 1.05$ K_u, the transistor is in the *On* state when $D > 3.5$ mJ/m² or $j > 5$ MA/cm², and in the *Off* state when $D = 3.3 \sim 3.5$ mJ/m² and $j = 5$ MA/cm². Table 12.5 shows that, when $K_{uv} = 1.10$ K_u, the transistor is in the *On* state for $D = 3.7$ mJ/m² and $j = 10$ MA/cm², while when $D < 3.7$ mJ/m² or $j < 10$ MA/cm², the transistor has a stable *Off* state.

The results show that the skyrmion may be pinned by the barrier induced by different PMAs in the nanotrack. The energy and the radius of the skyrmion strongly depend on the magnetic anisotropy. The energy is given by $E_{Sk} = -\dfrac{D^2\pi^4}{4K\pi + \dfrac{16}{\pi}B} + 38.7A$ (Zhang et al. 2015d), where D is the magnitude of the DMI, A is the exchange constant, B is the magnetic field, and K is the PMA constant. If $K_{uv} = K_u$, there is no potential barrier in the nanotrack, and the skyrmion can move from the left to right side of the nanotrack smoothly driven by the spin current. If $K_{uv} < K_u$, a potential well is formed at the left border of the gate region, while a potential barrier is formed at the right border of the gate region. In this case, the skyrmion can pass the left border but cannot pass the right border. If $K_{uv} > K_u$, a potential barrier is formed at the left border of gate region, while a potential well is formed at the right border of the gate region. In this case, the skyrmion cannot pass the left border. However, when the spin current density j is larger than a threshold value, the driving force on the skyrmion is strong enough to overcome the potential barrier.

Table 12.3   Working Status of Voltage-Gated Skyrmion Transistor at Different DMIs and Spin Current Densities with Fixed Voltage-Controlled PMA $K_{uv} = 0.95\ K_u$

D (mJ/m²)	j (MA/cm²)					
	5	6	7	8	9	10
3.7	*On*	*On*	*On*	*On*	*On*	*On*
3.6	*On*	*On*	*On*	*On*	*On*	*On*
3.5	*On*	*On*	*On*	*On*	*On*	*On*
3.4	*Off*	*On*	*On*	*On*	*On*	*On*
3.3	*Off*	*On*	*On*	*On*	*On*	*On*

Source: Zhang, X., et al., *Sci. Rep.*, 5, 11369, 2015d. With permission.

Table 12.4   Working Status of Voltage-Gated Skyrmion Transistor at Different DMIs and Spin Current Densities with Fixed Voltage-Controlled PMA $K_{uv} = 1.05\ K_u$

D (mJ/m²)	j (MA/cm²)					
	5	6	7	8	9	10
3.7	*On*	*On*	*On*	*On*	*On*	*On*
3.6	*On*	*On*	*On*	*On*	*On*	*On*
3.5	*Off*	*On*	*On*	*On*	*On*	*On*
3.4	*Off*	*On*	*On*	*On*	*On*	*On*
3.3	*Off*	*On*	*On*	*On*	*On*	*On*

Source: Zhang, X., et al., *Sci. Rep.*, 5, 11369, 2015d. With permission.

Table 12.5   Working Status of Voltage-Gated Skyrmion Transistor at Different DMI and Spin Current Densities with Fixed Voltage-Controlled PMA $K_{uv} = 1.10\ K_u$

D (mJ/m²)	j (MA/cm²)					
	5	6	7	8	9	10
3.7	*Off*	*Off*	*Off*	*Off*	*Off*	*On*
3.6	*Off*	*Off*	*Off*	*Off*	*Off*	*Off*
3.5	*Off*	*Off*	*Off*	*Off*	*Off*	*Off*
3.4	*Off*	*Off*	*Off*	*Off*	*Off*	*Off*
3.3	*Off*	*Off*	*Off*	*Off*	*Off*	*Off*

Source: Zhang, X., et al., *Sci. Rep.*, 5, 11369, 2015d. With permission.

Chapter 12

12.5.2 The Size Effect of the Skyrmion Transistor

As shown in the above section, the working state of this skyrmion-based gate can be controlled by the gate voltage through the modulation of PMA in the gate region. Besides, the working state also can be tuned with the driving spin current as well as other parameters. The size effect of the skyrmion transistor has also been investigated by Zhang et al. (2015d), as shown in Figure 12.14. In Figure 12.14a, the system size is $600 \times 100 \times 1$ nm^3. As the size of the skyrmionic transistor decreases to $300 \times 100 \times 1$ nm^3 (Figure 12.14b), the skyrmionic transistor works the same as in the case of $600 \times 100 \times 1$ nm^3. When the local electric field and the spin current are turned on, the PMA in voltage-gated region $K_{uv} = 1.10\ K_u$, the skyrmion is pinned at the left border of the gate region. When the local electric field is turned off, $K_{uv} = K_u$, the skyrmion driven by spin current moves from the left side to the right side of the nanotrack smoothly. When $K_{uv} = 0.90\ K_u$, the skyrmion passes the left border but is pinned at the right border of the voltage-gated region. When the system size further reduces to $150 \times 50 \times 1$ nm^3 (Figure 12.14c), the working conditions match well with the cases in Figure 12.14a and b, indicating the stability of the skyrmionic transistor model.

12.5.3 Two Strategies Employing Voltage Gates in the Skyrmion–Based Racetrack Memory

The pinning/depinning characteristics of a skyrmion in a nanotrack with the voltage-controlled magnetic anisotropy (VCMA) effect also can be used to build a skyrmion-based RM electrical model (Kang et al. 2016). Two strategies can be employed in the skyrmion-based RM, as shown in Figure 12.15.

Firstly, the state (*On* or *Off*) can be directly controlled by the voltage of the VCMA-gated region. In this case, when the voltage is on, the driving force is not enough to overcome the energy barrier of the VCMA-gated region. Figure 12.15a shows the trajectory of the skyrmion motion along the nanotrack. The skyrmion stops at the left boundary of the VCMA-gated region when the voltage of the VCMA gate is on and then continues to move when the voltage is off. By placing the VCMA gates evenly along the nanotrack, this strategy is rather suitable for skyrmion-based RM applications with step-by-step motion based on the clock frequency. Secondly, the state (*On* or *Off*) can be

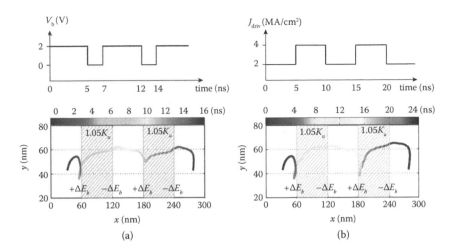

FIGURE 12.15 Trajectory of the skyrmion motion along the nanotrack: (a) the case of controlling the on/off voltage of the VCMA gate, and (b) the case of modulating the driving current configuration. (From Kang, W., et al., *Sci. Rep.*, 6, 23164, 2016. With permission.)

controlled by modulating the driving current with the unchanged K_{uv} of the VCMA-gated region. As shown in Figure 12.15b, the total driving current is composed of a DC component and an AC pulsed current. The current density of the DC component is limited. Therefore, the skyrmion stops at the left boundary of the VCMA-gated region without the AC current pulse. However, when the AC current is on, the total driving current is large enough to drive the skyrmion to pass through the voltage gate. This strategy offers more freedom for skyrmion-based RM design by dynamically configuring the driving current.

12.6   Magnetic Skyrmions in SAF Racetracks

Skyrmions can be driven by the spin-polarized current. However, a skyrmion cannot move in a straight line along the driving current direction due to the Magnus force, which is referred as SkHE in recent researches. Recently, one approach is proposed by Zhang et al. (2016a) to suppress the SkHE with SAF racetracks. This idea provides a guideline for designing realistic ultradense and ultrafast information processing, storage, and logic computing devices based on magnetic skyrmions.

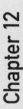

12.6.1 Magnetic Skyrmions in a Bilayer SAF Racetrack

The mode of the bilayer SAF racetrack is shown in Figure 12.16. Considering two perpendicularly magnetized FM sublayers, they are strongly coupled via the antiferromagnetic (AFM) exchange interaction with a heavy-metal layer beneath the bottom FM layer. One skyrmion in the top FM layer and the other skyrmion in the bottom FM

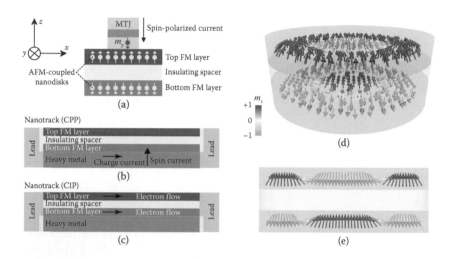

FIGURE 12.16 Schematics of the antiferromagnetically exchange-coupled bilayer systems and the bilayer-skyrmion. (a) The AFM-coupled bilayer nanodisk with a diameter of 100 nm for bilayer-skyrmion creation. The spin-polarized current (polarization direction $\mathbf{p} = -z$) is injected into the top ferromagnetic (FM) layer in the central circle with a diameter of 40 nm, which can be realized by means of a magnetic tunnel junction (MTJ) injector above the nanodisk. (b) The AFM-coupled bilayer nanotrack ($500 \times 50 \times 3$ nm³) for the study of the motion of a bilayer-skyrmion driven by the current perpendicular to the plane (CPP). The charge current flows through the heavy-metal substrate along the x-direction, which gives rise to a spin current ($\mathbf{p} = +y$) perpendicularly injected to the bottom FM layer because of the spin Hall effect. The skyrmion in the bottom FM layer is driven by the spin current, whereas the skyrmion in the top FM layer moves accordingly due to the interlayer AFM exchange coupling. (c) The AFM-coupled bilayer nanotrack ($500 \times 50 \times 3$ nm³) for the study of the motion of a bilayer-skyrmion driven by the in-plane current (CIP). The electrons flow towards the right in both the top and bottom FM layers, that is, the corresponding charge currents flow along the $-x$ direction. The skyrmions in both the top and bottom FM layers are driven by the current. In all the models, the thickness of both the top FM layer, the bottom FM layer and the insulating spacer are equal to 1 nm. The initial state of the top FM layer is almost spin-up (pointing along $+z$) and that of the bottom FM layer is almost spin-down (pointing along $-z$). (d) Illustration of a pair of skyrmions, (that is, the bilayer-skyrmion) in an AFM-coupled nanodisk. (e) Side view of the bilayer-skyrmion along the diameter of (d). The color scale represents the out-of-plane component of the magnetization. (From Zhang, X., et al., *Nat. Commun.*, 7, 10293, 2016a. With permission.)

layer can be simultaneously created with the injection of spin polarized current into the top layer. Such a pair of AFM-coupled magnetic skyrmions as a magnetic bilayer-skyrmion can travel over arbitrarily long distances driven by spin currents without touching nanotrack edges when the interlayer AFM exchange coupling is strong enough, as shown in Figure 12.17. The skyrmion in the top layer follows the motion of the skyrmion in the bottom layer even when the current is injected only into the bottom FM layer. The Magnus forces acting on the skyrmions between the top and bottom FM layers are cancelled when they are strongly bound by the interlayer AFM exchange coupling. The SkHEs are completely suppressed with the sufficient interlayer AFM coupling strength. When the top and bottom FM layers are decoupled or coupled with small interlayer AFM exchange coupling, the skyrmions in the top and bottom FM layers move left-handed and right-handed, respectively. That means the SkHE is active, which leads to the destruction of skyrmions in the top FM layer and/or bottom FM layer by touching the nanotrack edges as shown in the decoupled case of Figure 12.17.

12.6.2   Magnetic Skyrmions in a Multilayer SAF Racetrack

The motion of magnetic skyrmions in multilayer SAF racetracks has been investigated by Zhang et al. (2016b). The mode is shown in Figure 12.18. The monolayer racetrack contains one FM layer and a

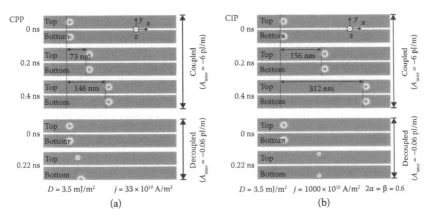

FIGURE 12.17   Current-induced motion of skyrmions in the top and bottom FM layers of an AFM-coupled bilayer nanotrack. Top views of the motion of skyrmions at selected interlayer exchange coupling constants and times driven by spin currents with (a) the current-perpendicular-to-plane (CPP) injection geometry and (b) the current-in-plane (CIP) injection geometry. (From Zhang, X., et al., *Nat. Commun.*, 7, 10293, 2016a. With permission.)

Chapter 12

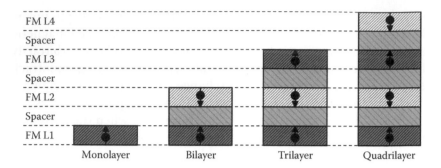

FIGURE 12.18 Schematics of the simulation models including the FM monolayer, bilayer SAF, trilayer SAF, and quadrilayer SAF racetracks. (From Zhang, X., et al., *arXiv*, 1601.03893, 2016b. With permission.)

heavy-metal substrate underneath the FM layer. N-layer SAF racetrack includes N FM layers which are separated by insulating spacer layers. In N-layer SAF racetracks ($N \geq 2$), the adjacent FM layers are antiferromagnetically exchange-coupled through the FM/Space/FM interface. The magnetization in each FM layer is perpendicular to the racetrack plane due to the high PMA, while the magnetization in neighboring FM layers is antiparallel due to the interlayer AFM exchange coupling.

Figure 12.19 shows trajectories of N-layer skyrmions in N-layer racetracks. The trajectory of a monolayer skyrmion in an FM monolayer racetrack ($N = 1$) at a driving current shows an obvious transverse shift of the monolayer skyrmion due to the SkHE. The trajectory of a bilayer-skyrmion in a bilayer racetrack ($N = 2$) at a driving current. The bilayer-skyrmion moves along the central line of the racetrack as a result of the suppression of the SkHE. The trajectory of a trilayer skyrmion in a trilayer racetrack ($N = 3$) also shows a transverse shift, which indicates that the skyrmion, the trilayer skyrmion, experiences the SkHE. On the other hand, the quadrilayer skyrmion moves along the central line of the racetrack without SkHE.

The SkHEs at $N = 1$ and $N = 3$ have been compared in Figure 12.20a. The skyrmion Hall angle v_y/v_x are shown as a function of time t. It is almost a constant for $N = 1$ and $N = 3$. The skyrmion angle is antiproportional to N. It means that the SkHE for $N = 3$ is three times smaller than that for the case of $N = 1$. At the same time, the velocity v_x is almost antiproportional to N as shown in Figure 12.20b and c. This because that the driving current is only applied to the bottom FM layer.

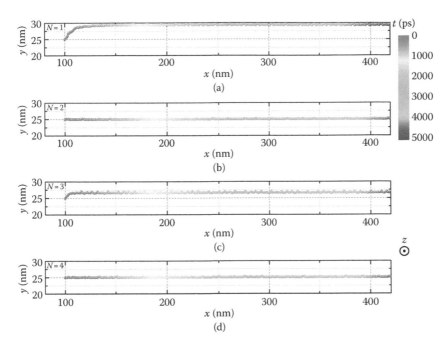

FIGURE 12.19 Typical trajectories of SAF *N*-layer skyrmions in N-layer SAF race-tracks. (a) The trajectory of a FM monolayer skyrmion in a monolayer FM racetrack (*N* = 1). The transverse shift of the monolayer skyrmion due to the SkHE is obvious. (b) The trajectory of a SAF bilayer-skyrmion in a bilayer SAF racetrack (*N* = 2). The SAF bilayer skyrmion moves along the central line (*y* = 25 nm) of the racetrack. (c) The trajectory of a SAF trilayer skyrmion in a trilayer SAF racetrack (*N* = 3). (d) The trajectory of a SAF quadrilayer skyrmion in a quadrilayer SAF racetrack (*N* = 4). The SAF quadrilayer skyrmion moves along the central line (*y* = 25 nm) of the racetrack. The dot denotes the center of the skyrmion. (From Zhang, X., et al., *arXiv*, 1601.03893, 2016b. With permission.)

12.7 Summary

In this chapter, we have presented the application of skyrmions to the RM. We have dealt with the details which are important to realize the skyrmion-based RM, including the repulsion between skyrmions, the reliable distance of the skyrmions, and the elimination of skyrmions in the racetrack. The reliable distance of the skyrmions is important for the writing process, the reading process, and the storage density. The notch at the ending of the racetrack proposed in this chapter provides a good solution for the skyrmion clogging when they are eliminated, which makes the skyrmion-based memory more feasible. Secondly, the motion of a skyrmion driven by SWs instead of electrical current has been proposed and demonstrated. Skyrmions can be driven

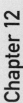

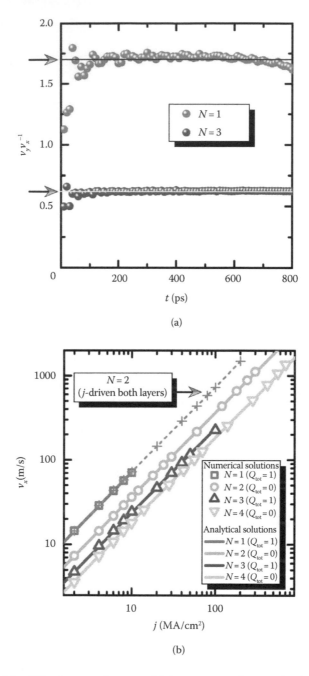

(a)

(b)

FIGURE 12.20 Velocity v as a function of driving current density j for the motion of N-layer skyrmions. (a) The skyrmion Hall angle v_y/v_x as a function of time. (b) Open symbols indicate the velocity v_x of N-layer skyrmions. *(Continued)*

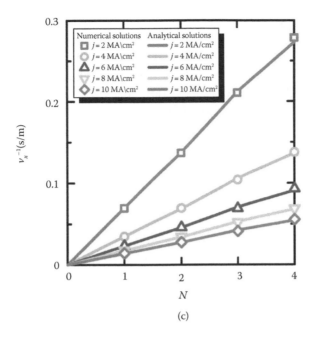

(c)

FIGURE 12.20 (Continued) (c) The inverse velocity $1/v_x$ versus the total FM layer number N. (From Zhang, X., et al., *arXiv*, 1601.03893, 2016b. With permission.)

by SWs, which gives an alternative means to control skyrmions when they are used as a basic building block of the skyrmion-based memory. Furthermore, the application of magnetic skyrmions in the logic devices has been overviewed. The conversion, duplication, and merging can be realized on the designed racetrack, which means the feasibility of conversion, duplication, and merging of the information encoded by the magnetic skyrmion on the designed racetrack. The investigation of these processes is essential to the real application of magnetic skyrmions. In addition, the magnetic skyrmion can be functioned as a transistor, where the PMA of the gate region is controlled by the applied voltage. Lastly, the latest researches focused on the skyrmions in SAF racetracks demonstrated that stable motion of skyrmions can be realized with the bilayer or multilayer structure. The diversity and feasibility of the skyrmion application show that the electronics aspects of the magnetic skyrmion deserves more investigation.

Acknowledgments

X.Z. was supported by JSPS RONPAKU (Dissertation PhD) Program. G.Z. acknowledges the support by National Natural Science Foundation

Chapter 12

of China (Grants No. 11074179 and No. 10747007) and the Construction Plan for Scientific Research Innovation Teams of Universities in Sichuan (Project No. 12TD008). M.E. thanks the support by the Grants-in-Aid for Scientific Research from JSPS KAKENHI (Grants No. JP25400317 and No. JP15H05854). Y.Z. acknowledges the support by Shenzhen Fundamental Research Fund (Grant No. JCYJ20160331164412545).

References

Bance S, et al. (2008). Micromagnetic calculation of spin wave propagation for magneto-logic devices. *J Appl Phys* 103:07E735.

Beg M, et al. (2015). Ground state search, hysteretic behaviour, and reversal mechanism of skyrmionic textures in confined helimagnetic nanostructures. *Sci Rep* 5:17137.

Butenko AB, Leonov AA, Rößler UK, and Bogdanov AN (2010). Stabilization of skyrmion textures by uniaxial distortions in noncentrosymmetric cubic helimagnets. *Phys Rev B* 82:052403.

Daichi C, et al. (2010). Control of multiple magnetic domain walls by current in a Co/Ni nano-wire. *Appl Phys Expr* 3:073004.

Du HF, et al. (2014). Highly stable skyrmion state in helimagnetic MnSi nanowires. *Nano Lett* 14:2026–2032.

Ezawa M (2010). Giant skyrmions stabilized by dipole-dipole interactions in thin ferromagnetic films. *Phys Rev Lett* 105:197202.

Fert A, Cros V, and Sampaio J (2013). Skyrmions on the track. *Nat Nanotech* 8:152–156.

Harriott LR (2001). Limits of lithography. *Proc IEEE* 89:366–374.

Hayashi M, Thomas L, Moriya R, Rettner C, and Parkin SSP (2008). Current-controlled magnetic domain-wall nanowire shift register. *Science* 320:209–211.

Heinze S, et al. (2011). Spontaneous atomic-scale magnetic skyrmion lattice in two dimensions. *Nat Phys* 7:713–718.

Hillebrands B and Thiaville A (2006). *Spin Dynamics in Confined Magnetic Structures III*. Berlin: Springer.

Iwasaki J, Beekman AJ, and Nagaosa N (2014). Theory of magnon-skyrmion scattering in chiral magnets. *Phys Rev B* 89:064412.

Iwasaki J, Mochizuki M, and Nagaosa N (2013a). Current-induced skyrmion dynamics in constricted geometries. *Nat Nanotech* 8:742–747.

Iwasaki J, Mochizuki M, and Nagaosa N (2013b). Universal current-velocity relation of skyrmion motion in chiral magnets. *Nat Commun* 4:1463.

Jiang W, et al. (2015). Blowing magnetic skyrmion bubbles. *Science* 349:283–286.

Jonietz F, et al. (2010). Spin transfer torques in MnSi at ultralow current densities. *Science* 330:1648–1651.

Kang W, et al. (2016). Voltage controlled magnetic skyrmion motion for racetrack memory. *Sci Rep* 6:23164.

Kim J-S, et al. (2014). Voltage controlled propagating spin waves on a perpendicularly magnetized nanowire. *arXiv* 1401.6910. Available at http://arxiv.org/abs/1401.6910, accessed on June 7, 2015.

Koshibae W, et al. (2015). Memory functions of magnetic skyrmions. *Jpn J Appl Phys* 54:053001.

Lin S-Z, Reichhardt C, Batista CD, and Saxena A (2013). Particle model for skyrmions in metallic chiral magnets: Dynamics, pinning, and creep. *Phys Rev B* 87:214419.

Miron IM, et al. (2011). Fast current-induced domain-wall motion controlled by the Rashba effect. *Nat Mater* 10:419–423.

Mühlbauer S, et al. (2009). Skyrmion lattice in a chiral magnet. *Science* 323:915–919.

Münzer W, et al. (2010). Skyrmion lattice in the doped semiconductor $Fe_{1-x}CO_xSi$. *Phys Rev B* 81:041203.

Nagaosa N and Tokura Y (2013). Topological properties and dynamics of magnetic skyrmions. *Nat Nanotech* 8:899–911.

Parkin S and Yang SH (2015). Memory on the racetrack. *Nat Nanotech* 10:195–198.

Parkin SSP, et al. (1999). Exchange-biased magnetic tunnel junctions and application to nonvolatile magnetic random access memory. *J Appl Phys* 85:5828–5833.

Parkin SSP, et al. (2003). Magnetically engineered spintronic sensors and memory. *Proc IEEE* 91:661–680.

Parkin SSP (2004). *Shiftable Magnetic Shift Register and Method of Using the Same.* US 6,834,005.

Parkin SSP (2009). *Unidirectional Racetrack Memory Device.* US 7,551,469 (Office, U. S. P. a. T., ed), p 5 US: International Business Machines Corporation.

Parkin SSP, Hayashi M, and Thomas L (2008). Magnetic domain-wall racetrack memory. *Science* 320:190–194.

Pavel E, et al. (2014). Quantum optical lithography from 1 nm resolution to pattern transfer on silicon wafer. *Opt Laser Technol* 60:80-84.

Pfleiderer C, et al. (2010). Skyrmion lattices in metallic and semiconducting B20 transition metal compounds. *J Phys Condens Matter* 22:164207.

Romming N, et al. (2013). Writing and deleting single magnetic skyrmions. *Science* 341:636–639.

Ryu KS, Thomas L, Yang SH, and Parkin S (2013). Chiral spin torque at magnetic domain walls. *Nat Nanotech* 8:527–533.

Ryu K-S, Yang S-H, Thomas L, and Parkin SSP (2014). Chiral spin torque arising from proximity-induced magnetization. *Nat Commun* 5:3910.

Sampaio J, Cros V, Rohart S, Thiaville A, and Fert A (2013). Nucleation, stability and current-induced motion of isolated magnetic skyrmions in nanostructures. *Nat Nanotech* 8:839.

Schellekens AJ, van den Brink A, Franken JH, Swagten HJM, and Koopmans B (2012). Electric-field control of domain wall motion in perpendicularly magnetized materials. *Nat Commun* 3:847.

Seki S, Yu XZ, Ishiwata S, and Tokura Y (2012). Observation of skyrmions in a multiferroic material. *Science* 336:198–201.

Shiota Y, et al. (2011). Quantitative evaluation of voltage-induced magnetic anisotropy change by magnetoresistance measurement. *Appl Phys Express* 4:043005.

Tchoe Y and Han JH (2012). Skyrmion generation by current. *Phys Rev B* 85:174416.

Tomasello R, et al. (2014). A strategy for the design of skyrmion racetrack memories. *Sci Rep* 4:6784.

Upadhyaya P, Yu G, Amiri PK, and Wang KL (2015). Electric-field guiding of magnetic skyrmions. *Phys Rev B* 92:134411.

Verba R, Tiberkevich V, Krivorotov I, and Slavin A (2014). Parametric excitation of spin waves by voltage-controlled magnetic anisotropy. *Phys Rev Appl* 1:044006.

Vlaminck V and Bailleul M (2008). current-induced spin-wave doppler shift. *Science* 322:410–413.

Yang SH, Ryu KS, and Parkin S (2015). Domain-wall velocities of up to 750 ms(−1) driven by exchange-coupling torque in synthetic antiferromagnets. *Nat Nanotech* 10:221–226.

Yi SD, Onoda S, Nagaosa N, and Han JH (2009). Skyrmions and anomalous Hall effect in a Dzyaloshinskii-Moriya spiral magnet. *Phys Rev B* 80:054416.

Chapter 12

Yu XZ, et al. (2010). Real-space observation of a two-dimensional skyrmion crystal. *Nature* 465:901–904.

Yu XZ, et al. (2011). Near room-temperature formation of a skyrmion crystal in thin-films of the helimagnet FeGe. *Nat Mater* 10:106–109.

Yu XZ, et al. (2012). Skyrmion flow near room temperature in an ultralow current density. *Nat Commun* 3:988.

Yu XZ, et al. (2013). Observation of the magnetic skyrmion lattice in a MnSi nanowire by Lorentz TEM. *Nano Lett* 13:3755–3759.

Yu XZ, et al. (2014). Biskyrmion states and their current-driven motion in a layered manganite. *Nat Commun* 5:3198.

Zhang X, et al. (2015a). All-magnetic control of skyrmions in nanowires by a spin wave. *Nanotechnology* 26:225701.

Zhang X, et al. (2015c). Skyrmion-skyrmion and skyrmion-edge repulsions on the skyrmion-based racetrack memory. *Sci Rep* 5:7643.

Zhang X, Ezawa M, and Zhou Y (2015b). Magnetic skyrmion logic gates: Conversion, duplication and merging of skyrmions. *Sci Rep* 5:9400.

Zhang X, Ezawa M, and Zhou Y (2016b). Thermally stable magnetic skyrmions in multilayer synthetic antiferromagnetic racetracks. *arXiv*1601.03893. Available at http://arxiv.org/abs/1601.03893.

Zhang X, Zhou Y, and Ezawa M (2016a). Magnetic bilayer-skyrmions without skyrmion Hall effect. *Nat Commun* 7:10293.

Zhang X, Zhou Y, Ezawa M, Zhao GP, and Zhao W (2015d). Magnetic skyrmion transistor: Skyrmion motion in a voltage-gated nanotrack. *Sci Rep* 5:11369.

Zhou Y and Ezawa M (2014). A reversible conversion between a skyrmion and a domain-wall pair in junction geometry. *Nat Commun* 5:4652.

13. Magnetic Skyrmion Channels

Guided Motion in Potential Wells

Wen Siang Lew, Wei Liang Gan,
Indra Purnama, and Fusheng Ma
Nanyang Technological University, Singapore

In this chapter, the dynamics of a skyrmion under the application of external current as well as the methods to guide the ensuing current-induced motion is discussed. The discussion is started by first deriving the equation of motion of the current-driven skyrmion from the more general Landau–Lifshitz–Gilbert equation with the addition of the spin-transfer torque

Chapter 13

(STT) terms. The calculations show that the motion of a skyrmion deviates from the direction of the conduction electron flow due to the presence of Magnus force, which itself originates from the fact that the torques that the skyrmion received from the electrons are all directed perpendicular to each other. In application perspective, the presence of the Magnus force is a bane to the realization of skyrmion-based devices, as the skyrmions will be pushed aside and get annihilated at the edge of the ferromagnetic film. To solve this issue, we will discuss several methods to confine the skyrmions in the device to prevent the skyrmion from being annihilated when current is applied. We show that the skyrmions can be confined in the center region of the device and can be guided to follow the shape of the device by creating a curbed nanowire or by tuning the magnetic anisotropy along the width of the nanowire. We show that the skyrmion motion can even be guided along a curbed U-shape without getting annihilated. We also show that by reducing the width between the curbs in the nanowire, the size of the skyrmions can be reduced which results in smaller separation distances and significant increase in the packing density of the skyrmions.

13.1 Introduction

Magnetic skyrmion, which is characterized by the in-plane curling magnetization and out-of-plane magnetization at the core, has attracted much attention because of its promising applications in high-density storage devices. Recently, they have been observed in extended lattices of bulk noncentrosymmetric magnetic materials such as MnSi, (FeCo)Si, or FeGe [1–3]. The skyrmions have also been observed via spin-polarized scanning tunneling microscopy (STM) at zero applied fields in the Fe monolayer that is grown on Ir(111) [4]. The presence of skyrmions in such ultrathin film is attributed to the Dzyaloshinskii–Moriya (DM) interaction [5,6].

Due to the topological configuration of skyrmions, they can be driven under low applied current and they are also shown to be far less hindered by defects [7–10], which makes them ideal as information carriers in future magnetic information storage and memory devices [11]. However, when the skyrmion is driven by a current, a Magnus force is generated which leads to the skyrmion moving away from the direction of the conduction electron flow [12,13]. In this case, the Magnus force is induced in the system due to the perpendicular relationship between the nonadiabatic and the adiabatic STTs. The deflection poses a serious problem to the realization of skyrmion-based devices, as it leads to skyrmion annihilation at the film edges [14].

From an application point of view, the challenge to realize skyrmion-based devices then lies in finding ways to overcome the Magnus force and prevent the induced skyrmion annihilation at the film edges. In this chapter, we will discuss how the Magnus force is derived and propose several methods to mitigate the Magnus force in order to prevent the skyrmions from being annihilated during their current-induced motions. In general, the skyrmions can be prevented from being annihilated by creating potential barriers to confine the skyrmions in the nanowire [15,16]. The creation of potential barriers can then be achieved by either increasing the perpendicular magnetic anisotropy (PMA), adding additional ferromagnetic layers, or changing the damping constant along the nanowire edges. Additionally, the method can be extended to create a magnetic skyrmion waveguide. Similar to an optical waveguide that guides electromagnetic wave, the magnetic skyrmion waveguide can confine the motion of skyrmions within a designed region and avoids possible loss of skyrmions and the information carried by them.

13.2   Skyrmion Dynamics

We start our discussion by understanding the skyrmion motion as derived from a basic magnetization dynamic equation. Firstly, under the application of current, the magnetization dynamics of the ferromagnet is expressed by the modified Landau–Lifshitz–Gilbert equation, which includes the adiabatic and the nonadiabatic contribution from the STT [17–22]:

$$\frac{d\mathbf{M}}{dt} = -\gamma\mathbf{M}\times\mathbf{H}_{\text{eff}} + \frac{\alpha}{M_s}\mathbf{M}\times\frac{d\mathbf{M}}{dt} - (\mathbf{u}\cdot\nabla)\mathbf{M} + \frac{\beta}{M_s}\mathbf{M}\times((\mathbf{u}\cdot\nabla)\mathbf{M}). \quad (13.1)$$

Here, $\mathbf{M}$ is the magnetization vector, γ is the gyromagnetic constant, $\mathbf{H}_{\text{eff}}$ is the effective field from the minimization of the ferromagnet's internal energies, α is the Gilbert damping constant, β is the nonadiabatic constant of the STT, and $\mathbf{u}$ represents the drift speed of the conduction electrons.

$$\mathbf{u} = \frac{g\mu_B P}{2eM_s}\mathbf{J}, \quad (13.2)$$

where P is the polarization rate of the ferromagnetic film and $\mathbf{J}$ is the current density vector. The LLG equation is then rearranged such that it can be read as "the time rate of change of angular momentum is equal to the torque due to linear dissipative effects, the torque due to reversible effects, as well as the torque from the spin transfer effect:"

$$-\frac{1}{\gamma}\frac{d\mathbf{M}}{dt}+\frac{1}{\gamma}\frac{\alpha}{M_s}\mathbf{M}\times\frac{d\mathbf{M}}{dt}-\mathbf{M}\times\mathbf{H}_{\text{eff}}-\frac{1}{\gamma}(\mathbf{u}\cdot\nabla)\mathbf{M}$$

$$+\frac{1}{\gamma}\frac{\beta}{M_s}\mathbf{M}\times\left((\mathbf{u}\cdot\nabla)\mathbf{M}\right)=0. \tag{13.3}$$

We can then create a set of magnetic fields (**H**) for each of the above torques, such that:

$$\mathbf{M}\times\mathbf{H}_g+\mathbf{M}\times\mathbf{H}_d+\mathbf{M}\times\mathbf{H}_{\text{int}}+\mathbf{M}\times\mathbf{H}_{\text{adia}}+\mathbf{M}\times\mathbf{H}_{\text{nonadia}}=0, \tag{13.4}$$

$$\mathbf{M}\times\mathbf{H}_g=-\frac{1}{\gamma}\frac{d\mathbf{M}}{dt}, \tag{13.5}$$

$$\mathbf{M}\times\mathbf{H}_d=\frac{1}{\gamma}\frac{\alpha}{M_s}\mathbf{M}\times\frac{d\mathbf{M}}{dt}, \tag{13.6}$$

$$\mathbf{M}\times\mathbf{H}_{\text{int}}=-\mathbf{M}\times\mathbf{H}_{\text{eff}}, \tag{13.7}$$

$$\mathbf{M}\times\mathbf{H}_{\text{adia}}=-\frac{1}{\gamma}(\mathbf{u}\cdot\nabla)\mathbf{M}, \tag{13.8}$$

$$\mathbf{M}\times\mathbf{H}_{\text{nonadia}}=\frac{1}{\gamma}\frac{\beta}{M_s}\mathbf{M}\times\left((\mathbf{u}\cdot\nabla)\mathbf{M}\right). \tag{13.9}$$

where $\mathbf{H}_g$, $\mathbf{H}_d$, $\mathbf{H}_{\text{int}}$, $\mathbf{H}_{\text{adia}}$, and $\mathbf{H}_{\text{nonadia}}$ represent the gyroscopic field, the dissipative field, the internal field, the adiabatic spin transfer field, and the nonadiabatic spin transfer field, respectively.

In equilibrium, the skyrmion can be expected to move at a constant velocity (v), and thus the magnetization configuration of the system as a function of time can be expressed as:

$$\mathbf{M}=\mathbf{M}(x-vt), \tag{13.10}$$

which also gives us:

$$\frac{d\mathbf{M}}{dt}=\frac{d\mathbf{M}}{dt}-(v\cdot\nabla)\mathbf{M}=-\begin{pmatrix}v_x\dfrac{\partial M_x}{\partial x}+v_y\dfrac{\partial M_x}{\partial y}+v_z\dfrac{\partial M_x}{\partial z}\\[2mm]v_x\dfrac{\partial M_y}{\partial x}+v_y\dfrac{\partial M_y}{\partial y}+v_z\dfrac{\partial M_y}{\partial z}\\[2mm]v_x\dfrac{\partial M_z}{\partial x}+v_y\dfrac{\partial M_z}{\partial y}+v_z\dfrac{\partial M_z}{\partial z}\end{pmatrix}. \tag{13.11}$$

In addition, the equilibrium state means that it is possible to write a force equilibrium equation on the skyrmion whereby:

$$0 = \mathbf{F}_g + \mathbf{F}_d + \mathbf{F}_{int} + \mathbf{F}_{adia} + \mathbf{F}_{nonadia}, \tag{13.12}$$

where $\mathbf{F}_g$, $\mathbf{F}_d$, $\mathbf{F}_{int}$, $\mathbf{F}_{adia}$, and $\mathbf{F}_{nonadia}$ correspond to the gyroscopic force, dissipative force, the internal force, the adiabatic spin transfer force, and the nonadiabatic spin transfer force, respectively. The forces are obtained by incorporating the magnetic fields that are obtained before into the following mathematical operation:

$$\mathbf{F} = \int \mathbf{f} dV = -\int \nabla_M (H \cdot M) dV. \tag{13.13}$$

For instance, the gyroscopic effective field is expressed as:

$$\mathbf{H}_g = \begin{pmatrix} H_x \\ H_y \\ H_z \end{pmatrix}_g = \frac{1}{\gamma M_s^2} \mathbf{M} \times \frac{d\mathbf{M}}{dt} - \frac{-1}{\gamma M_s^2} \begin{pmatrix} M_y \dfrac{dM_z}{dt} - M_z \dfrac{dM_y}{dt} \\[2mm] M_z \dfrac{dM_x}{dt} - M_x \dfrac{dM_z}{dt} \\[2mm] M_x \dfrac{dM_y}{dt} - M_y \dfrac{dM_x}{dt} \end{pmatrix}. \tag{13.14}$$

Which can then be used to calculate the gyroscopic force:

$$\mathbf{F}_g = \int \mathbf{f}_g dV = -\int \nabla_M \left(H_g \cdot M \right) dV, \tag{13.15}$$

$$\begin{pmatrix} f_x \\ f_y \\ f_z \end{pmatrix}_g = -\begin{pmatrix} (H_x)_g \dfrac{\partial M_x}{\partial x} + (H_y)_g \dfrac{\partial M_y}{\partial x} + (H_z)_g \dfrac{\partial M_z}{\partial x} \\[2mm] (H_x)_g \dfrac{\partial M_x}{\partial y} + (H_z)_g \dfrac{\partial M_z}{\partial y} + (H_y)_g \dfrac{\partial M_y}{\partial y} \\[2mm] (H_x)_g \dfrac{\partial M_x}{\partial z} + (H_y)_g \dfrac{\partial M_y}{\partial z} + (H_z)_g \dfrac{\partial M_z}{\partial z} \end{pmatrix}. \tag{13.16}$$

Chapter 13

Substituting Equations 13.11 and 13.14 in Equation 13.16 will give us:

$$(f_i)_g = (g_{ji}v_j + g_{ki}v_k),$$
(13.17)

where

$$g_{ji} = \frac{1}{\gamma M_s^2} \left(\begin{array}{l} M_x \left(\dfrac{\partial M_y}{\partial j} \dfrac{\partial M_z}{\partial i} - \dfrac{\partial M_z}{\partial j} \dfrac{\partial M_y}{\partial i} \right) \\[2mm] + M_y \left(\dfrac{\partial M_z}{\partial j} \dfrac{\partial M_x}{\partial i} - \dfrac{\partial M_x}{\partial j} \dfrac{\partial M_z}{\partial i} \right) \\[2mm] + M_z \left(\dfrac{\partial M_x}{\partial j} \dfrac{\partial M_y}{\partial i} - \dfrac{\partial M_y}{\partial j} \dfrac{\partial M_x}{\partial i} \right) \end{array} \right).$$
(13.18)

Or in a vector form:

$$g_{ji} = \frac{1}{\gamma M_s^2} \mathbf{M} \cdot \left(\frac{\partial \mathbf{M}}{\partial j} \times \frac{\partial \mathbf{M}}{\partial i} \right).$$
(13.19)

The force density can then be expressed as:

$$\mathbf{f}_g = \begin{bmatrix} 0 & g_{yx} & g_{zx} \\ g_{xy} & 0 & g_{zy} \\ g_{xz} & g_{yz} & 0 \end{bmatrix} \begin{bmatrix} v_x \\ v_y \\ v_z \end{bmatrix}.$$
(13.20)

However, we see that

$$g_{ji} = -g_{ij},$$
(13.21)

This allows us to write the gyro force as:

$$\mathbf{f}_g = \begin{bmatrix} 0 & -g_{xy} & g_{zx} \\ g_{xy} & 0 & -g_{yz} \\ -g_{zx} & g_{yz} & 0 \end{bmatrix} \begin{bmatrix} v_x \\ v_y \\ v_z \end{bmatrix} = \begin{bmatrix} 0 & -g_Z & g_Y \\ g_Z & 0 & -g_X \\ -g_Y & g_X & 0 \end{bmatrix} \begin{bmatrix} v_x \\ v_y \\ v_z \end{bmatrix}.$$
(13.22)

In the vector form, this can be written as:

$$\mathbf{f}_g = \mathbf{g} \times \mathbf{v},$$
(13.23)

where

$$g_X = \frac{1}{\gamma M_s^2} \mathbf{M} \cdot \left(\frac{\partial \mathbf{M}}{\partial y} \times \frac{\partial \mathbf{M}}{\partial z} \right) \quad g_Y = \frac{1}{\gamma M_s^2} \mathbf{M} \cdot \left(\frac{\partial \mathbf{M}}{\partial z} \times \frac{\partial \mathbf{M}}{\partial x} \right),$$
(13.24)

$$g_Z = \frac{1}{\gamma M_s^2} \mathbf{M} \cdot \left(\frac{\partial \mathbf{M}}{\partial x} \times \frac{\partial \mathbf{M}}{\partial y} \right).$$

The gyroscopic force is then equal to

$$\mathbf{F}_g = \int \mathbf{f}_g \, dV = \mathbf{G} \times \mathbf{v} = \left(\int \mathbf{g}_x + \mathbf{g}_y + \mathbf{g}_z \, dV \right) \times \alpha \mathbf{v} = 4\pi \left(\hat{\mathbf{z}} \times \mathbf{v} \right).$$ (13.25)

Similar treatment is given to the dissipative force. The dissipative field is given by:

$$\mathbf{H}_d = \frac{\alpha}{\gamma M_s} \frac{d\mathbf{M}}{dt}.$$
(13.26)

If we substitute dM/dt:

$$\mathbf{H}_d = -\frac{\alpha}{\gamma M_s} (\mathbf{v} \cdot \nabla) \mathbf{M} = \begin{pmatrix} H_x \\ H_y \\ H_z \end{pmatrix}_d = -\frac{\alpha}{\gamma M_s} \begin{pmatrix} v_x \dfrac{\partial M_x}{\partial x} + v_y \dfrac{\partial M_x}{\partial y} + v_z \dfrac{\partial M_x}{\partial z} \\ v_x \dfrac{\partial M_y}{\partial x} + v_y \dfrac{\partial M_y}{\partial y} + v_z \dfrac{\partial M_y}{\partial z} \\ v_x \dfrac{\partial M_z}{\partial x} + v_y \dfrac{\partial M_z}{\partial y} + v_z \dfrac{\partial M_z}{\partial z} \end{pmatrix}.$$
(13.27)

The dissipative force is expressed as:

$$\begin{pmatrix} f_x \\ f_y \\ f_z \end{pmatrix}_d = -\begin{pmatrix} (H_x)_d \dfrac{\partial M_x}{\partial x} + (H_y)_d \dfrac{\partial M_y}{\partial x} + (H_z)_d \dfrac{\partial M_z}{\partial x} \\ (H_x)_d \dfrac{\partial M_x}{\partial y} + (H_z)_d \dfrac{\partial M_z}{\partial y} + (H_y)_d \dfrac{\partial M_y}{\partial y} \\ (H_x)_d \dfrac{\partial M_x}{\partial z} + (H_y)_d \dfrac{\partial M_y}{\partial z} + (H_z)_d \dfrac{\partial M_z}{\partial z} \end{pmatrix}.$$ (13.28)

If we substitute Equation 13.27 in 13.28, we see that the components of the dissipative force can be written as:

$$(f_i)_d = \alpha(d_{ii}v_i + d_{ji}v_j + d_{ki}v_k), \qquad (13.29)$$

where

$$d_{ij} = \frac{1}{\gamma M_s}\left(\frac{\partial M_x}{\partial i}\frac{\partial M_x}{\partial j} + \frac{\partial M_y}{\partial i}\frac{\partial M_y}{\partial j} + \frac{\partial M_z}{\partial i}\frac{\partial M_z}{\partial j}\right). \qquad (13.30)$$

And thus the dissipative force can be simplified as:

$$\mathbf{F}_d = \int \mathbf{f}_d\, dV = D\,\alpha\mathbf{v}. \qquad (13.31)$$

where D is the dissipative tensor with components as written in Equation 13.30.

For the internal force, we can consider it as 0 when no external magnetic field is applied and the skyrmion is driven through the application of the current:

$$F_{in} = \int \nabla_M\left(\mathbf{H}_{int}\cdot\mathbf{M}\right)dV = 0. \qquad (13.32)$$

For the adiabatic spin torque, we see that the adiabatic torque expression shown in Equation 13.8 is very much similar to the gyroscopic torque shown in Equation 13.5 after the substitution of $dM/dt = -\mathbf{v}\cdot\nabla\mathbf{M}$, but with a negative sign and different speed vector (u instead of v). Hence, we can immediately deduce that the corresponding force from the adiabatic STT can be expressed as:

$$\mathbf{F}_{adia} = \int \mathbf{f}_{adia}\, dV = -\mathbf{G}\times\mathbf{u} = -4\pi\left(\hat{\mathbf{z}}\times\mathbf{u}\right). \qquad (13.33)$$

For the nonadiabatic part of the STT, the effective field can be expressed as:

$$\mathbf{H}_{nonadia} = \frac{1}{\gamma}\frac{\beta}{M_s}(\mathbf{u}\cdot\nabla)\mathbf{M}. \qquad (13.34)$$

Similarly, we see that it is very similar to the dissipative field shown in Equation 13.6, but with a different sign and constant. Hence, we can immediately deduce that the nonadiabatic force takes the form of:

$$\mathbf{F}_{nonadia} = -D\beta\mathbf{u}. \qquad (13.35)$$

The force equation, which is also known as the Thiele equation, now becomes:

$$0 = \mathbf{G} \times (\mathbf{v} - \mathbf{u}) + D(\alpha\mathbf{v} - \beta\mathbf{u}), \tag{13.36}$$

which shows that the skyrmion speed (v) can be directly related to the electron drift velocity (u) by:

$$v_x = \left(\frac{\beta}{\alpha} + \frac{\alpha - \beta}{\alpha^3 (D/G)^2 + \alpha} \right) u, \tag{13.37}$$

$$v_y = \left(\frac{(\alpha - \beta)(D/G)}{\alpha^2 (D/G)^2 + 1} \right) u \tag{13.38}$$

The calculated relation between v and u shows that even when the current is only flown in the x-direction, the skyrmion will have a speed component in the y-direction when $\alpha \neq \beta$ due to the gyroscopic/Magnus force.

13.2.1   Micromagnetic Simulation

The skyrmion motion can be visualized by making use of micromagnetic simulation softwares such as the Mumax program [23,24] and OOMMF [25]. The material parameters for the simulations can be chosen to correspond to Co/Pt multilayers [26]. The values are as follows: exchange stiffness 15×10^{-12} J/m, saturation magnetization (M_{sat}) = 580×10^3 A/m, Gilbert damping (α) = 0.1, DMI strength (D) = 3×10^{-3} J/m^3, nonadiabaticity of STT (β) = 0.35, magnetocystalline anisotropy constant = 6×10^5 J/m^3, and spin polarization (P) = 0.7. The track thickness and width is t_{nano} = 0.4 nm and w = 60 nm, respectively. The results that are shown here are obtained with curb thickness of t_{curb} = 0.4 nm, and the results remain the same even when thicker curbs are considered. It is also possible to include a spacer layer underneath the curbs to separate the curb and the skyrmion layer while still maintaining the overall results.

13.2.2   Skyrmion in Thin Film

The inset in Figure 13.1a shows the movement of a skyrmion in a wide plane under the application of an in-plane current for various Gilbert damping (α) and nonadiabatic STT constants (β). In the in-plane

Chapter 13

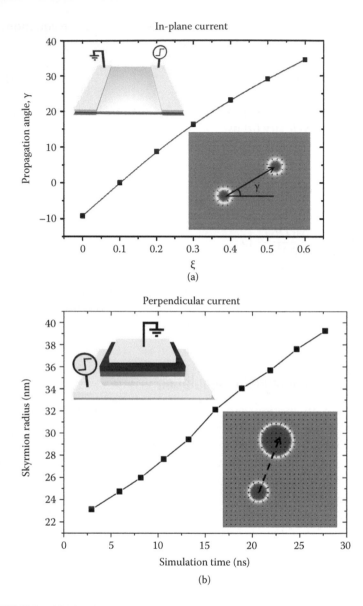

FIGURE 13.1   (a) The change in the propagation angle of the skyrmion as a function of the nonadiabatic constant. Top inset is a schematic of a setup for skyrmion driving with in-plane current. Bottom inset is a snapshot of a skyrmion driving simulation with in-plane current. (b) The change in the size of the skyrmion as a function of time. Top inset is a schematic of a setup for skyrmion driving with perpendicular current. Bottom inset is a snapshot of a skyrmion driving simulation with perpendicular current.

driving case, the magnetization dynamics is expressed by the modified Landau–Lifshitz–Gilbert (LLG) equation (see Equation 13.1), and the skyrmion is driven mostly by the field-like torque from the STT [18]. The simulation results show that under the application of the in-plane current, the skyrmion moves at an angle with respect to the conduction electron flow when $\alpha \neq \beta$ [27]. Figure 13.1a shows the change in the propagation angle as a function of the nonadiabatic constant.

The micromagnetic simulations also show that the skyrmion is deflected from the intended direction when it is driven under perpendicular current injection. In the case of perpendicular current injection [28,29], the magnetization dynamics is expressed by:

$$\tau = \frac{\gamma}{1+\alpha^2}\left(\mathbf{m}\times\mathbf{H}_{\text{eff}} +\alpha\left(\mathbf{m}\times\left(\mathbf{m}\times\mathbf{H}_{\text{eff}}\right)\right)\right)+\tau_{\text{SL}}, \tag{13.39}$$

$$\tau_{\text{SL}} = \frac{j_z\hbar}{M_{\text{sat}}ed}\frac{\varepsilon-\alpha\varepsilon'}{1+\alpha^2}\left(\mathbf{m}\times\left(\mathbf{m}_{\text{p}}\times\mathbf{m}\right)\right)-\frac{j_z\hbar}{M_{\text{sat}}ed}\frac{\varepsilon'-\alpha\varepsilon}{(1+\alpha^2)}\mathbf{m}\times\mathbf{m}_{\text{p}}, \tag{13.40}$$

$$\varepsilon = \frac{P\Lambda^2}{(\Lambda^2+1)+(\Lambda^2-1)(\mathbf{m}\cdot\mathbf{m}_{\text{p}})} \tag{13.41}$$

where j_z is the current density along the z-axis, d is the skyrmion layer thickness, $\mathbf{m}_{\text{p}}$ is the fixed layer magnetization, P is the spin polarization, Λ is the Slonczewski parameter which characterizes the spacer layer, and ε' is the secondary spin-torque parameter.

The inset in Figure 13.1b shows the schematic and the snapshot of the micromagnetic simulations of the skyrmion driving with perpendicular current. The perpendicular current injection can be achieved by using the setup of magnetic tunnel junction (MTJ) or spin Hall devices [30,31], with the magnetization of the reference layer fixed along the x-or y-axis. Under the injection of perpendicular current, the skyrmion is now driven mostly by the torque from the angular momentum transfer as compared to the field-like torque of the STT [32]. As a result, the skyrmion is able to move with a considerably higher speed due to the angular momentum transfer torque being much stronger than the field-like torque. The LLG equation which describes the magnetization dynamics under the application of perpendicular current is expressed in Equation 13.39. The skyrmion is also shown to increase in size as it moves along the plane. Figure 13.1b shows the change in the skyrmion size as a function of simulation time as it moves along the ultrathin magnetic layer.

13.3 Guided Skyrmion Motion

13.3.1 Skyrmion Guiding in Geometrically Curbed Nanowires

In this section, we show that it is possible to guide the movement of the skyrmion and prevent it from annihilating by surrounding and compressing the skyrmion with strong local potential barriers. The compressed skyrmion receives higher contribution from the STT, which results in the significant increase of the skyrmion speed.

13.3.1.1 Skyrmion Motion in 1D Potential Wells

As discussed above, when the skyrmion is driven on a track, it deviates from the intended propagation direction after a short travelling distance because of the Magnus force. The deflection towards the edge of the track eventually leads to skyrmion annihilation. The deviation of the skyrmion motion from the conduction electron flow also prevents reversible two-way operation, as the skyrmion tends to move to the opposite edge of the track when the applied current is reversed. Figure 13.2a shows the schematic of a curbed track that will impose a stronger edge potential barrier on the skyrmion [15,16,33]. The curbed track is formed by creating a rectangular groove on the surface of the perpendicular magnetization anisotropy (PMA) nanowire, leaving the two edges of the nanowire with limited height and width. The curb structures function to confine the skyrmion within the PMA grove. Figure 13.2b and c shows snapshots of simulation with a skyrmion within the uncurbed device and curbed device, respectively. If we consider each of the curbs to have a width of s, the diameter (d) of the skyrmion is then forcibly shrunk into $d = w - 2s$, where w is the width of the ferromagnetic layer. At $t = 0$, an in-plane current is applied to a skyrmion in the relaxed state on both the uncurbed and curbed tracks. At $t = 0.7$ ns, the skyrmion on the uncurbed track is shown to be pushed towards the edge of the track, while the skyrmion on the curbed track is shown to propagate further. At $t = 1.3$ ns, the skyrmion on the uncurbed track is approaching the edge of the track and it starts to get annihilated. Eventually, the skyrmion on the uncurbed track is annihilated completely at $t = 1.9$ ns. In contrast, the skyrmion on the curbed track is stable and the propagation is well guided without any sign of deflection.

Figure 13.3a shows the speed of the skyrmion in the proposed curbed device. The results show that the skyrmion speed is increased due to the inclusion of the curbs, for both in-plane driving setup and also perpendicular driving setup. For instance, when the skyrmion is driven in the in-plane driving setup with a current density of $J = 10 \times 10^{11}$ A m^2

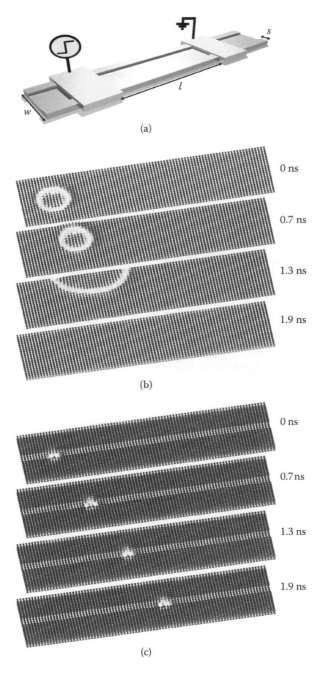

FIGURE 13.2   (a) Schematic of the proposed curbed skyrmion track. (b) Snapshot of the skyrmion driving simulation on a track without curbs which results in skyrmion annihilation. (c) Snapshot of the skyrmion driving simulation on a curbed track which results in fast skyrmion motion.

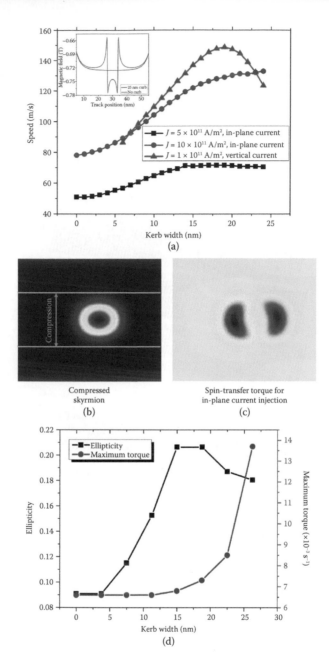

FIGURE 13.3 (a) The skyrmion speed as a function of the curb width for various applied current density. Inset is a cross-sectional plot of the curbed track to show the stray field that is generated by the curbs. (b) Snapshot of the compressed skyrmion on the curbed track. (c) The associated spin-transfer torque that is acting on the skyrmion when an in-plane current is applied. (d) The ellipticity of the skyrmion and the maximum torque that the skyrmion receives as functions of curb width.

and curbs of $s = 15$ nm wide, the speed of the skyrmion is found to be increased by almost 50%, from 75 to 110 m/s. The speed increment is found to be more prevalent in the case of the perpendicular driving setup. For instance, the speed can be increased to as much as 130 m/s by using the same curbs at just a tenth of the current density that is used in the in-plane driving setup ($J = 1 \times 10^{11}$ A/m^2). The inset in Figure 13.3a shows the stray magnetic field that the curbs exert to the skyrmion. The stray field from the curbs is directed in the same direction as the skyrmion magnetization; hence, the stray field acts to increase the stability of the skyrmion against thermal fluctuation [34].

The increase in skyrmion driving speeds can be attributed to the increased STT that the skyrmion receives when the width of the curbed track is less than the original diameter of the skyrmion. Figure 13.3b shows a snapshot of a skyrmion that is compressed by the curbs at the two edges as the skyrmion can only exist within the area that is not covered by the curb. As a result, the skyrmions become elliptical and elongated along the nanotrack axis. Figure 13.3c shows the STT that acts on a skyrmion in the case of in-plane driving. The calculated torque is shown to be most prominent at the front and the back sides of the skyrmion, where the front torque is directed at $-z$ while the back torque is directed at $+z$. The two torques are equal in value with a maximum value of $\tau_{drive} = 6.5 \times 10^{-3}$ s^{-1} when the skyrmion is driven on an uncurbed track. Figure 13.3d shows the ellipticity of the skyrmion and the maximum torque (τ_{drive}) as functions of the curb width (s). The maximum torque that acts on the skyrmion is found to increase rapidly with increasing curb width, which results in the increased skyrmion speed. The increased torque can be attributed to the fact that the compressed skyrmion possesses a higher magnetization divergence, which directly affects the STT, as shown in Equation 13.1.

13.3.1.2   Skyrmion Motion in U-Shape Potential Wells

It is possible to extend the skyrmion guiding technique to drive the skyrmion on a curved track. Figure 13.4a shows the snapshots of the simulations when the skyrmion is driven on a curved track which has an 180° turning angle. The results show that without the curb (i), the skyrmion is annihilated within the curvature even though the current is distributed to follow the shape of the track. However, with the inclusion of the curbs to the track (ii), the annihilation is prevented and the skyrmion is guided to make the 180° turn. Figure 13.4b shows the position of the skyrmion within the curbed curved track as a function of time in the case of clockwise and anti clockwise driving. In contrast to the straight track where the skyrmion maintains a constant velocity

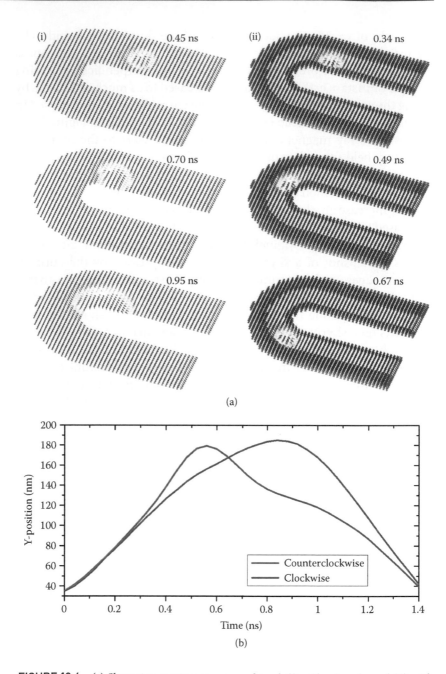

FIGURE 13.4   (a) Skyrmion motion on a curved track (i) without curbs and (ii) with curbs. (b) Skyrmion position along the y-axis as a function of time for both clockwise driving and counterclockwise driving.

throughout its motion, the skyrmion on the curved track is shown to be slowed down with different deceleration depending on the direction of the applied current. The slowing down of the skyrmion can be attributed to the uneven current distribution that the skyrmion receives within the curved track, while the difference in the behavior between the clockwise and counterclockwise driving can be attributed to the Magnus force. As discussed before, during the application of current, the skyrmion is pushed to one side of the track due to the Magnus force. Therefore, with the application of clockwise and counterclockwise currents, the skyrmion is pushed to the outer and inner arc, respectively. Since the turning radius for the inner arc is much smaller, it is more difficult for the skyrmion to complete the turn when it is driven in the counterclockwise direction.

The reduced size of a skyrmion can also confer a significant increase in data storage density to a potential skyrmion-based memory device. As skyrmions in a magnetic thin film are of the same magnetic charge, two skyrmions experience repulsion from each other, and the strength of the repulsion is related to the size of the skyrmions, as shown by Figure 13.5a. The minimum interskyrmion distance whereby two skyrmions can remain undisturbed from each other therefore shall also depend on the skyrmion size. The micromagnetic simulations show that the interskyrmion distance (d_e) and the radius of the skyrmion (r_s) varies almost linearly, as shown in Figure 13.5b. This means that the curbed track can also be utilized to pack the skyrmions closer to each other. At curb widths of around 5 nm, the micromagnetic simulations show that the skyrmion was unexpectedly enlarged. This behavior can be attributed to the presence of the stray magnetic field from the curb as explained previously. However, the size increment from the stray magnetic field is immediately overcome by the skyrmion compression from the curbs as the distance between the curbs is decreased.

13.3.2   Skyrmion Guiding in Nanowires with Tunable Perpendicular Anisotropy

By tuning the magnetic anisotropy along the width of the nanowire, barriers can be formed to confine the skyrmions residing in the nanowire to within the center region. The effective field from nanowire with a low anisotropy constant K in the center region is shown in Figure 13.6a. The effective field is much lower at the center region of the nanowire, allowing the magnetization to flip more easily. The lower magnetic anisotropy at the center therefore represents a path of lower resistance compared to the higher magnetic anisotropy at the edges.

Chapter 13

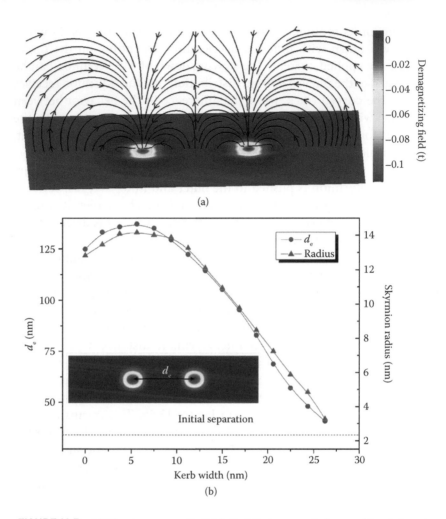

(a)

(b)

FIGURE 13.5   (a) The stray magnetic fields that that are created around two interacting skyrmions. (b) Plot showing the relation between the equilibrium distance (d_e),the skyrmion radius, and the curb width.

Figure 13.6b shows simulation snapshots of skyrmion motion in a nanowire with uniform magnetic anisotropy. At $t = 0.5$ ns, the skyrmion moves towards the edge due to Magnus force and at $t = 1.0$ ns, it starts to get annihilated. Figure 13.6c shows snapshots of skyrmion motion in a nanowire with lower magnetic anisotropy at the center and higher magnetic anisotropy at the outer regions. Annihilation at the nanowire edge is prevented as the skyrmion is confined in the center region and moves along the intended direction.

The PMA of Co/Pt multilayers can be patterned by light ion irradiation and accurately controlled by the influence of the irradiation [35].

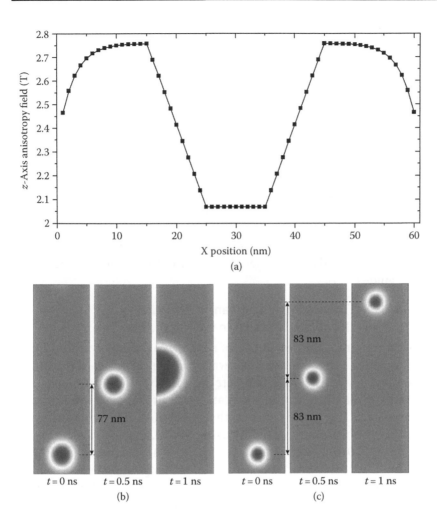

FIGURE 13.6 (a) z-axis anisotropy field (T) along the width (nm) of the nanowire with a higher PMA of $8{\times}10^5$ J/m^3 at the outer region of the nanowire and a lower PMA of 6×10^5 J/m^3 at the 10-nm wide center region, the gradient between the higher and lower PMA spans over 10 nm. The z-axis anisotropy field near the edges are lower due to the spin tilting near the nanowire edge. The lower PMA at the center region confines skyrmions residing in the nanowire. (b) Snapshots of simulation at times $t = 0$, 0.5, $t = 1.0$ ns of a skyrmion driven by current in a nanowire with uniform PMA of 6×10^5 J/m^3. (c) Snapshots of simulations of a skyrmion driven by current in a nanowire with tuned PMA to confine the skyrmion in the center region.

He$^+$ irradiation lowers the PMA of Co/Pt multilayers and when combined with high-resolution lithography, magnetic anisotropy patterning at nanometer scale can be achieved [36]. The velocity of current-driven skyrmions is doubled by lowering the magnetic anisotropy of the center region from 8×10^5 to 6×10^5 J/m^3.

13.3.3 Skyrmion Guiding in Nanowires with Tunable Damping Constant

In this section, the self-focusing of magnetic skyrmions in magnonic waveguides with periodic property is demonstrated theoretically and through micromagnetic simulations. In the proposed magnetic skyrmion waveguide, the motion of skyrmions is self-focused without annihilation at the waveguide edges, which is similar to the waveguides of electromagnetic waves. It is shown numerically that a magnonic waveguide of a properly designed periodicity can serve as a magnetic skyrmion waveguide by periodically modulating the damping constant of the waveguide in either the longitudinal or the transverse direction. Moreover, the simultaneous propagation of skyrmions along multiple channels is also investigated in both the straight and the curved waveguides.

13.3.3.1 The Damping Constant Is Periodically Modulated Parallel to the Nanowire

To demonstrate the feasibility of such a magnetic skyrmion guide, we consider a special magnonic waveguide in which α–β has almost the same magnitude but opposite sign in the two unit blocks. The lengths of block A and block B are 1000 nm so that the skyrmion moves near the medial axis of the waveguide. The micromagnetic simulations were carried out by using MUMAX3 package, the waveguide is 520 nm wide and 1 nm thick. The magnetic parameters of the Co/Pt wires are used as: the exchange constant $A = 1.5 \times 10^{-11}$ J/m, the saturation magnetization $M_s = 8 \times 10^5$ A/m, the PMA $K_u = 8 \times 10^5$ J/m^3, and the DMI constant $D = 3 \times 10^{-3}$ J/m^2. The mesh size is $2 \times 2 \times 1$ nm^3. The nonadiabatic STT coefficient $\beta = 0.2$ is used. The damping coefficients in blocks A and B are 0.1 and 0.3, respectively.

The time dependence of the skyrmion core position under a current density of 2.76×10^{12} A/m^2, corresponding to $u = 200$ m/s, is shown in Figure 13.7. After the current is switched on, the skyrmion starts to propagate and enters the waveguide. For $\alpha = 0.2$ and $\beta = 0.2$, the skyrmion is driven along the waveguide center. For $\alpha = 0.1$ and $\beta = 0.2$, the skyrmion moves upward and is finally annihilated at the top edge of the waveguide. For $\alpha = 0.3$ and $\beta = 0.2$, the skyrmion moves downward and is finally annihilated at the bottom edge of the waveguide. For $\alpha_A = 0.1$, $\alpha_B = 0.3$, and $\beta = 0.2$, the skyrmion moves upward by 16 nm for about 2.2 ns in block A and then downward by 16 nm for about 2.2 ns in block B alternatively, as expected. As shown in Figure 13.8, the longitudinal velocity

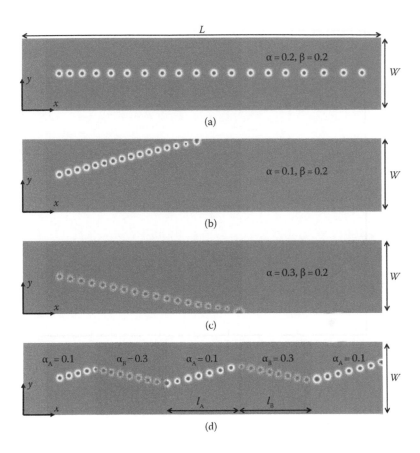

FIGURE 13.7 Sketches of moving direction of a skyrmion in magnonic waveguide with $\beta = 0.2$ and (a) $\alpha = 0.2$, (b) $\alpha = 0.1$, (c) $\alpha = 0.3$, and (d) α_A (α_B) = 0.1 (0.3). The x-, y-, and the z-axis are along the length direction, the width direction, and the thickness direction, respectively. The shading is coded by the value of m_z as indicated by the right bar. (d) The magnonic waveguide of a magnetic skyrmion: block A: $\alpha < \beta$; block B: $\alpha > \beta$. The zig-zag line sketches the trajectory of a skyrmion in the waveguide.

of the skyrmion v_x is independent of the waveguide width, but the transverse velocity of the skyrmion v_y is quite dependent on the waveguide width. The current density dependence of skyrmion agrees well with the theoretical prediction as shown in Figure 13.8b.

13.3.3.2 The Damping Constant Is Periodically Modulated Perpendicular to the Nanowire

Obviously, a perfect skyrmion waveguide as proposed in the previous section requires a complete compensation of transverse motion of a skyrmion inside blocks A and B. This is not easy to achieve because the motion of a skyrmion is sensitive to many material parameters

Chapter 13

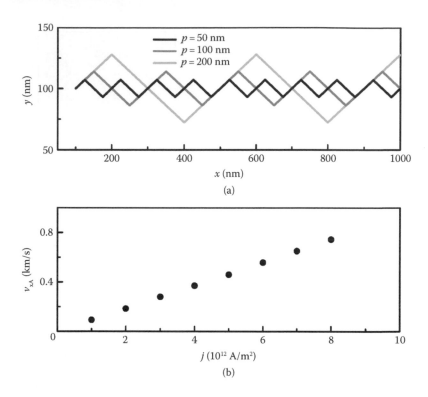

FIGURE 13.8   (a) Skyrmion core position as the functions of time t under a current density of 2.76×10^{12} A/m², corresponding to $u = 200$ m/s for different widths of the blocks. (b) Current density dependence of skyrmion core velocity. The longitudinal velocity v_x is linearly dependent on the current density.

including both α and β. Any fluctuation or uncertainty in these parameters shall lead to an overall shift in the transverse direction, no matter how small it might be. After a long time and/or a long travel distance, the skyrmion might eventually collide with the waveguide edges and annihilate there. Furthermore, our current understanding of current-driven skyrmion motion does not give us the power to predict the exact compensatory length for blocks A and B for given sets of material parameters. Mostly, using micromagnetic simulations to provide certain guidance if one wants to build a long skyrmion waveguide. Thus, an ideal way is that one can make a self-focusing waveguide for skyrmions as what an optical fiber does for light. Interestingly, if we periodically vary the damping of the waveguide in the transverse direction as shown in Figure 13.9, a skyrmion can be confined near the interface if the top and bottom layers (block B) are made of material with $\alpha > \beta$ and the center layer (block A) is a material with $\alpha < \beta$. According to the above formula for the transverse motion of a skyrmion, the skyrmion moves

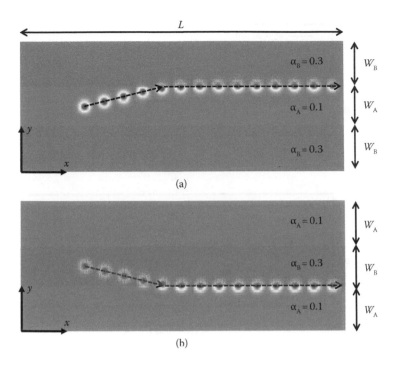

FIGURE 13.9 The trajectory of a skyrmion in the second type of magnonic waveguide with the damping constant varied along the transverse direction: (a) $\alpha_A = 0.1$, $\alpha_B = 0.3$, and (b) $\alpha_A = 0.3$, $\alpha_B = 0.1$. The skyrmion core position as a function of time t under a current density of 2.76×10^{12} A/m^2, corresponding to $u = 200$ m/s. The shading is coded by the value of m_z as indicated by the right bar.

downward in layer B and upward in layer A. As a result, the skyrmion shall move around the interface during its lateral propagation along the interface of the layered structures.

To demonstrate such a self-focused waveguide is capable of confining the skyrmion near the interface of the waveguide, we consider a waveguide with $\beta = 0.2$ in both blocks A and B, and $\alpha = 0.1$ in block A and $\alpha = 0.3$ in block B. Both blocks A and B have the same width of 260 nm and the same thickness of 1 nm. Figure 13.10a shows the position of a skyrmion under an electric current density of 2.76×10^{12} A/m^2, corresponding to $u = 200$ m/s. An initially created skyrmion enters the waveguide, and then it is confined around the interface. The longitudinal motion of the skyrmion is at a constant speed of $v_x \simeq 474$ m/s.

The proposed waveguides can be extended from single channel to multichannels. As shown in Figure 13.11, we presented the double-channel waveguide as an instance. It numerically demonstrates the simultaneous propagation of skyrmions along multiple channels in both the straight and the curved skyrmion waveguides.

Chapter 13

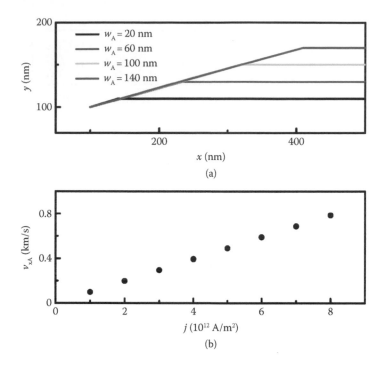

FIGURE 13.10   (a) Skyrmion core position as the functions of time t under a current density of 2.76×10^{12} A/m^2, corresponding to $u = 200$ m/s in the second type of skyrmion waveguide with different widths of block A. (b) Current density dependence of skyrmion core velocity. The longitudinal velocity v_x is linearly dependent on the current density.

13.4   Conclusion

In conclusion, we have shown that it is possible to guide the skyrmion motion and prevent the skyrmion from being annihilated at the same time by modifying the physical or the magnetic properties of nanostructures. The modification can be done by either modifying the edges of the nanostructure or by creating a periodic adjustment in the nanostructure to mimic a waveguide. When the edge modification is considered, we show that the technique can be extended to drive the skyrmion on a curved track, with different speeds depending on the direction of the applied current. We also show that the speed of a current-driven skyrmion is increased significantly when the skyrmion is compressed by the curbs. The increase in the current-driven speed can be attributed to the higher torque that the compressed skyrmion receives from the conduction electron. For the wave guides, two types of skyrmion waveguide structures are proposed and numerically studied. One is with the

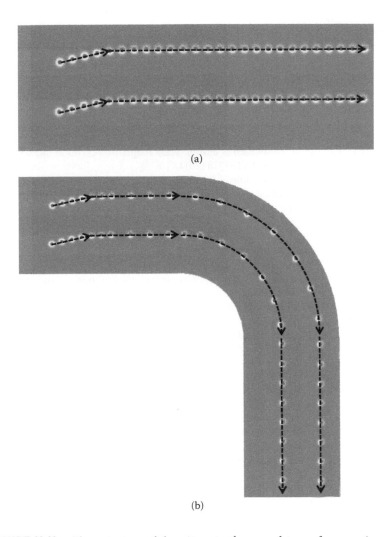

(a)

(b)

FIGURE 13.11 The trajectory of skyrmions in the second type of magnonic wave-guides with two channels for $\alpha_A = 0.1$ and $\alpha_B = 0.3$: (a) straight and (b) curved wave-guide, respectively. The skyrmion core position as a function of time t under a current density of 2.76×10^{12} A/m², corresponding to $u = 200$ m/s. The shading is coded by the value of m_z as indicated by the right bar.

periodic damping constant along the waveguide, whose two unit blocks have opposite signs of α–I. Such a guide requires an exact compensation of the transverse motion of a skyrmion in blocks A and B, and high demands in the accurate design of two basic blocks is needed. The second type of skyrmion waveguide is the periodic damping constant transverse to the waveguide, in which α–β has opposite signs in two layers.

Chapter 13

References

1. Muhlbauer, S., et al. Skyrmion lattice in a chiral magnet. *Science* **323**, 915–919 (2009).
2. Yu, X. Z., et al. Real-space observation of a two-dimensional skyrmion crystal. *Nature* **465**, 901–904 (2010).
3. Huang, S. X. and Chien, C. L. Extended skyrmion phase in FeGe (111) thin films. *Phys. Rev. Lett.* **108**, 267201 (2012).
4. Heinze, S., et al. Spontaneous atomic-scale magnetic skyrmion lattice in two dimensions. *Nat. Phys.* **7**, 713–718 (2011).
5. Fert, A. and Levy, P. M. Role of anisotropic exchange interactions in determining the properties of spin-glasses. *Phys. Rev. Lett.* **44**, 1538–1541 (1980).
6. Fert, A. Magnetic and transport properties of metallic multilayers, *Metal. Multilayers* **59–60**, 439 (1990).
7. Nagaosa, N. and Tokura, Y. Topological properties and dynamics of magnetic skyrmions. *Nat. Nanotechnol.* **8**, 889–911 (2013).
8. Sampaio, J., Cros, V., Rohart, S., Thiaville, A., and Fert, A. Nucleation, stability and current-induced motion of isolated magnetic skyrmions in nanostructures. *Nat. Nanotechnol.* **8**, 839–844 (2013).
9. Fert, A., Cros, V., and Sampaio, J. Skyrmions on the track. *Nat. Nanotechnol.* **8**, 152–156 (2013).
10. Yu, X. Z., et al. Skyrmion flow near room temperature in an ultralow current density. *Nat. Commun.* **3**, 988 (2012).
11. Tomasello, R., et al. A strategy for the design of skyrmion racetrack memories, *Sci. Rep.* **4**, 6784 (2014)
12. Iwasaki, J., Mochizuki, M., and Nagaosa, N. Universal current-velocity relation of skyrmion motion in chiral magnets. *Nat Commun.* **4**, 1463 (2013).
13. Schultz, T., et al. Emergent electrodynamics of skyrmion in a chiral magnet. *Nat. Phys.* **8**, 301–304 (2012).
14. Iwasaki, J., Mochizuki, M., and Nagaosa, N. Current-induced skyrmion dynamics in constricted geometries. *Nat. Nanotechnol.* **8**, 742–747 (2013).
15. Purnama, I., Gan, W. L., Wong, D. W. and Lew, W. S. Guided current-induced skyrmion motion in 1D potential well. *Sci. Rep.* **5**, 10620 (2015).
16. Fook, H. T., Gan, W. L., Purnama, I. and Lew, W. S. Mitigation of magnus force in current-induced skyrmion dynamics, *IEEE Trans. Magn.* **51**, 1500204 (2015)
17. Brataas, A., Kent, A. D., and Ohno, H. Current-induced torques in magnetic materials. *Nat. Mater.* **11**, 372 (2012).
18. Thiaville, A., Nakatani, Y., Miltat, J., and Suzuki, Y. Micromagnetic understanding of current-driven domain wall motion in patterned nanowires. *Europhys. Lett.* **69**, 990 (2005).
19. Haazen, P. P. J., et al. Domain wall depinning governed by the spin Hall effect. *Nat. Mater.* **12**, 299 (2013).
20. Ryu, J., Lee, K. J., and Lee, H. W. Current-driven domain wall motion with spin Hall effect: Reduction of threshold current density. *Appl. Phys. Lett.* **102**, 172404 (2013).
21. Miron, I. M., et al. Fast current-induced domain-wall motion controlled by the Rashba effect. *Nat. Mater.* **10**, 419–423 (2011).
22. Ryu, J., Seo, S. M., Lee, K. J., and Lee, H. W. Rashba spin–orbit coupling effects on a current-induced domain wall motion. *J. Magn. Magn. Mater.* **324**, 1449 (2012).

23. Vansteenkiste, A., et al. The design and verification of mumax3. *AIP Adv.* **4**, 107133 (2014).
24. Najafi, M., et al. Proposal for a standard problem for micromagnetic simulations including spin-transfer torque. *J. Appl. Phys.* **105**, 113914 (2009).
25. Donahue, M. J. and Porter, D. G. *"OOMMF User's Guide, Version 1.0," Interagency Report No. NISTIR 6376*, National Institute of Standards and Technology, Gaithersburg, MD (1999).
26. Metaxasm, P. J., et al. Creep and flow regimes of magnetic domain-wall motion in ultrathin Pt/Co/Pt films with perpendicular anisotropy. *Phys. Rev. Lett.* **99**, 217208 (2007).
27. Zhou, Y. and Ezawa, M. A reversible conversion between a skyrmion and a domain-wall pair in a junction geometry. *Nat. Commun.* **5**, 4652 (2014).
28. Boone, C.T., et al. Rapid domain wall motion in permalloy nanowires excited by a spin-polarized current applied perpendicular to the nanowire. *Phys. Rev. Lett.* **104**, 097203 (2010).
29. Theodonis, I., Kioussis, N., Kalitsov, A., and Chshiev, M., Butler, W. H. Anomalous bias dependence of spin torque in magnetic tunnel junctions. *Phys. Rev. Lett.* **97**, 237205 (2006).
30. Khvalkovskiy, A. V., et al. Matching domain-wall configuration and spin–orbit torques for efficient domain-wall motion. *Phys. Rev. B* **87**, 020402 (2013).
31. Khvalkovskiy, A., et al. High domain wall velocities due to spin currents perpendicular to the plane. *Phys. Rev. Lett.* **102**, 067206 (2009).
32. Slonczewski, J. C. Currents, torques, and polarization factors in magnetic tunnel junctions. *Phys. Rev. B* **71**, 024411 (2005).
33. Rohart, S. and Thiaville, A. Skyrmion confinement in ultrathin film nanostructures in the presence of Dzyaloshinskii-Moriya interaction. *Phys. Rev. B*, 184422 (2013).
34. Romming, N., Hannekan, C., Menzel, M., Bickel, J. E., Wolter, B., von Bergmann, K., Kubetzka, A. and Wiesendanger, R. Writing and deleting single magnetic skyrmions. *Science* **341**, 6146 (2013).
35. Chappert, C., et al. Planar patterned magnetic media obtained by ion irradiation. *Science* **280**, 1919 (1998).
36. Devolder, T., et al. Sub-50 nm planar magnetic nanostructures fabricated by ion irradiation. *App. Phys. Lett.*, **74**, 3383–3385 (1999).

Chapter 13

14. Skyrmions in Other Condensed Matter Systems

Guoping Zhao, Nian Ran, and Ping Lai
Sichuan Normal University, Chengdu, People's Republic of China

Guoping Zhao, François Jacques Morvan
Chinese Academy of Sciences, Ningbo, People's Republic of China

Jing Xia, Xichao Zhang, and Yan Zhou
The Chinese University of Hong Kong, Shenzhen,
People's Republic of China

Chapter 14

Skyrmions can exist in various condensed matter systems, which is reviewed in this chapter, including the antiferromagnetic (AFM) materials, ferromagnetic bilayers with AFM interface coupling, multiferroic insulators, quantum Hall systems, liquid crystals, and Bose–Einstein condensate (BEC) systems. The basic theory, advantage, and weakness for AFM skyrmions and skyrmions in metallic bilayers with AFM coupling have been examined, which can overcome the skyrmion Hall effect (SkHE) in metallic skyrmion racetrack. Creation and motion of mulitiferroic skyrmion crystals in the magnetic insulator Cu_2OSeO_3 have also been demonstrated, with the mechanism, nucleation, and dynamic properties summarized, which have no SkHE and can be manipulated by electric potentials. Skyrmions in quantum Hall systems can also be manipulated by electric fields due to the electrical charge carried, whose existence have been confirmed by NMR and other experimental methods. Such unstable skyrmions are mostly found in GaAs and related quantum wells. A 1/4 spiral skyrmion is formed in liquid crystals, where double-twist cylinders are energetically more favorable than the uniaxial helical structures. On the other hand, a full spiral skyrmion is predicted in cold-atom systems, which can be deteriorated due to strong spin–orbit (SO) coupling. While the so-called skyrmions in liquid crystals are far away from a full skyrmion, much work needs to be done before the skyrmions in quantum Hall and cold-atom systems have real applications.

14.1   Introduction

Skyrmions were first proposed by the British physicist Tony Hilton Royle Skyrme in early 1960s as a theoretical model to account for the stability of hadrons in nuclear physics (Skyrme 1962). Such a topological model assumes that the universe is in some ferromagnetic state, which turns out to be a ubiquitous model figuring almost all branches of physics (Girvin 2000).

In condensed-matter physics, magnetic skyrmions are quasiparticles (Sondhi et al. 1993) with hedgehog or vortex-like spin configurations (Bogdanov and Rossler 2001; Dupe et al. 2014; Iwasaki et al. 2013; Rossler et al. 2006; Sondhi et al. 1993), which have been observed experimentally (Mühlbauer et al. 2009; Romming et al. 2013) in bulk semiconductor B20 alloys, such as MnSi (Mühlbauer et al. 2009),

FeCoSi, FeGe, and MnFeGe. They are chiral by nature and may exist both as dynamic excitations (Sondhi et al. 1993) or stable and metastable states (Romming et al. 2013), as elaborated in previous chapters.

Since their discovery, magnetic skyrmions attracted more and more attention (Dai et al. 2013; Du et al. 2014, 2015; Fert et al. 2013; Miao et al. 2014; Mühlbauer et al. 2009; Nagaosa and Tokura 2013; Sun et al. 2013; Yu et al. 2010; Zhang et al. 2015a), and most of the researches made since demonstrate that magnetic skyrmions are promising for building next-generation magnetic memories and spintronic devices due to their stability, small size, and the extremely low currents needed to move them (Fert et al. 2013; Nagaosa and Tokura 2013).

The small size of magnetic skyrmions makes them a good candidate for future data storage solutions. The topological charge, representing the existence and nonexistence of skyrmions, can represent the bit states "1" and "0." Physicists at the University of Hamburg have even managed to read and write skyrmions using scanning tunneling microscopy (Romming et al. 2013). However, despite their small size, a repulsion force between them must be considered. Zhang et al. (2015a) demonstrated that an average distance of 60 nm between skyrmions is necessary to prevent repulsion in a thin film system with typical magnetic material parameters. This repulsion does deteriorate the data storage density, but even if physicists do not surmount it, storage devices as small as microSD cards could theoretically reach the maximum capacity of 10 TB (such a result is being calculated as 1/10 of the space being composed of skyrmions, 9/10 mainly being composed of the architectural parts controlling the reading and writing of skyrmions): it is 50 times the actual highest capacity available in microSD cards (200 GB ["Sandisk SD card C10," 2015]). If we imagine that one day this repulsion is partially overcome, an average distance of 30 nm between skyrmions is coherent, and storage devices as small as microSD cards could theoretically reach the maximum capacity of 80 TB, namely, 400 times the actual highest capacity available, and 10 times the highest capacity (8 TB ["Ultrastar He8," 2015]) among 3.5″ hard disk drives (HDD) available on the market. However, this is not the only way skyrmions can revolutionize the high-technology industry and our life.

Magnetic skyrmions have a latency of 0.5 ns, with a theoretical bandwidth of a few TB/s, so that they have a very fast storage speed. Moreover, in comparison to solid-state drives (SSD) and HDD, magnetic skyrmions are more stable and can last longer. Further, magnetic skyrmions can be used as logic gates and transistors, which can be controlled electronically through the electric field and spin current (Kim et al. 2014; Verba et al. 2014; Zhang et al. 2015b).

Chapter 14

The aforementioned skyrmions in B20 alloys are mainly Bloch types. Recently, the attention has shifted to ultrathin ferromagnetic/heavy metal films, where Néel-type skyrmions can be nucleated with wider applications. Particular interest has been paid to multilayers deposited by sputtering (Boulle et al. 2016; Monso et al. 2002) because the anisotropy, Dzyaloshinskii–Moriya interaction (DMI), and exchange interaction (Bogdanov and Hubert 1994; Rohart and Thiaville 2013) can be easily tuned by changing the nature and thickness of the ultrathin films (Boulle et al. 2016). The interface DMI (Belmeguenai et al. 2015; Di et al. 2015; Emori et al. 2014; Freimuth et al. 2014; Pizzini et al. 2014; Nembach et al. 2015; Stashkevich et al. 2015; Yang et al. 2015) is larger in these ultrathin films, resulting in chiral Néel domain walls (DWs) (Chen et al. 2013; Tetienne et al. 2015). In addition, the current induced SO torques (Garello et al. 2013; Miron et al. 2011a) in these multilayers are great, leading to fast current-induced DW motion (Miron et al. 2011b; Thiaville et al. 2012). Further, the fast and spatially homogeneous deposition by sputtering is compatible with standard spintronics devices such as magnetic tunnel junctions, which make the industrial integration straightforward (Boulle et al. 2016). Stable chiral skyrmions in sputtered ultrathin Pt/Co/MgO nanostructures at room temperature and zero applied magnetic field have been observed recently, where the chiral Néel internal structure has been demonstrated by photoemission electron microscopy combined with X-ray magnetic circular dichroism (XMCD-PEEM) (Boulle et al. 2016).

One big drawback of the metallic skyrmions in constricted geometry is that skyrmions can drift from the direction of electron flow because of the Magnus force and annihilate at the magnetic film edge (Barker and Tretiakov 2015; Purnama et al. 2015; Zhang et al. 2016b, 2016c), which is called the SkHE (Zang et al. 2011; Jiang et al. 2016). Many efforts have been made to weaken the SkHE, i.e., by turning the perpendicular magnetic anisotropy (PMA) along the nanotrack width or by creating a curbed nanotrack (Fook et al. 2015; Purnama et al. 2015). Recently, two novel solutions, i.e., AFM skyrmions (Barker and Tretiakov 2015; Zhang et al. 2016b) and skyrmions in two-layer magnetic metals with AFM coupling (Zhang et al. 2016c), have been proposed to completely inhibit the SkHE, which are addressed in Sections 14.2 and 14.3 of this chapter. In the former case, the Magnus force disappears because the net local magnetization is zero. On the other hand, the Magnus forces sensed by the skyrmions in the two-metal layer system can be exactly cancelled for the latter case. Skyrmion crystals can form in the magnetic insulator Cu_2OSeO_3 (Adams et al. 2012; Seki et al. 2012a, 2012b), where SkHE is out of problem. The mechanism, creation, and dynamic properties of such multiferroic skyrmions are given in Section 14.4, where manipulation of skyrmions by electric fields is possible. The multifaceted skyrmions in quantum Hall systems, liquid crystals, and BEC are addressed in the last three sections of this chapter.

14.2 Creation and Motion of Skyrmions in AFM Materials

AFM materials have aroused a lot of interests recently, which found many applications (Biswas et al. 2013; Nascimento and Bernhard 2013; Gomonay and Loktev 2014; Zherlitsyn et al. 2014; Jungwirth et al. 2016; Wadley et al. 2016). AFM skyrmions (AFMSk) have been proposed simultaneously by two groups in 2015, they can be nucleated and driven by spin currents in typical AFM materials and have various advantages over ferromagnetic skyrmions (FMSk) (Fert et al. 2013). They have no demagnetizing field and are not sensitive to the stray fields. They will not experience the Magnus force so that they can move parallel to the current without any disturbance. The longitudinal velocity of AFMSk induced by current can reach a few km/s.

Both research teams from The University of Hong Kong (HKU), the University of Tokyo (UTokyo), and Tohoku University (Tohoku U) have proposed an AFMSk model and studied the spin texture as well as the motion of skyrmions using this model. Zhang et al. (2016b) from HKU and UTokyo focused on the creation of AFMSk and proposed two methods for the nucleation of AFMSk. Zhang et al. (2016b) also studied the motion dynamics of AFMSk driven by perpendicular-to-plane current. On the other hand, Barker and Tretiakov (2015) from Tohoku U studied the Brownian motion of a single AFMSk and found that the diffusion coefficient is larger than that for FMSk. Barker and Tretiakov (2015) derived the formula for the AFMSk radius depending on the temperature and found that the fluctuations of the AFMSk radius around its mean value are larger than those of the corresponding FMSk. Barker and Tretiakov (2015) also investigated the motion dynamics of AFMSk driven by in-plane current. Wadley et al. (2016) demonstrated room-temperature electrical switching between stable configurations in AFM CuMnAs thin film devices by applied current with magnitudes of order 10^6 A/cm^2.

The following discusses the basic theory, nucleation, and motion of AFMSk based on Zhang et al.'s and Barker et al.'s works.

14.2.1 Theory

The theory for AFMSk is similar to that of the FMSk introduced in the previous chapters, by solving the Landau–Lifshitz–Gilbert–Slonczewski

Chapter 14

(LLGS) equation of the system. First, the Hamiltonian of the AFM system is given by:

$$H_{AFM} = J \sum_{\langle i,j \rangle} \boldsymbol{m}_i \cdot \boldsymbol{m}_j + \sum_{\langle i,j \rangle} \boldsymbol{D} \cdot \left(\boldsymbol{m}_i \times \boldsymbol{m}_j \right) - K \sum_i (m_i^z)^2 \tag{14.1}$$

The three terms on the right-hand side of Equation 14.1 represent the AFM exchange interaction, the DMI, and the PMA, respectively. J, D, and K are the AFM exchange stiffness, the DMI vector, and the anisotropic constant, respectively. m_i represents the normalized magnetization ($|m_i| = 1$), and $\langle i, j \rangle$ runs over all the nearest neighbor sites. The dynamics of the magnetization m_i at the lattice site i can be obtained by solving the LLGS equation (see Equation 14.2).

The Hamiltonian in Equation 14.1 is different from the FM case in that J can be negative while a positive J holds for all FM. For example, the square lattice AFM system contains A sites with $\boldsymbol{m}_A(i)$ and B sites with $\boldsymbol{m}_B(i)$. Because of the AFM exchange interaction, $\boldsymbol{m}_A(i) \approx -\boldsymbol{m}_B(i)$ for all sites, i.e., the directions are opposite between the neighboring sites.

When a perpendicular-to-plane spin current is applied, the LLGS equation is defined as:

$$\frac{d\boldsymbol{m}_i}{dt} = -|\gamma| \boldsymbol{m}_i \times \boldsymbol{H}_i^{\text{eff}} + \alpha \boldsymbol{m}_i \times \frac{d\boldsymbol{m}_i}{dt} + |\gamma| \beta (\boldsymbol{m}_i \times \boldsymbol{p} \times \boldsymbol{m}_i) - |\gamma| \beta' (\boldsymbol{m}_i \times \boldsymbol{p}),$$

$$\tag{14.2}$$

where γ is the gyromagnetic ratio, α is the Gilbert damping coefficient originating from spin relaxation, β is the Slonczewski-like spin-transfer torque (STT) coefficient, β' is the out-of-plane STT coefficient, and p is the spin polarization direction. Here, $\beta = \left| \dfrac{\hbar}{\mu_0 e} \right| \dfrac{j|p|}{2dM_S}$ with μ_0, j, d, and M_S representing the vacuum magnetic permittivity, the current density, the film thickness, and the saturation magnetization, respectively. $\boldsymbol{H}_i^{\text{eff}} = -\dfrac{\partial H_{AFM}}{\partial m}$ is the effective magnetic field induced by the Hamiltonian Equation 14.1.

14.2.2   Nucleation of AFMSk

Figure 14.1a illustrates an AFMSk in a square-lattice AFM thin film. There are two methods to create an AFMSk: one is injecting a spin-polarized current perpendicularly in a circular region of an AFM nanodisk to flip the magnetization of the current injection region, which is similar to the creation of a FMSk in an FM nanodisk, as shown in Figure 14.1b. The other method is analogous to the conversion from an FM DW pair to a FMSk (Zhou and Ezawa 2014). An AFM DW pair can convert to an AFMSk in a junction geometry. The specific conversion process is given in Figure 14.1c. Firstly, an AFM DW pair in the narrow nanotrack is driven by the spin-polarized current towards the right. At $t = 30$ ps, both ends of the DWs are pinned by the junction, but the central parts continue moving rightward driven by the current. Finally, both ends of the DWs will break away from the junction and form an AFMSk, which will continue moving rightward in the wide nanotrack.

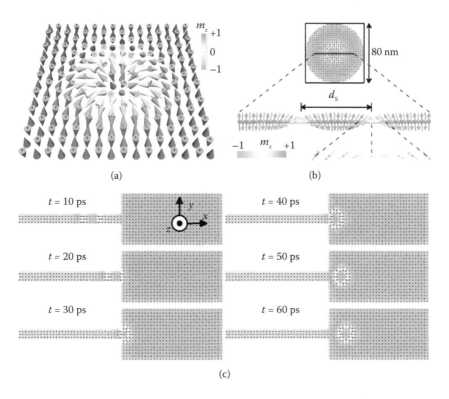

(a)   (b)   (c)

FIGURE 14.1   (a) Illustration of an AFMSk. The in-plane and out-of-plane components of magnetization are represented by the arrows and color scale, respectively. (b) Creation of an AFMSk in a nanodisk by injecting spin-polarized current perpendicularly to the nanodisk. (c) Creation of an AFMSk in a nanotrack from an AFM DW pair in a nanotrack. (Reprinted from Zhang, X.C., et al., *Sci. Rep.*, 6, 24795, 2016b. With permission.)

Chapter 14

14.2.3　Advantages and Limitations

Compared with FMSk, the current-induced motion of AFMSk in a nanotrack has nontrivial advantages. Firstly, the AFMSk has no transverse motion, which ensures its straight motion without annihilation on the edge, as illustrated in Figure 14.2a. In addition, as shown in Figure 14.2b and c, the stable motion velocity of an AFMSk can reach

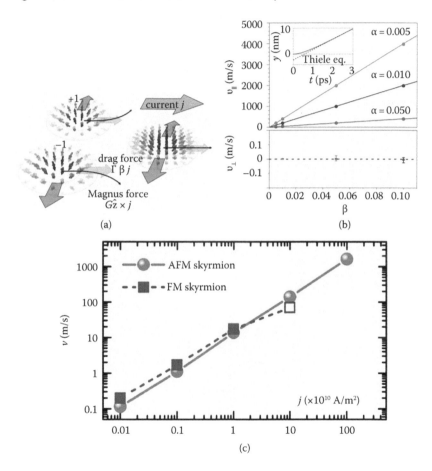

(a)

(b)

(c)

FIGURE 14.2 (a) Schematic illustration of AFMSk composed of two topological objects with opposite topological charge, hence the Magnus force acts in opposite directions. Because of the strong coupling between the sublattices, the AFMSk has no transverse motion due to the cancellation of the Magnus forces towards opposite directions. (Reprinted from Barker, J., and Tretiakov, O.A., *Phys Rev Lett.* 116: 147203, 2015. With permission.) (b) The longitudinal and transverse velocities of AFMSk driven by in-plane spin-polarized current. (Reprinted from Barker, J., and Tretiakov, O.A., *Phys Rev Lett.* 116: 147203, 2015. With permission.) (c) The longitudinal velocities of AFMSk and FMSk driven by perpendicular-to-plane spin-polarized current. (Reprinted from Zhang, X.C., et al., *Sci. Rep.*, 6, 24795, 2016b. With permission.)

the level of km/s, either driven by in-plane or out-of-plane current. Such an ultrahigh motion speed, which is an important benefit for the fabrication of ultrafast spintronic devices, cannot be reached by the FMSk.

On the other hand, the AFMSk has its limitations although it has great potential to be an ideal information carrier. Antiferromagnets are hard to be controlled by external magnetic fields because of the alternating directions of magnetic moments on individual atoms, which leads to zero net magnetization. The fluctuations of the AFMSk radius about the mean value and the diffusion coefficient are much larger than those of the FMSk. Moreover, it is still a big challenge to search for AFMSk experimentally at room temperature.

14.3   Magnetic Bilayer-Skyrmions with AFM Interface Coupling

An entirely different way of suppressing the SkHE was proposed and investigated, based on a bilayer metallic system with AFM interface coupling, where a pair of SkHE-free skyrmions can be created (Zhang et al. 2016c). It is shown that such a magnetic skyrmion pair, i.e., a bilayer-skyrmion, can be nucleated by vertical spin-polarized current or converted from a bilayer DW pair. Such a bilayer-skyrmion pair can travel over an arbitrary long distance driven by either the in-plane or perpendicular-to-plane spin-polarized current. The concept is highly promising for practical applications such as the ultra-dense storage devices as well as ultra-fast information processing devices.

14.3.1   Creation of AFM-Coupled Bilayer-Skyrmions

The bilayer-skyrmion can be nucleated by a vertical spin-polarized current in a bilayer system with interlayer AFM coupling, as shown in Figure 14.3a. Spins can flip in both FM layers separated by an insulating spacer when the spin-polarized current is only injected onto the top FM layer. This way, two skyrmions with antiparallel spin orientations can be nucleated in the bilayers. Such a bilayer-skyrmion as illustrated in Figure 14.3d and e, can be driven by the perpendicular-to-plane spin-polarized current (see Figure 14.3b) or by the in-plane current (see Figure 14.3c) in the AFM-coupled bilayer nanotrack.

Zhang et al. (2016c) have illustrated the role of the interlayer AFM coupling, as shown in Figure 14.4. In the absence of the AFM interlayer exchange coupling, as shown in Figure 14.4a, skyrmions can be formed only in the top FM layer. On the other hand, the other skyrmions can be formed in the bottom FM layer due to the interlayer AFM coupling

Chapter 14

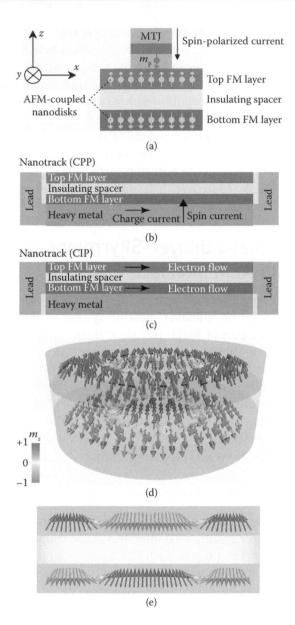

FIGURE 14.3 Schematics of the bilayer nanodisk, nanotrack, and the bilayer-skyrmion. (a) The AFM-coupled bilayer nanodisk used for the creation of the bilayer-skyrmion, its diameter being 100 nm. The spin current (polarized along −*z*) is injected into the top layer in the central circle region (diameter of 40 nm). (b) and (c) show the AFM-coupled bilayer nanotrack (500 nm × 50 nm × 3 nm) for the motion of the bilayer-skyrmion driven by the current perpendicular to plane (CPP) and in-plane current (CIP), respectively. (d) Illustration of the bilayer-skyrmion in a nanodisk, which is a pair of antiferromagnetically exchange-coupled skyrmions. (e) Side view of the bilayer-skyrmion along the diameter of (d). (Reprinted from Zhang, X., et al., *Nat. Commun.*, 7, 10293, 2016c. With permission.)

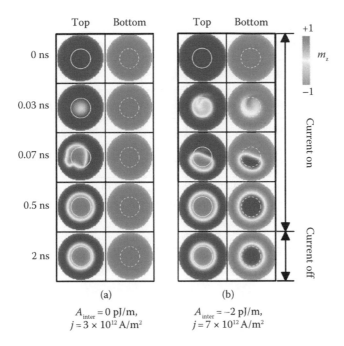

FIGURE 14.4 Creation of a bilayer-skyrmion in the bilayer nanodisk through vertical spin-polarized current injection. (a) Top-layer-skyrmion due to the absence of AFM coupling. (b) Bilayer-skyrmion due to the AFM coupling. (Reprinted from Zhang, X., et al., *Nat. Commun.*, 7, 10293, 2016c. With permission.)

which favors antiparallel orientations of the spins in the two layers, as shown in Figure 14.4b.

The other method for creating a bilayer-skyrmion is to convert a bilayer–DW pair to a bilayer-skyrmion, as illustrated in Figure 14.5. Due to the presence of the interlayer AFM coupling, a bilayer-DW pair can be created at $t = 50$ ps through the injection of vertical spin-polarized current to the top FM layer. The bilayer-DW pair is driven by perpendicular-to-plane current applied to the bottom FM layer. Finally, the bilayer-skyrmion can be created, converted from a bilayer-DW pair by passing the junction interface at $t = 190$ ps. Spins at both ends of the DW are pinned at the junction, whereas the central part of the DW is not pinned due to the spin-transfer torques in the wide part of the nanotrack. As a result, the bilayer-DW pair is deformed into a curved shape and a bilayer-skyrmion texture thus forms.

14.3.2   Motion of Bilayer-Skyrmions

Figure 14.6 shows the motion of bilayer-skyrmions driven by per-pendicular-to-plane and in-plane spin-polarized currents. It can be

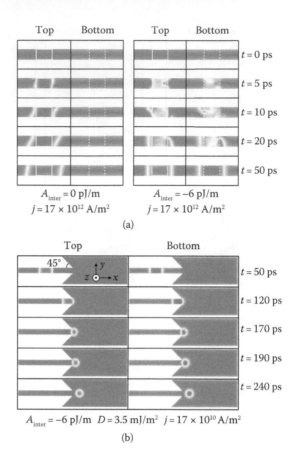

FIGURE 14.5 Creation of a bilayer-DW pair and its conversion into a bilayer-skyrmion. (a) Creation of a bilayer-DW Pair and (b) its conversion into a bilayer-skyrmion. (Reprinted from Zhang, X., et al., *Nat. Commun.*, 7, 10293, 2016c. With permission.)

seen that there is no SkHE for a magnetic bilayer-skyrmion so that the bilayer-skyrmion can move along the central line of the nano-track at a high speed of a few hundred meters per second, which is much larger than that of a monolayer-skyrmion. Zhang et al. (2016c) have demonstrated that the relation between the velocity and the driving current density for a bilayer-skyrmion is almost the same as that for a monolayer-skyrmion when it is driven by in-plane current. In addition, a different DMI constant in the top and bottom layers does not much change the velocity of the bilayer-skyrmion motion as DMI only changes the size of the bilayer-skyrmion. Similar ideas can be extended to a multilayer or a superlattice where the skyrmions are strongly coupled for a better manipulation of skyrmions (Zhang et al. 2016a).

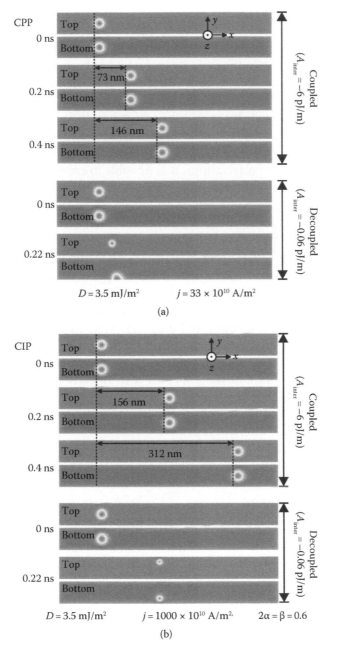

FIGURE 14.6 Top views of the motion of bilayer-skyrmions at selected interlayer exchange coupling and times driven by (a) the perpendicular-to-plane spin-polarized current and (b) the in-plane spin-polarized current. (Reprinted from Zhang, X., et al., *Nat. Commun.*, 7, 10293, 2016c. With permission.)

14.4 Multiferroic Skyrmion

Experimentally, the observation of skyrmion phases was limited to the metallic B20 alloys in the early stages and it has been extended to ultra-thin metallic films very recently (Boulle et al. 2016). A skyrmion crystal was discovered in the magnetic insulator Cu_2OSeO_3 in 2012 (Adams et al. 2012; Seki et al. 2012a, 2012b), which has no SkHE and exhibits a multiferroic nature with spin-induced ferroelectricity. In addition, due to the strong magnetoelectric coupling between noncollinear skyrmion spins and electric polarizations, it is possible to create and manipulate skyrmions by an application of electric fields instead of the injection of electric currents so that there are no Joule-heating losses.

The theories for multiferroic skyrmions include the first-principle calculations (Chizhikov and Dmitrienko 2015; Janson et al. 2014; Romhányi et al. 2014; Yang et al. 2012), the simplified classical spin model, and the spin-wave mode. Microscopic analyses based on the first-principle calculations combined with electron spin resonance (ESR) experiments (Ozerov et al. 2014) can catch significant quantum nature and predict complicated magnetic systems well. On the other hand, the simplified classical spin model can describe quite accurately the magnetic properties of Cu_2OSeO_3. Onose et al. measured the absorption and transmission spectra with linearly polarized electromagnetic waves, confirming the predicted skyrmion resonance modes (Schwarze et al. 2015) based on the spin-wave modes of the skyrmion crystal.

In the following, we review briefly the simplified classical spin model, the origin of multiferroics in Cu_2OSeO_3, as well as the nucleation and dynamic properties of multiferroic skyrmions.

14.4.1 Simplified Classical Spin Model

The continuum spin model used to describe multiferroic skyrmions is similar to the model used in the previous chapters, which was proposed by Bak and Jensen in 1980 for sufficiently slow spatial and temporal varied magnetic order:

$$H = \int dr \left[\begin{array}{l} \dfrac{J}{2a}(\nabla m)^2 + \dfrac{D}{a^2} m \cdot (\nabla \times m) - \dfrac{g\mu_B\mu_0}{a^3} H \cdot m \\[2mm] + \dfrac{A_1}{a^3}\left(m_x^4 + m_y^4 + m_z^4\right) - \dfrac{A_2}{2a}\left[(\nabla_x m_x)^2 + \left(\nabla_y m_y\right)^2 + \left(\nabla_z m_z\right)^2\right] \end{array} \right]$$

(14.3)

In this model, four types of energy compete with each other and determine the phase diagrams of the magnetic order. The first two terms are the exchange energy and the DM interaction, whereas the last two terms are the fourth-order and the second-order anisotropy energy. Similar expressions of these energies have been given in Equation 14.1. The third term is the Zeeman energy, which denotes the interaction between the applied field and the spins. The exchange interaction favors the parallel orientation of neighboring spins, whereas the DMI prefers the perpendicular orientation. The Zeeman energy and the two types of magnetic anisotropies drive the spins to the applied field or the easy axis direction. The anisotropy terms are usually very small in realistic multiferroic materials so that they can be ignored. In order to treat this continuum spin model numerically, the space occupied by the material concerned is divided into cubic meshes and the energy terms in Equation 14.3 can be rewritten as (Mochizuki and Seki 2015):

$$H = -J \sum_{i,\gamma} m_i \cdot m_{i+\gamma} - D \sum_{i,\gamma} \left(m_i \times m_{i+\gamma} \cdot \gamma \right) - g \mu_B \mu_0 H \cdot \sum_i m_i$$

$$+ A_1 \sum_i \left[\left(m_i^x \right)^4 + \left(m_i^y \right)^4 + \left(m_i^z \right)^4 \right] \tag{14.4}$$

$$- A_2 \sum_i \left(m_i^x m_{i+x}^x + m_i^y m_{i+y}^y + m_i^z m_{i+z}^z \right).$$

The calculated theoretical diagrams based on the above simplified model fit experimental diagrams, justifying the model (Mochizuki and Seki 2015).

14.4.2 Origin of Multiferroics in Cu_2OSeO_3

The noncollinear skyrmion spin textures in Cu_2OSeO_3 can induce electric polarizations via the so-called spin-dependent metal–ligand hybridization mechanism, resulting in the multiferroic nature of the material. The schematics of the spin-dependent metal–ligand hybridization mechanism (Mochizuki and Seki 2015) are shown in Figure 14.7.

As can be seen from Figure 14.7, local electric polarizations P_{ij} can be induced along the bond vector e_{ij}, connecting the ith magnetic metal ion Cu^{2+} and the jth ligand ion O^{2-}. The magnitude of the local polarization

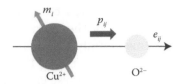

FIGURE 14.7   Schematics of the spin-dependent metal–ligand hybridization mechanism as an origin of magnetism-induced electric polarizations. (Reprinted from Mochizuki, M., and Seki, S., *J. Phys. Condens. Matter*, 27, 503001, 2015. With permission.)

depends on the relative direction of magnetization m_i against the bond, given by (Mochizuki and Seki 2015):

$$P_{ij} \propto \left(e_{ij} \cdot m_i \right)^2 e_{ij} \,. \tag{14.5}$$

For a Cu_2OSeO_3 with a skyrmion spin configuration as shown in Figure 14.8a, the local electric polarization $p(r)$ can be obtained through one-by-one correspondence based on Equation 14.3, which is illustrated in Figure 14.8b.

Further, the magnitude of the polarization p can be modulated by changing the direction of the applied magnetic field, as illustrated in Figure 14.8c through e.

14.4.3   Nucleation

The coupling between magnetism and electricity offers an opportunity to manipulate magnetic skyrmions electrically by modulating the spatial distributions of their electric polarizations. Mochizuki and Watanabe (2015) simulated the dynamic process of the creation of skyrmions by a local application of the electric field.

A skyrmion on a thin specimen of multiferroic chiral magnet can be created very rapidly by applying a local electric field via an electrode tip where the number of skyrmions created can be tuned by the duration of an E-field application, as shown in Figure 14.9. The application of the E-field to a small spot is shown in Figure 14.9b, with a uniform ferromagnetic state leading to the local reversal of the magnetization at the point within the E-field area (see Figure 14.9c). This reversed spot grows into a line-shaped structure (see Figure 14.9d) with a finite skyrmion number $Q = -1$ after the local application of the E-field for 0.22 ns. The local shutdown of the E-field at this time results in the formation of a single skyrmion after relaxation, as shown in the process of Figure 14.9g and h. If the duration of the applied field increases to 2.2 ns, the line-shaped

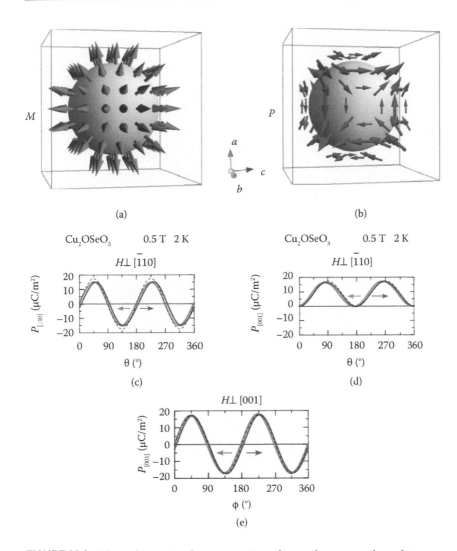

FIGURE 14.8 Three-dimensional representation of general correspondence between (a) **M**- and (b) **P**-directions in the collinear spin state where arrows at the same position in (a) and (b) represent the **M**-vector and the corresponding induced **P**-vector, respectively. (c) [110] and (d) [001] components of ferroelectric polarization **P** simultaneously measured under a magnetic field **H** rotating around the [$\bar{1}$10] axis. (e) [001] component of **P** under **H** rotating around the [001] axis. Both measurements are performed for the collinear ferrimagnetic state at 2 K with H = 0.5 T. Dashed lines indicate the theoretically expected behaviors from Equation 14.3, and arrows denote the direction of **H**-rotation. (Reprinted from Mochizuki, M., and Seki, S., *J. Phys. Condens. Matter*, 27, 503001, 2015. With permission.)

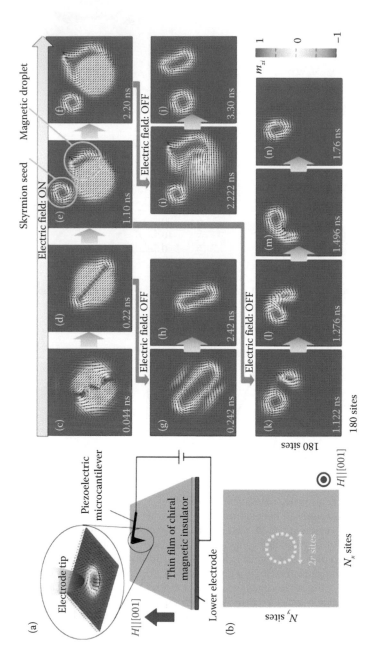

FIGURE 14.9 (a) Schematic of the skyrmion creation via local application of an electric field using an electrode tip. (b) An electric field $E = (0, 0, E_z)$ is applied to sites within the area indicated by the dashed circle with a diameter of $2r$. (c–n) Simulated spatiotemporal dynamics of the magnetizations m_i for the electrical skyrmion creation process. (Reprinted from Mochizuki, M., and Watanabe, Y., *Appl. Phys. Lett.*, 107, 082409, 2015. With permission.)

structure will break into two parts, as shown in Figure 14.9e through h, which will evolve into two skyrmions after shutdown of the E-field, as shown in Figure 14.9i and j. If the duration of the applied field is 1.1 ns, which is in between the two cases discussed above, a complicate process of separation and combination will occur and only a single skyrmion is obtained, as shown in Figure 14.9k through n.

14.4.4   Dynamical Magnetoelectric Phenomena of Skyrmions in the Microwave-Frequency Regime

The coupled dynamics of local magnetizations m_i (left panels) and local polarizations p_i (right panels) have been studied, where both the counterclockwise rotational mode and the breathing mode are examined and shown in Figure 14.10. The simultaneous electric and magnetic activities of the resonant modes illustrated in Figure 14.10 are quite different for the counterclockwise rotational mode and the breathing mode, due to different angles between the dc and ac fields applied (Mochizuki and Seki 2015). This will cause some special dynamical magnetoelectric phenomena of skyrmions in the microwave-frequency regime, i.e., the microwave absorption density changes a lot for different incident directions (Mochizuki and Seki 2015).

In summary, we have provided an overview on isolated magnetic skyrmions that can be created electrically in multiferroic chiral magnets through local electric-field application via an electrode tip on thin-film samples. The simplified spin models for skyrmions and skyrmion crystals have been introduced, respectively. In addition, the mechanisms of spin-dependent metal–ligand hybridization and some dynamical magnetoelectric phenomena of skyrmions in the microwave-frequency regime have been studied.

14.5   Quantum Hall Skyrmions

Skyrmion topological spin textures can also form in quantum Hall (QH) systems due to the competition between the Column exchange energy and the Zeeman coupling (Girvin 2000).

The exchange energy is large which strongly favors locally parallel spins and hence the ferromagnetic order. On the other hand, the Zeeman splitting is small in QH systems. Therefore, it is energetically cheap for the spins to partially turn over and form a skyrmion spin structure. When the filling factor $v = 1$, which is primarily discussed in the QH systems, a significant energy of about 30 K is needed to excite a skyrmion or antiskyrmion, which will vanish at lower temperature (Girvin 2000).

Chapter 14

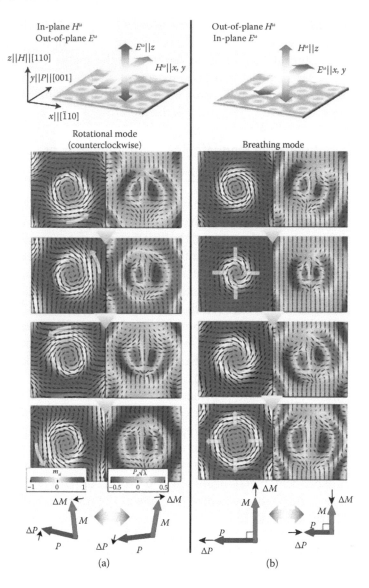

FIGURE 14.10 Spatiotemporal dynamics of magnetizations m_i (left panels) and polarizations p_i (right panels) for electromagnon excitations in a skyrmion crystal under the static magnetic field $\mathbf{H}\|[110]$ ($\|z$). (a) Counterclockwise rotational mode activated by in-plane ac magnetic field $H^\omega\perp[110]$ or by out-of-plane ac electric field $E^\omega\|[110]$. (b) Breathing mode activated by out-of-plane ac magnetic field $H^\omega\|[110]$ or by in-plane ac electric field $E^\omega\perp[110]$. Dynamical behaviors of net magnetization M and ferroelectric polarization P are shown for each mode in the lowest panel. The angle between M and P becomes larger and smaller in an oscillatory manner in the rotational mode, while in the breathing mode, elongation and shrinkage of M and P in length occur synchronously, keeping the angle at $90°$. (Reprinted from Mochizuki, M., and Seki, S., *J. Phys. Condens. Matter*, 27, 503001, 2015. With permission.)

Such a skyrmion excitation will carry charge and can be manipulated by electrostatic potentials (Girvin 2000), unlike the skyrmions in magnetic systems. Skyrmions can also be excited by a small amount of defects in principle and form skyrmion crystals in a QH system. Recent experiments have evidenced the existence of the skyrmion crystal in this system, although this spin configuration is not a ground state and requires very harsh conditions to be created. To understand the skyrmions in a QH system well, a brief introduction of the QH effect and QH ferromagnet is needed.

14.5.1   Quantum Hall Effect

In 1980, K.V. Klitzing, G. Dorda, and M. Pepper found that the two-dimensional electron gas formed in an artificial semiconductor quantum well demonstrated the quantization of the platform subjected to low temperatures and strong magnetic fields (about 10 T). This ideal behavior, called the QH effect or integer QH effect, which can only occur in imperfect samples, is now used by standards laboratories to maintain ohm (Girvin 2000).

In essence, the QH effect is a quantum-mechanical version of the Hall effect, observed in two-dimensional electron systems. The Hall conductance $\sigma = I_{channel}/V_{Hall}$ turns out to be universal, which takes the following quantized values: $\sigma = ve^2/h$, where $I_{channel}$ is the channel current, V_{Hall} is the Hall voltage, e is the elementary charge, and h is Planck's constant. The prefactor v, known as the "filling factor," can take on either integer ($v = 1, 2, 3, ...$) or fractional ($v = 1/3, 2/5, 3/7, ...$) values with an odd denominator, corresponding to the integer or fractional QH effects, respectively.

Four kinds of QH effect have been discovered so far, where the integer QH effect is first observed and very well understood. The integer QH effect can be simply attributed to an excitation gap associated with the filling of discrete kinetic energy levels (Landau levels), which is based on the single-particle orbital for an electron in the imposed magnetic field. The fractional QH effect occurs when one of the Landau levels is fractionally filled, which is more complicated as its existence relies fundamentally on electron–electron interactions. Discoverers of integer and fractional QH effects won the Nobel Prize in Physics in 1985 and 1998, respectively. The quantum anomalous Hall (QAH) was proposed in 1988, which refers to the QH effect without Landau levels. On the other hand, quantum spin Hall effect is an analogue of the QH effect, where spin currents flow instead of charge currents (Girvin 2000). In the following, the QH effect refers to the integer QH effect.

Chapter 14

14.5.2   Quantum Hall Ferromagnet and Quantum Hall Skyrmions

QH systems can exhibit spontaneous magnetic order as the lower spin state is more energy favorable due to Zeeman splitting. For the ground state corresponding to $v = 1$, only the lowest spin state of the lowest orbital Landau level is completely filled with spin up electrons (Girvin 2000).

The strong Column interaction would exact a large exchange-energy penalty if a spin in a QH ferromagnet were reversed so that the ferromagnet is 100% polarized at zero temperature. However, forming a topological spin structure, i.e., a skyrmion, by smoothly distorting the ferromagnetic order in a vortex-like spin texture (Figure 14.11), is energetically cheaper (Girvin 2000) at a high temperature (>30 K). Such a skyrmion configuration in QH systems holds exactly one extra unit of charge, which is directly proportional to the "topological charge" and is given by: $Q = h\, Q_t \sigma_{xy}/e$, where σ_{xy} is the Hall conductivity and Q_t is the "topological charge" of the magnetic order parameter for a 2D system given in the previous chapters. For larger filling factors, skyrmions can exist even at a zero temperature, where the ferromagnetic order is not collinear anymore and the formation of a finite density of skyrmions is the cheapest way to add or subtract a charge (Girvin 2000). Excitation of skyrmions in a QH system has been confirmed by NMR and some optical and transport measurements, where most experimental evidence is associated with $v = 1$, as given below.

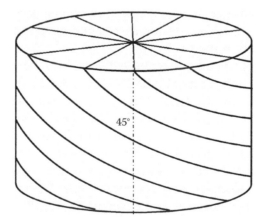

FIGURE 14.11   Perspective view on the double-twist cylinder. The lines on the outside are supposed to indicate a 45° rotation of the director at that distance from the center line. For a better understanding of the structure, only five helical axes are demonstrated; however, theoretically, there are an infinite number. © CC BY-SA 3.0. (Available at https://en.wikipedia.org/wiki/File:DoubleTwist-Cylinder-SV.png.)

14.5.3   Evidence for Quantum Hall Skyrmions

Skyrmions in a QH system were first predicted in 1993 by Karlhede et al. (1993), which aroused lots of interest in the past 20 years. Skyrmions and associated spin flips were observed 2 years later in GaAs quantum wells based on NMR shifts (Dabbagh et al. 1995; Tycko et al. 1995), and the tilted-field magnetotransport measurements (Eisenstein et al. 1995). These experiments suggest that skyrmions are the lowest-lying charged excitations of the fully polarized QH fluid for $v = 1$. In 1996, Aifer et al. observed skyrmion excitations about $v = 1$ in n-modulation-doped GaAs single quantum wells based on interband optical transmission (Aifer and Goldberg 1996).

Unlike skyrmions in magnetic materials, skyrmions in QH systems are not stable, as evidenced by direct observation of spin-wave excitations well below the Zeeman energy by inelastic light scattering (Yann et al. 2008).

On the other hand, skyrmion size increases in a QH system as the Zeeman energy decreases, which can be realized by application of hydrostatic pressure or confining the 2D electron system to an AlGaAs quantum well. Large skyrmions were found in multiple (30) AlGaAs quantum wells from an NMR study, which might be more stable due to its very small electronic polarization (Maude and Portal 1996; Horvatić et al. 2007).

In the meantime, skyrmion crystals in a QH system have also aroused much attention, which was predicted theoretically by Brey et al. (1995). Bayot et al. (1996) and Grivei et al. (1997) found that specific heat is greatly enhanced by the presence of skyrmions in GaAs/AlGaAs heterostructures, which shorten the nuclear spin-lattice relaxation time and bring the nuclei into thermal equilibrium.

Evidence for skyrmion crystallization was found in GaAs/AlGaAs quantum well from NMR relaxation experiments (Gervais et al. 2005; Yann et al. 2008) by Gervais et al., while microwave pinning mode resonance of a skyrmion crystal was later illustrated by Zhu et al., based on the broadband microwave conductivity spectroscopy (Sambandamurthy 2010).

In summary, existence of skyrmions and skyrmion crystals in QH systems has been confirmed by many experiments, where an NMR is a powerful tool providing valuable information on skyrmion spin structures. Skyrmion excitations near $v = 1$ have a rich microscopic nature and dynamics for the many-body electron state, which is still a hot topic in condensed matter physics. Moving away from $v = 1$, novel phenomena such as a "canted antiferromagnetic" state might occur (Girvin 2000), producing even richer physics to be explored.

Chapter 14

14.6 Skyrmions in Liquid Crystals

Topological electric dipole structures, considered as skyrmions, have been observed in liquid crystals, more precisely, in blue phases (Wright and Mermin 1989; Nagaosa and Tokura 2013). Liquid crystals can occur in thousands of substances that exhibit one (or more) intermediate-state displaying properties between those of conventional liquid (the molecules are free to tumble) and those of solid crystal (the molecules are rigidly locked into place). In these intermediate phases, the molecules have an ordered arrangement and yet can still flow like a liquid; therefore, these substances are called "liquid crystals." Many proteins and cell membranes are liquid crystals, and most contemporary electronic displays use liquid crystals (Gray 1962; Chandrasekhar 1992; Sluckin et al. 2004).

The different types of liquid crystals can be characterized by the type of ordering of the mesophase, i.e., the liquid-crystal phase. Among the three main types of liquid crystals, i.e., thermotropic, lyotropic, and metallotropic phases, the thermotropic phase (in which skyrmions have been predicted) is a type of ordering that can be divided into four main subtypes: the nematic, smectic, chiral, and blue phases. Skyrmion structure can only occur in blue phases in the temperature range between those of a typical chiral nematic phase and an isotropic liquid phase. Blue phases have a regular three-dimensional cubic structure and are naturally occurring in crystals whose periodicity is on a scale of hundreds of nanometers.

Although the double-twisted cylinder structures of blue phases (Figure 14.11) were known for decades (Wright and Mermin 1989), it is only recently that a link has been made with skyrmions (Nagaosa and Tokura 2013). In order to understand which kind of skyrmion the cylinder structure is, a clarification is needed: two main kinds of skyrmion exist, the hedgehog and spiral skyrmions, which can be seen in Figure 14.12. In the case of the "hedgehog skyrmion" (Figure 14.12a), the progression of magnetization across the diameter is cycloidal, whereas in the case of the "spiral skyrmion" (Figure 14.12b), also known as "vortex skyrmion," the progression of magnetization is helical.

The structures observed in blue phases are topological electric dipole textures whose dipole distributions are helical, which make them spiral skyrmions. It should be noted that in some papers the words helical and spiral are interchangeable, so that helical skyrmions and spiral skyrmions actually refer to the same concept.

Skyrmions are not only divided between hedgehog and spiral skyrmions, they are also divided between full and incomplete skyrmions.

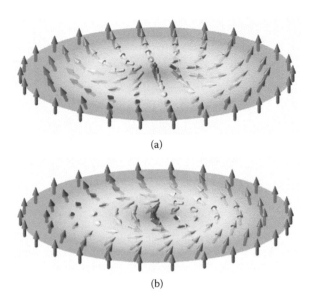

(a)

(b)

FIGURE 14.12 The vector field of two 2D magnetic skyrmions: (a) a hedgehog skyrmion and (b) a spiral skyrmion. © CC BY-SA 3.0. (Available at https://commons. wikimedia.org/wiki/File:2skyrmions.PNG.)

A full skyrmion occurs when the angle between the spin at the center and those at the edge is 180°, as shown in Figure 14.12. In the case of the blue phase, the director has no specific sense, it is only an axis, i.e., a headless arrow. Indeed, if the director had a sense associated with it, the line singularities would have a very different character and there would be no blue phases. The angle between the central director and the outermost directors being 45°, we only have a 1/4 skyrmion.

Figure 14.13 represents the internal structure of blue phases, where many cylinder structures arranged cubically can be observed. Each cylinder is a double-twisted cylinder structure, also known as a spiral skyrmion. As the temperature is cooled, the structure will lose its complexity to become a single helical structure, signaling that the chiral phase is in place. The double-twisted structure is only a 1/4 skyrmion.

Double-twisted cylinders are a majority over single helical axis structures in blue phases because their energy is lower than that of the uniaxial helical structures. After derivation, the final equation of the gradient free energy difference between them is as follows:

$$\Phi^{dt}(r) - \Phi^{uh} = \frac{3}{4}\lambda_0^2 \left[\left(\frac{\sin\left(\frac{1}{2}r\right)}{\frac{1}{2}r} \right) - \frac{2\sin(r)}{r} \right], \qquad (14.6)$$

Chapter 14

FIGURE 14.13 Cubic structures of double-twisted cylinders in liquid crystal blue phase I (left) and II (right). © CC BY-SA 3.0. (Available at https://en.wikipedia.org/wiki/File:Structures_of_double_twisted_cylinders_in_liquid_crystal_blue_phase_I_(left)_and_II_(right)..jpg.)

where ϕ represents the gradient free energy, the superscripts dt and uh stand for double-twisted cylinder and uniaxial helical structures, respectively, λ_0 represents a temperature-dependent tensor, and r is the distance from the central director. The result is negative, i.e., the double-twisted structure is the energy minimum, until $\tan\left(\dfrac{r}{2}\right) = r$.

Although blue phases are of interest for fast light modulators or tunable photonic crystals, they were long believed to exist only in a very narrow temperature range, usually less than a few kelvins. However, recently the stabilization of blue phases over a temperature range of more than 60 K including room temperature (260–326 K) has been demonstrated (Coles and Pivnenko 2005).

Thanks to their spiral skyrmion-based structure, blue phases stabilized at room temperature allow electro-optical switching with response times of the order of 10^{-4} s which is coherent with the very low latencies of skyrmions. Such a structure allows display devices to run at a few thousands hertz (up to ten thousand hertz), one order of magnitude higher than the fastest devices available on the market (Kikuchi et al. 2002, 2007). In May 2008, Samsung Electronics announced that it has developed the world's first Blue Phase LCD panel (Samsung has developed blue phase technology 2008), which can be operated at an unprecedented refresh rate of 240 Hz. It is worth noting that the driving voltage of blue phase LCs in IPS structures is still a little bit too high. To reduce the voltage, material engineering for developing high Kerr constant mixtures is critically important. Moreover, device design is also an effective way, with proper device structure design, that the driving voltage can be largely reduced.

If the control of the blue phase, and the skyrmions within, is enhanced, skyrmion-based display devices could revolutionize the high technology industry the same as skyrmion-based storage devices would.

14.7   Skyrmions in Cold-Atom Systems

Skyrmions have also been predicted in cold-atom systems. In a gas, each atom has its own energy and is able to move freely. When these atoms are bosons, they are allowed to have the same energy at the same time. Besides, when the gas is cooled down, the atoms slow down and their energies decrease. Because of their quantum nature, the atoms behave as waves that increase in size as temperature decreases, as shown in Figure 14.14a. At very low temperature, the size of the waves becomes larger than the average distance between two atoms, as shown in Figure 14.14b. Then, when the temperature is very close to absolute zero (namely, less than 10^{-6} K), all of the bosons are able to be at the very same energy in the same quantum state: they all form a single collective quantum wave called a BEC, as shown in Figures 14.14c and 14.15 (Bose 1924; Einstein 1925).

As in many other systems, vortices can exist in BECs (Nagaosa and Tokura 2013; Senthil et al. 2004). These can be created, for example, by stirring the condensate with lasers, or rotating the confining trap. The vortex created will be a quantum vortex. The link with skyrmions is that coreless vortices in BCEs have been referred to as skyrmions (Ho 1998). Indeed, the structures observed in BECs are topological spin textures whose spin distributions are vortices, which make them spiral skyrmions. Their characteristic is that they can only appear in ultracold atoms, under a temperature close to the absolute zero (10^{-6} K). Their size is about 20 μm, and their level of stability is the same as that in superfluids. Their energy consumption is high as we need to reach a temperature close to absolute zero to make them appear.

In his paper, Ho (1998) shows that in an optical trap, the ground states of spin-1 bosons such as ^{23}Na, ^{39}K, and ^{87}Rb can be either ferromagnetic or "polar" states. While ordinary vortices are stable in the polar state, only those with unit circulation are stable in the ferromagnetic state. He also demonstrates that spin variations in the ferromagnetic states in general lead to superflows. This phenomenon is identical to that of superfluid ^{3}He-A, which has exactly the same spin texture, superfluid velocity, and topological instability (Vollhardt and Wölfle 1990). In the case of ^{3}He-A, it is known that external rotations can distort the texture

Chapter 14

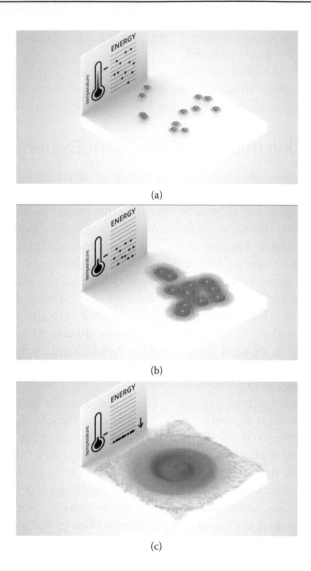

(a)

(b)

(c)

FIGURE 14.14   Schematic of a Bose–Einstein condensate with (a) showing the influence of low temperatures, (b) the influence of very low temperatures, and (c) the influence of temperatures near absolute zero. © CC BY-SA 3.0. (Available at http:// toutestquantique.fr/en/bose-einstein-condensate/.)

so as to generate a velocity field to mimic the external rotation as closely as possible. Such textural distortions will occur here for the same energetic reasons. The aforesaid textural spin structures are skyrmions.

To understand better how they occur, we must first determine the different contributions within the calculation of BECs energy. The Gross–Pitaevskii equation (named after Eugene P. Gross [1961] and

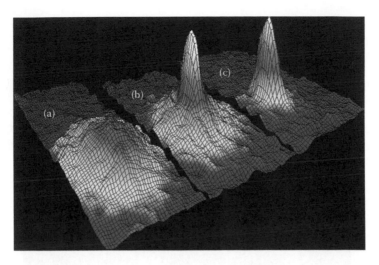

FIGURE 14.15 Velocity-distribution data (three views) for a gas of rubidium atoms, confirming the discovery of a new phase of matter, the Bose–Einstein condensate. (a) Just before the appearance of a Bose–Einstein condensate. (b) Just after the appearance of the condensate. (c) After further evaporation, leaving a sample of nearly pure condensate (public domain).

Lev Petrovich Pitaevskii [1961]), an approximation limited to the case of ultracold temperatures (fitting most experiments), is the most common energy equation of BECs. The equation is as follows:

$$E = \int d\vec{r} \left[\frac{\hbar}{2m} |\nabla\Psi(\vec{r})|^2 + V(\vec{r})|\Psi(\vec{r})|^2 + \frac{1}{2}U_0|\Psi(\vec{r})|^4 \right], \tag{14.7}$$

with the first term being the mass of the bosons, the second term the external potential, and the third term the interparticle interactions.

Quantum vortices within BECs rely on the following equation:

$$E_{qv} = \pi n \frac{\hbar}{m} \ln\left(1.464\left(\frac{b}{\xi} \right) \right), \tag{14.8}$$

where m is the mass of the bosons, n is the boson density, b is the farthest distance from the vortex considered, and ξ is the healing length of the condensate. We can see from both equations that the mass and interparticle interactions (which are related to the boson density and the temperature) are important for both BECs and quantum vortices to occur.

Xu and Han (2011) investigated the combined effects of Rashba SO coupling and rotation on trapped spinor BECs by analytical and numerical means. It was found that at strong rotation, increasing SO coupling strength can favor a triangular vortex lattice and weaken the skyrmion crystal order. The equation is as follows:

$$\rho_s(\mathbf{r}) = \frac{1}{4\pi} \frac{\mathbf{n} \cdot (\partial_x \mathbf{n} \times \partial_y \mathbf{n})}{|\mathbf{n}|^3}. \tag{14.9}$$

As shown in Figure 14.16, as the relative SO coupling constant γ increases from 0 to 0.3, the total skyrmion number decreases from

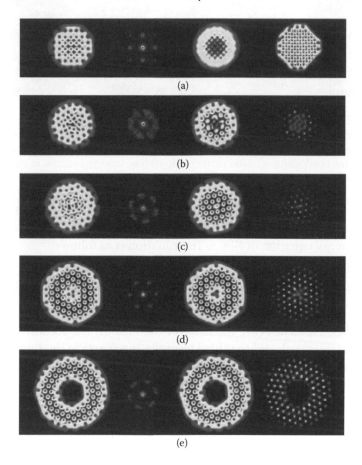

(a)

(b)

(c)

(d)

(e)

FIGURE 14.16   Ground state evolution with the relative spin–orbit coupling constant γ for large rotation $b = 0.998$. From left to right: density of up-component $|\psi_\uparrow|^2$, magnitude of its Fourier transformation $\left|\int \exp[-i\mathbf{k} \cdot \mathbf{r}] |\psi_\uparrow|^2\right|$, total density of two components $(|\psi_\uparrow|^2 + |\psi_\downarrow|^2)$, and the skyrmion density ρ_s. (a–e) $\gamma = 0.0, 0.002, 0.1, 0.2$, and 0.3, while the total skyrmion number, defined as the spatial integral of Equation 14.9, is 40.9143, 19.2982, 12.8454, 8.1136, and 3.7456, respectively. (Reprinted from Xu, X.Q., and Han, J.H., *Phys. Rev. Lett.*, 107, 200401, 2011. With permission.)

41 to 3.7, signifying the significant deterioration of the skyrmions due to strong SO coupling.

The primary application of atomic BEC systems is in basic research areas at the moment and will probably remain so for the foreseeable future. If we reduce the potential applications to those linked with skyrmions, quantum vortices within BECs are currently the subject of analogue gravity research, studying the possibility of modeling black holes and their related phenomena in such environments in the laboratory. No commercial applications are currently being developed, and even theoretically, no commercial applications using skyrmions in BECs were thought at all. All their potential applications are in the fundamental research areas, and it seems that it will remain so at least for the following decade.

14.8 Summary

In this chapter, skyrmions in various condensed matter systems have been reviewed, including the AFM materials, ferromagnetic bilayers with AFM interface coupling, multiferroic insulators, quantum Hall systems, liquid crystals, and BEC systems. Nucleation and motion of AFM skyrmions and skyrmions in metallic bilayers with AFM coupling have been examined, which can inhibit totally the SkHE in a metallic skyrmion racetrack. The basic theory, advantage, and weakness of these novel solutions for SkHE are summarized. Skyrmion crystals in the magnetic insulator Cu_2OSeO_3 have been demonstrated, focusing on the mechanism, creation, and dynamic properties of such multiferroic skyrmions. Such multiferroic skyrmions can be manipulated by electric potentials, where SkHE has not any influence. Skyrmions in quantum Hall systems carry charge, which can also be manipulated by electric fields. Existence of skyrmions and skyrmion crystals have been confirmed by NMR and other experimental methods, which are unstable and mostly found in GaAs and related quantum wells. Vortex-like skyrmions have been observed in liquid crystals and predicted in cold-atom systems. For the former, double-twisted cylinders are energetically more favorable than the uniaxial helical structures in blue phases, where a 1/4 spiral skyrmion is formed. For the latter, a full spiral skyrmion can be formed, which can be deteriorated due to strong SO coupling. While the so-called skyrmion in liquid crystal is far away from a full skyrmion, there is still a long way to go before the skyrmions in quantum Hall and cold-atom systems will have real applications.

Chapter 14

Acknowledgment

This work was supported by the National Natural Science Foundation of China (Grant No. 11074179 and No. 10747007) of China and the Scientific Research Fund of Sichuan Provincial Education Department (Project No. 16ZA0372).

References

Adams T, et al. (2012). Long-wavelength helimagnetic order and skyrmion lattice phase in Cu_2OSeO_3. *Phys Rev Lett* 108: 237204.

Aifer E H and Goldberg B B (1996). Evidence of skyrmion excitations about $v = 1$ in n-modulation-doped single quantum wells by interband optical transmission. *Phys Rev Lett* 76: 680–683.

Barker J and Tretiakov O A (2015). Static and dynamical properties of antiferromagnetic skyrmions in the presence of applied current and temperature. *Phys Rev Lett* 116: 147203.

Bayot V, Grivei E, Melinte S, Santos M B and Shayegan M (1996). Giant low temperature heat capacity of GaAs quantum wells near landau level filling. *Phys Rev Lett* 76: 4584.

Belmeguenai M, et al. (2015). Interfacial Dzyaloshinskii-Moriya interaction in perpendicularly magnetized Pt/Co/AlOx ultrathin films measured by Brillouin light spectroscopy. *Phys Rev B* 91: 180405.

Biswas A, et al. (2013). The universal behavior of inverse magnetocaloric effect in antiferromagnetic materials. *J Appl Phys* 113: 17A902.

Bogdanov A and Hubert A (1994). Thermodynamically stable magnetic vortex states in magnetic crystals. *J Magn Magn Mater* 138: 255–269.

Bogdanov A N and Rossler U K (2001). Chiral symmetry breaking in magnetic thin films and multilayers. *Phys Rev Lett* 87: 037203.

Bose S N (1924). Plancks Gesetz und Lichtquantenhypothese. *Zeitschrift Für Physik A Hadrons & Nuclei* 26: 178–181.

Boulle O, et al. (2016). Room-temperature chiral magnetic skyrmion in ultrathin magnetic nanostructures. *Nat. Nanotech* 11: 449–454.

Brey L, Fertig H A, Côté R and MacDonald A H (2015). Skyrme Crystal In A Two-Dimensional Electron Gas. *Phys Rev Lett* 75: 2562–2565.

Chandrasekhar S (1992). *Liquid Crystals* (2nd ed.). Cambridge: Cambridge University Press.

Chen G, et al. (2013). Novel chiral magnetic domain wall structure in Fe/Ni/Cu (001) films. *Phys Rev Lett.* 110: 177204.

Chizhikov V A and Dmitrienko V E (2015). Microscopic description of twisted magnet Cu_2OSeO_3. *J Magn Magn Mater* 382: 142.

Coles H J and Pivnenko M N (2005). Liquid crystal 'blue phases' with a wide temperature range. *Nature.* 436: 997.

Dabbagh G, Pfeiffer L N, West K W, Barrett S E, and Tycko R (1995). Optically pumped NMR evidence for finite-size skyrmions in GaAs quantum wells near Landau level filling. *Phys Rev Lett.* 74: 5112.

Dai Y Y, et al. (2013). Skyrmion ground state and gyration of skyrmions in magnetic nanodisks without the Dzyaloshinsky-Moriya interaction. *Phys Rev B.* 88: 054403.

Di K, et al. (2015). Asymmetric spin-wave dispersion due to Dzyaloshinskii-Moriya interaction in an ultrathin Pt/CoFeB film. *Appl Phys Lett.* 106: 052403.

Du H F. et al. (2014). Highly skyrmion state in helimagnetic MnSi nanowires. *Nano Lett.* 14: 2026–2032.

Du H F, et al. (2015). Electrical probing of field-driven cascading quantized transitions of skyrmion cluster states in MnSi nanowires. *Nat Commun.* 6: 8637.

Dupe B, Hoffmann M, Paillard C, and Heinze S (2014). Tailoring magnetic skyrmions in ultra-thin transition metal films. *Nat. Commun.* 5: 4030.

Einstein A (1925). Quantentheorie des einatomigen idealen Gases. *Sitzungsberichte der Preussischen Akademie der Wissenschaften* 1: 3.

Eisenstein L N P J P, Schmeller A, and West K W (1995). Evidence for skyrmions and single spin flips in the integer quantized hall effect. *Phys Rev Lett.* 75: 4290.

Emori S, et al. (2014). Spin hall torque magnetometry of Dzyaloshinskii domain walls. *Phys Rev B.* 90: 184427.

Fert A, Cros V, and Sampaio J (2013). Skyrmions on the track. *Nat Nanotech.* 8: 152–156.

Fook H T, Gan W L, Purnama I, and Lew W S (2015). Mitigation of magnus force in current-induced skyrmion dynamics. *IEEE Trans on Magn.* 51: 1–4.

Freimuth F, Blügel S, and Mokrousov Y (2014). Berry phase theory of Dzyaloshinskii-Moriya interaction and spin–orbit torques. *J Phys Condens Matter.* 26: 104202.

Garello K, et al. (2013). Symmetry and magnitude of spin-orbit torques in ferromagnetic heterostructures. *Nat Nanotech.* 8: 587–593.

Gervais et al. (2005). Evidence for Skyrmion Crystallization from NMR Relaxation Experiments. *Phys Rev Lett.* 94: 196803.

Grivei S M E, Beuken J M, Bayot V, and Shayegan M (1997). Critical behavior of nuclear-spin diffusion in GaAsyAlGaAs heterostructures near landau level filling. *Phys Rev Lett.* 79: 1718.

Girvin S M (2000). Spin and isospin: Exotic order in quantum hall ferromagnets. *Phys Today.* 53: 39–45.

Gomonay E V and Loktev V M (2014). Spintronics of antiferromagnetic systems. *Low Temp Phys.* 40: 17–35.

Gray G W (1962). *Molecular Structure and the Properties of Liquid Crystals.* Academic Press.

Gross E P (1961). Structure of a quantized vortex in boson systems. *Il Nuovo Cimento.* 20: 454–477.

Ho T L (1998). Spinor Bose condensates in optical traps. *Phys Rev Lett.* 81: 742.

Horvatić S A L M, Berthier C, Mitrović V F, and Shayegan M (2007). NMR study of large skyrmions in Al0.13Ga0.87As quantum wells. *Phys Rev B.* 76: 115335.

Iwasaki J, Mochizuki M, and Nagaosa N (2013). Current-induced skyrmion dynamics in constricted geometries. *Nat Nanotech.* 8: 742–747.

Janson O, et al. (2014). The quantum nature of skyrmions and half-skyrmions in Cu_2OSeO_3. *Nat Commun* 5: 5376.

Jiang W, et al. (2016). Direct observation of the skyrmion Hall effect. *arXiv* 1603.07393: https://arxiv.org/abs/1603.07393.

Jungwirth T, Marti X, Wadley P, and Wunderlich J (2016). Antiferromagnetic spintronics. *Nat Nanotech.* 11: 231–241.

Karlhede S A K A, Sondhi S L, and Rezayi E H (1993). Skyrmions and the crossover from the integer to fractional quantum Hall effect at small Zeeman energies. *Phys Rev B.* 47: 16419.

Kikuchi H, Higuchi H, Haseba Y, and Iwata T (2007). Fast electro-optical switching in polymer-stabilized liquid crystalline blue phases for display application. *SID07 Dig.* 38: 1737.

Chapter 14

Kikuchi H, Yokota M, Hisakado Y, Yang H, and Kajiyama T (2002). Polymer-stabilized liquid crystal blue phases. *Nat Mater.* 1: 64.

Kim J-S, et al. (2014). *Voltage controlled propagating spin waves on a perpendicularly magnetized nanowire. arXiv* 1401.6910: http://arxiv.org/abs/1401.6910.

Maude M P D K and Portal J C (1996). Spin excitations of a two-dimensional electron gas in the limit of vanishing Landég factor. *Phys Rev Lett.* 77: 4604.

Miao B F, et al. (2014). Experimental realization of two-dimensional artificial skyrmion crystals at room temperature. *Phys Rev B.* 90: 174411.

Miron I M, et al. (2011a). Perpendicular switching of a single ferromagnetic layer induced by in-plane current injection. *Nature.* 476: 189–193.

Miron I M, et al. (2011b). Fast current-induced domain-wall motion controlled by the Rashba effect. *Nat Mater.* 10: 419–423.

Mochizuki M and Seki S (2015). Dynamical magnetoelectric phenomena of multiferroic skyrmion. *J Phys Condens Matter.* 27: 503001.

Mochizuki M and Watanabe Y (2015). Writing a skyrmion on multiferroic materials. *Appl Phys Lett.* 107: 082409.

Monso S, et al. (2002). Crossover from in-plane to perpendicular anisotropy in Pt/CoFe/AlOx sandwiches as a function of Al oxidation: A very accurate control of the oxidation of tunnel barriers. *Appl Phys Lett.* 80: 4157–4159.

Mühlbauer S, et al. (2009). Skyrmion lattice in a chiral magnet. *Science* 323: 915–919.

Nagaosa N and Tokura Y (2013). Topological properties and dynamics of magnetic skyrmions. *Nat Nanotech.* 8: 899–911.

Nascimento D and Bernhard B H (2013). The magnetocaloric effect in itinerant antiferromagnetic materials. *Solid State Commun.* 167: 40–45.

Nembach H T, Shaw J M, Weiler M, Jué E, Silva and T J (2015). Linear relation between Heisenberg exchange and interfacial Dzyaloshinskii-Moriya interaction in metal films. *Nat Phys.* 11: 825–829.

Ozerov M, et al. (2014). Establishing the fundamental magnetic interactions in the chiral skyrmionic mott insulat or Cu_2OSeO_3 by terahertz electron spin resonance. *Phys Rev Lett.* 113: 157205.

Pitaevskii L P (1961). Vortex lines in an imperfect Bose gas. *Sov Phys JETP.* 13: 451.

Pizzini S, et al. (2014). Chirality-induced asymmetric magnetic nucleation in Pt/Co/AlOx ultrathin microstructures. *Phys Rev Lett* and 113: 047203.

Purnama I, Gan W L, Wong D W and Lew W S (2015). Guided current-induced skyrmion motion in 1D potential well. *Sci Rep.* 5: 10620.

Rohart S and Thiaville A (2013). Skyrmion confinement in ultrathin film nanostructures in the presence of Dzyaloshinskii-Moriya interaction. *Phys Rev B.* 88: 184422.

Romhányi J, van den Brink J, and Rousochatzakis I (2014). Entangled tetrahedron ground state and excitations of the magnetoelectric skyrmion material Cu_2OSeO_3. *Phys Rev B.* 90: 140404.

Romming N, et al. (2013). Writing and deleting single magnetic skyrmions. *Science.* 341: 636–639.

Rossler U K, Bogdanov A N, and Pfleiderer C (2006). Spontaneous skyrmion ground states in magnetic metals. *Nature Lett.* 442: 797–801.

Sambandamurthy L N P G (2010). Pinning-mode resonance of a skyrme crystal near Landau-level filling factor. *Phys Rev Lett.* 104: 226801.

Samsung has developed blue phase technology, 2008, Available at http://phys.org/news/2008-05-samsung-worlds-blue-phase-technology.html.

Schwarze T, et al. (2015). Universal helimagnon and skyrmion excitations in metallic, semiconducting and insulating chiral magnets. *Nat Mater.* 14: 478

Seki S, Yu X Z, Ishiwata S, and Tokura Y (2012a). Observation of skyrmion in a multiferroic material. *Science.* 336: 198–201.

Seki S, et al. (2012b). Formation and rotation of skyrmion crystal in the chiral-lattice insulator Cu_2OSeO_3. *Phys Rev B*. 85: 220406.

Senthil T, Vishwanath A, Balents L, Sachdev S, and Fisher M P A (2004). Deconfined quantum critical points. *Science*. 303: 1490–1494.

Skyrme T H R (1962). A unified field theory of mesons and baryons. *Nucl Phys* 31: 556–569.

Sluckin T J, Dunmur D A, and Stegemeyer H (2004). *Crystals That Flow. Classic Papers from the History of Liquid Crystals*. London: Taylor & Francis.

Sondhi S L, Karlhede A, and Kivelson S A (1993). Skyrmions and the crossover from the integer to fractional quantum Hall effect at small Zeeman energies. *Phys Rev B*. 47: 16419–16426.

Stashkevich A A, et al. (2015). Experimental study of spin-wave dispersion in Py/Pt film structures in the presence of an interface Dzyaloshinskii-Moriya interaction. *Phys Rev B*. 91: 214409.

Sun L, et al. (2013). Creating an artificial two-dimensional skyrmion crystal by nanopatterning. *Phys Rev Lett*. 110: 167201.

Tetienne J P, et al. (2015). The nature of domain walls in ultrathin ferromagnets revealed by scanning nanomagnetometry. *Nat Commun*. 6: 6733.

Thiaville A, Rohart S, Jué É, Cros V, and Fert A (2012). Dynamics of Dzyaloshinskii domain walls in ultrathin magnetic films. *Europhys Lett*. 100: 57002.

Tycko R, Barrett S E, Dabbagh G, Pfeiffer L N, and West K W (1995). Electronic states in gallium arsenide quantum wells probed by optically pumped NMR. *Science*. 268: 1460–1463.

Verba R, Tiberkevich V, Krivorotov I, and Slavin A (2014). Parametric excitation of spin waves by voltage-controlled magnetic anisotropy. *Phys Rev Appl*. 1:044006.

Vollhardt D and Wölfle P (1990). *The Superfluid Phases of Helium 3*. London: Taylor & Francis.

Wadley P, et al. (2016). Electrical switching of an antiferromagnet. *Science*. 351: 587–590

Wright D C and Mermin N D (1989). Crystalline liquids: The blue phases. *Rev Mod Phys*. 61: 385.

Xu X Q and Han J H (2011). Spin-orbit coupled Bose-Einstein condensate under rotation. *Phys Rev Lett* 107: 200401.

Yang H, Thiaville A, Rohart S, Fert A and Chshiev M (2015). Anatomy of Dzyaloshinskii-Moriya interaction at Co/Pt interfaces. *Phys Rev Lett* 115: 267210.

Yang J H, et al. (2012). Strong Dzyaloshinskii-Moriya interaction and origin of ferroelectricity in Cu_2OSeO_3. *Phys Rev Lett* 109: 107203

Yann G, Jun Y, Aron P, Pfeiffer L N, and West K W (2008). Soft spin wave near ν = 1: Evidence for a magnetic instability in skyrmion systems. *Phys Rev Lett* 100: 086806.

Yu X Z, et al. (2010). Real-space observation of a two-dimensional skyrmion crystal. *Nature* 465: 901–904.

Zang J, mostovoy M, Han J H, Nagaosa N (2011). Dynamics of skyrmion crystals in Metallic thin films. *Phys Rev Lett* 107: 136804

Zhang X, et al. (2015a). Skyrmion-skyrmion and skyrmion-edge repulsions in skyrmion-based racetrack memory. *Sci Rep* 5: 7643.

Zhang X, Ezawa M, and Zhou Y (2016a). Thermally stable magnetic skyrmions in multilayer synthetic antiferromagnetic racetracks. *arXiv* 1601.03893: https://arxiv.org/abs/1601.03893.

Zhang X, Zhou Y, and Ezawa M (2016b). Antiferromagnetic skyrmion: Stability, creation and manipulation. *Sci Rep* 6, 24795.

Zhang X, Zhou Y, and Ezawa M (2016c). Magnetic bilayer-skyrmions without skyrmion Hall effect. *Nat Commun* 7: 10293.

Chapter 14

Zhang X, Zhou Y, Ezawa M, Zhao G P and Zhao W S (2015b). Magnetic skyrmion transistor: Skyrmion motion in a voltage-gated nanotrack. *Sci Rep* 5: 11369.

Zherlitsyn S, et al. (2014). Spin-lattice effects in selected antiferromagnetic materials. *J Low Temp Phys*. 40: 123–133.

Zhou Y and Ezawa M (2014). A reversible conversion between a skyrmion and a domain-wall pair in junction geometry *Nat Commun* 5: 4652.

Index